European Studies in Philosophy of Science

Volume 12

This new series results from the synergy of EPSA - European Philosophy of Science Association - and PSE - Philosophy of Science in a European Perspective: ESF Networking Programme (2008–2013). It continues the aims of the Springer series "The Philosophy of Science in a European Perspective" and is meant to give a new impetus to European research in the philosophy of science. The main purpose of the series is to provide a publication platform to young researchers working in Europe, who will thus be encouraged to publish in English and make their work internationally known and available. In addition, the series will host the EPSA conference proceedings, selected papers coming from workshops, edited volumes on specific issues in the philosophy of science, monographs and outstanding Ph.D. dissertations. There will be a special emphasis on philosophy of science originating from Europe. In all cases there will be a commitment to high standards of quality. The Editors will be assisted by an Editorial Board of renowned scholars, who will advise on the selection of manuscripts to be considered for publication.

More information about this series at http://www.springer.com/series/13909

Brigitte Falkenburg

Kant's Cosmology

From the Pre-Critical System to the Antinomy of Pure Reason

Springer

Brigitte Falkenburg
Department of Philosophy and Political Science
Faculty of Human Sciences and Theology
TU Dortmund
Dortmund, Germany

ISSN 2365-4228 ISSN 2365-4236 (electronic)
European Studies in Philosophy of Science
ISBN 978-3-030-52289-6 ISBN 978-3-030-52290-2 (eBook)
https://doi.org/10.1007/978-3-030-52290-2

This Springer imprint is published by the registered company Springer Nature Switzerland AG
The registered company address is: Gewerbestrasse 11, 6330 Cham, Switzerland

In memory of my brother

Preface

This book provides a systematic investigation of Kant's philosophical development from his pre-critical to his critical cosmology. It studies Kant's theoretical philosophy as it was expressed in his pre-critical metaphysics and cosmology, through the critical turn, to the antinomy of pure reason, focusing on his principles of metaphysical theory formation. The book has three parts. The first sets out the methodological approach that underlies Kant's pre-critical project to reconcile the principles of Wolff's system of metaphysics and Newtonian physics (Chaps. 1 and 2). The second part examines the way in which Kant transformed the foundations of his precritical metaphysical system in the course of his critical turn and explores his reasons for this transformation (Chaps. 3 and 4). The third part analyzes the cosmological antinomy of the *Critique of Pure Reason* and its relation to transcendental idealism (Chaps. 5 and 6). Finally, an extensive appendix sets out the crucial systematic concepts behind Kant's pre-critical and critical metaphysics of nature.

Kant's pre-critical work of the years 1755/1756 includes several writings on natural science and the foundations of metaphysics. From the systematic point of view taken in this book, the *Theory of the Heavens*, the *New Elucidation*, and the *Physical Monadology* are of decisive importance. Together, these three writings promised to lay the foundations of an all-encompassing metaphysics of nature. The *Prize Essay* published in 1764 defended the "analytic" method of metaphysics employed in this endeavour. But by this time, Kant had already begun to realize that the foundations of his pre-critical system were untenable. His development from the *Dreams of a Spirit Seer* of 1766 to the *Inaugural Dissertation* of 1770, and through his so-called silent period to the first *Critique*, is generally thought to be well understood. However, controversy remains concerning the genesis of the antinomy of pure reason, while, in addition, the structure and significance of the cosmological antinomy for Kant's critical philosophy has often been misunderstood. I hope that my attempt at a systematic reconstruction of these matters will shed more light on the specific role played by the doctrine of the antinomy within Kant's critical metaphysical project.

This book is an enlarged and completely revised English version of my *Kants Kosmologie* published 20 years ago. The motivation behind that book was to clarify the reasons why, in his critical period, Kant rejected his early work on the foundations of physics and metaphysics. The work he rejected included his impressive cosmology of 1755, which in a certain sense laid the foundations for modern physical cosmology as a well-established discipline of physics. This cosmology contained a theory of the formation of the solar system from a matter vortex, still considered essentially correct today, which he then extrapolated to an impressive cosmogony and added a physico-theological proof of the existence of God. In view of this theory, Kant's critical turn raises two questions. First, why did he reject not only this metaphysical addendum to that impressive cosmogony but the entire project of a physical cosmology itself, claiming that any theory of the whole universe must turn out to be contradictory? Second, does Kant's doctrine of the cosmological antinomy still provide lessons for modern physical cosmology, and if so, what?

Twenty years ago I could not give a definitive answer to the second question, nor do I attempt to do so in the present book. Instead, I focus on the first question, seeking to work out Kant's development more precisely and to clarify the significance of the antinomy for his critical project. Therefore, here I omit the topics of the last chapter of my previous book, which will have to be addressed in a separate book on Neo-Kantian approaches to modern physics. Other changes include the correction of several errors in my earlier book, a restructuring of the presentation in order to make the line of reasoning more transparent, and the addition of new material.

Parts I and II of the present book contain many new details concerning Kant's pre-critical "analytic" method of metaphysics, its analogy to Newton's inductive methodology, the collapse of the pre-critical system, and Kant's critical turn. Part I focuses on the structure and methods of the pre-critical system. The contents of Chap. 1 are now presented in a more systematic way, particularly concerning Kant's early attempt at reconciliation in the *True Estimation*, its consequences for his pre-critical metaphysical project, the structure of the 1755/1756 system, and the methodological challenges posed by the quest to unify the principles of Newtonian physics and metaphysics in Wolff's style. Chapter 2 studies the analytic and synthetic methods employed by Kant in his 1755/1756 trilogy that provided the foundations for his pre-critical system, the background of these methods in early modern science and philosophy, and what he made out of them in the *Prize Essay* of 1764. Particular emphasis is laid on the methodological analogy between metaphysics and Newtonian science, which also sheds new light on the proof structure employed in the *Only Possible Argument*.

Part II investigates the collapse of the pre-critical system and the critical turn. Chapter 3 investigates how the distinction between logical and real grounds affected the foundations of the pre-critical system, culminating in the *Dreams* of 1766 and the 1768 argument from incongruent counterparts. The latter gave rise to the puzzling result that Kant saw himself as left without any tenable concept of space as a real entity, be it absolute or relational. His way out was the 1770 theory of space and time as forms of pure intuition. From a modern logical point of view, however, his solution was sufficient but not necessary to resolve the puzzle of 1768. Chapter 4 takes a closer look at Kant's critical turn. From Neo-Kantianism to recent interpretations, it has been argued that Kant's 1770 *Dissertation* resolved the cosmological antinomy. This traditional view is rejected here, given that in 1770 Kant was far from stating the later antinomy of pure reason. Rather, he continued to attempt to reconcile the cosmological concept of traditional metaphysics with his new theory of space and time as pure forms of intuition. It was only after 1772, in the course of developing his critical theory of cognition, that he discovered the mathematical antinomy. His notes on metaphysics of 1773–1775 indicate that he then realized that in cosmology only the metaphysical concept of a potential infinite applies, in contrast to the mathematical concept of an actual infinite.

Part III examines Kant's critical cosmology, which to a large extent consists only in the critique of traditional cosmology. My logical reconstruction of the cosmological antinomy in Chap. 5 remains essentially the same as in my 2000 book, but I now hope to have presented Kant's arguments and their assessment in a clearer way. Detailed investigation again leads me to reject certain influential views concerning several alleged fallacies in Kant's reasoning, which miss the point that Kant consciously constructed the antinomies based on traditional cosmological arguments, some of which he himself adhered to in his pre-critical metaphysics, including the 1770 *Dissertation*, and which from his critical point of view turn out to be fallacious. Chapter 6 takes up some ideas from my 2000 book, but most of it is new. In particular, it contains a detailed analysis of the experiment of pure reason, which the preface to the second edition of the first *Critique* presents as an argument in favour of transcendental idealism. This thought experiment can be understood as a transcendental argument against transcendental realism. From a modern point of view, it seems weaker than Kant claimed: it proves only that transcendental idealism is sufficient, but not necessary, for avoiding the antinomy, just as Kant's way out of the 1768 puzzle was. However, his thought experiment does at least demonstrate that the doctrine of cosmological antinomy has an anti-naturalistic impact. Therefore, my main conclusion about the function of the antinomy of pure reason within Kant's critical project is that it is the critical correlate of his pre-critical physico-theology.

The foundations for this book were laid 25 years ago, when I wrote large parts of the underpinning German version during my stay at the Wissenschaftskolleg zu Berlin in the academic year 1995/1996. The remainder of my Heisenberg fellowship granted by the Deutsche Forschungsgemeinschaft subsequently enabled me to continue my discussions with Peter Mittelstaedt on the topics of this book and their relations to the philosophy of physics, at the Universität zu Köln.

After my appointment to the Technische Universität Dortmund in 1997, I was able to deepen my work on Kant's argument from incongruent counterparts, the cosmological antinomy, and the analytic-synthetic methods of early modern science and philosophy through several research projects: the project *Functions and Limitations of Intuition in Physics* (MWF of Nordrhein-Westfalen, Project IVA-6000, 1998–2001) with Renate Huber; a three-year cooperation on the *Functions of Intuition in Mathematics and Physics* with Michael Hallett and Emily Carson from McGill University (Montreal), which was supported by the German–American Academic Council (TransCoop Program 1998–2000); and the research project *Hypotheses non fingo: Newtons Methodenlehre*, supported by the Deutsche Forschungsgemeinschaft (DFG FA 261/5-1, 2002–2003) with Karl-Norbert Ihmig.

Over the next decade and a half, I was able to continue my work on Kant only sporadically, due to teaching and administrative obligations as well as research projects on other topics. Yet, over this period, on the occasion of talks at conferences or other universities, or when inviting colleagues to Dortmund, I nevertheless had many stimulating discussions about Kant's theory of nature. Of those over the years who gave me the opportunity to present my work on Kant, made critical comments on it, or provided me with important insights, I would like to mention Claus Beisbart, Silvia De Bianchi, Cinzia Ferrini, Tobias Häusler, Dietmar Heidemann, Katharina Kraus, Michela Massimi, Hernán Pringe, Helmut Pulte, Peter Rohs, Simon Saunders, Jürgen Stolzenberg, Thomas Sturm, Dieter Sturma, Violetta Waibel, and Brigitta von Wolff-Metternich, not to forget Michael Wolff who first instructed me about the logical structure of Kant's antithetic of pure reason, when I was a PhD student in Bielefeld many years ago.

In addition, I would like to thank Christian Feldbacher-Escamilla, Andreas Hüttemann, Oliver Scholz, Gerhard Schurz, and Ansgar Seide from our research group *Inductive Metaphysics*, supported by the Deutsche Forschungsgemeinschaft (DFG FOR 2495), for drawing my attention to the parallels and differences between the traditional analytic-synthetic method, inductive metaphysics from the late nineteenth to the early twentieth century, and creative abduction, as well as the parallels between Kant's experiment of pure reason and transcendental arguments. I would also like to thank Kristina Engelhard for carefully reading parts of the book and critically commenting on them, Benedict Young for his patient and constructive language corrections, and the Deutsche Forschungsgemeinschaft for supporting this work as well as two related conferences in our research project *Kant and Inductive Methods in Eighteenth Century Metaphysics* (DFG FA 261-15/1, which is part of FOR 2495).

To Dennis Dieks, Maria Carla Galavotti, and Wenceslao J. Gonzalez, the Editors of the European Studies in Philosophy of Science, I am grateful for the opportunity to publish the book in this series; I also express my gratitude to Christopher Wilby and Deepthi Vasudevan from Springer for their support in preparing the contract and guiding the publishing process, and to the publishers of my previous work on Kant and for the permission to reuse material. This gratitude extends of course to the publisher of the original German version of the book, Klostermann, as well as to the various publishers who gave permission to make use of several articles of mine written after that book: while some of these articles were based on the previous German book, others presented entirely new material that I again include here.

Finally, the translation tool *DeepL* (https://www.deepl.com/translator) helped me a great deal with the passages I took from previously published German material, and I hope in return to have helped train it on Kant's terminology.

Berlin, Germany
May 2020

Brigitte Falkenburg

Contents

Citations, Abbreviations, and Previously Published Material

Citations

References to Kant's works are given as follows: The *Critique of Pure Reason* is cited according to the pages of the first (A) edition and second (B) edition. All other works are cited according to the German standard edition *Kants gesammelte Schriften*, ed.: the Deutsche (formerly: Königlich-Preußische) Akademie der Wissenschaften, Vols. 1–29, Berlin 1900–, in the form (volume:page).

All translations are taken from the Cambridge Edition of the Works of Immanuel Kant, unless explicitly stated otherwise. In particular, the following volumes are quoted:

Theoretical Philosophy 1755–1770, transl. and ed.: D. Walford in collab. with R. Meerbote, Cambridge University Press, 1992.

Lectures on Logic, transl. and ed.: J. M. Young, Cambridge University Press, 1992.

Religion and Rational Theology, transl. and ed.: A. W. Wood and G. di Giovanni, Cambridge University Press, 1996.

Critique of Pure Reason, transl. and ed.: P. Guyer and A. Wood Cambridge University Press, 1998.

Critique of the Power of Judgment, ed.: P. Guyer, transl. by P. Guyer and E. Matthews, Cambridge University Press, 2000.

Theoretical Philosophy after 1781, ed.: H. Allison and P. Heath, transl. by G. Hatfield and M. Friedman, Cambridge University Press, 2002.

Notes and Fragments, ed.: P. Guyer, transl. by C. Bowman, P. Guyer, and F. Rauscher, Cambridge University Press, 2005.

Anthropology, History, and Education, ed.: G. Zöller and R. B. Louden, transl. by M. Gregor et al., Cambridge University Press, 2007.

Natural Science, ed.: E. Watkins, transl. by L. W. Beck et al., Cambridge University Press, 2012.

Abbreviations

For Kant's writings, I use the following abbreviations:

True Estimation Thoughts on the true estimation of living forces and assessment of the demonstration that Leibniz and other scholars of mechanics have made use of in this controversial subject, together with some prefatory considerations pertaining to the force of bodies in general (1749)

Theory of the Heavens Universal natural history and theory of the heavens or essay on the constitution and the mechanical origin of the whole universe according to Newtonian principles (1755)

On Fire Succinct exposition of some meditations on fire (1755)

New Elucidation A new elucidation of the first principles of metaphysical cognition (1755)

Physical Monadology The employment of metaphysics combined with geometry in natural philosophy, of which sample I contains the physical monadology (1756)

Plan and Announcement Plan and announcement of a series of lectures on physical geography with an appendix containing a brief consideration of the question: Whether the West winds in our regions are moist because they travel over a great sea (1757)

New Doctrine New doctrine of motion and rest and the conclusions associated with it in the fundamental principles of natural science while at the same time his lectures for this half-year are announced (1758)

Only Possible Argument The only possible argument in support of a demonstration of the existence of God (1763)

Negative Magnitudes Attempt to introduce the concept of negative magnitudes into philosophy (1763)

Prize Essay Inquiry concerning the distinctness of the principles of natural theology and morality (1764)

Dreams Dreams of a spirit-seer elucidated by dreams of metaphysics (1766)

Directions in Space Concerning the ultimate ground of the differentiation of directions in space (1768)

Dissertation On the form and principles of the sensible and the intelligible world (Inaugural Dissertation) (1770)

CPR Critique of Pure Reason (A: 1781, B: 1787)

Prolegomena Prolegomena to any future metaphysics that will be able to come forward as science (1783)

MFNS Metaphysical foundations of natural science (1786)

To Orient Oneself What does it mean to orient oneself in thinking? (1786)

CPJ Critique of the Power of Judgment (1790, 1793)

Progress What real progress has metaphysics made in Germany since the time of Leibniz and Wolff? (1793/1804)

Conflict of the Faculties The conflict of the faculties (1798)

Logic Immanuel Kant's Logic. A Manual for Lectures, edited by Gottlob Benjamin Jäsche (1800)

Previously Published Material

Material from the following publications of mine has been used in this book:

Kants Kosmologie. Die wissenschaftliche Revolution der Naturphilosophie im 18. Jahrhundert, Frankfurt am Main:Vittorio Klostermann 2000.

"Kants Naturalismus-Kritik", in: K. Engelhard and D. Heidemann (eds.), *Warum Kant heute?* Berlin: de Gruyter 2004, 177–206.

"Die Funktion der Naturwissenschaft für die Zwecke der Vernunft", in: V. Gerhardt and T. Meyer (eds.), *Kant im Streit der Fakultäten*, Berlin: de Gruyter 2006, 117–133.

"Intuition and Cosmology: The Puzzle of Incongruent Counterparts", in: E. Carson and R. Huber (eds.), *Intuition and the Axiomatic Method*, Dordrecht: Springer 2006, 157–180.

"From Kant's Early Cosmology to the Cosmological Antinomy", in: Silvia De Bianchi (ed.), *The Harmony of the Spheres*, Newcastle upon Tyne: Cambridge Scholars Publishing 2013, 48–70.

"Kant and the Scope of the Analytic Method", in: *Studies in History and Philosophy of Science* 71 (2018), 13–23.

"Was beweist Kants Experiment der reinen Vernunft?", in: Violetta L. Waibel et al. (eds.), *Natur und Freiheit: Akten des XII. Internationalen Kant-Kongresses*, Berlin: de Gruyter 2018, 641–659.

"The Function of Natural Science for the Ends of Reason", in: P. Órdenes and A. Pickhan (eds.), *Teleologische Reflexion in Kants Philosophie*, Berlin: Springer 2019, 93–109.

Part I
The Pre-critical System

The first part of this book focuses on the origins of Kant's systematic philosophy. It is well known that his pre-critical project was to reconcile the principles of Wolff's system of metaphysics with those of Newton's physics. Here, the methodological principles according to which Kant attempted to do so are examined in detail.

Chapter 1 reconstructs the systematic problems of Kant's pre-critical reconciliation project and the unifying principles he employed in order to resolve them. His first writing, the *True Estimation* published in 1749, remained in the eclectic tradition of his day; but with the writings of 1755/1756 he then sought to escape from eclecticism. Together, the *Theory of the Heavens*, the *New Elucidation*, and the *Physical Monadology* aimed at establishing the foundations of a system of metaphysics in Wolff's style. In view of the manifest conflict between the principles of Leibniz's metaphysics and the foundations of Newton's physics, this project was most ambitious. Kant faced a complex unification problem that involved various levels of theory formation, ranging from physics to metaphysics, from atoms to monads, from space, time, and matter to the principle of sufficient reason, from the assumption of divine intervention in the world to the system of pre-established harmony.

Chapter 2 investigates Kant's "analytic" method, its background in the analytic-synthetic methods of early modern science and philosophy, and the ways in which the *Theory of the Heavens*, the *New Elucidation*, and the *Physical Monadology* employed several variants of this method in order to bridge the discrepancies between Newtonianism and Wolffianism. His defence of the "analytic" method of metaphysics in the *Prize Essay* (1764) is of particular interest. There, Kant compares the "analytic" method of metaphysics to Newton's inductive method. But to talk in a one-sided way of his "analytic" method neglects the fact that Kant did indeed insist on a two-step methodology, in analogy to the analytic *and* the synthetic part of Newton's method of analysis and synthesis as explained by Newton in his *Opticks*, which itself traces back to ancient geometry (Pappus). In parallel to defending the use of this method in metaphysics, however, Kant also became increasingly aware of its limitations.

Chapter 1
Physics and Metaphysics

But how can metaphysics be married to geometry, when it seems easier to mate griffins with horses than to unite transcendental philosophy with geometry? (1:475)

This chapter investigates Kant's pre-critical writings until 1764 with regard to the unifying strategy that stands behind them. Kant scholars have long emphasized that the main goal of Kant's pre-critical philosophy was to reconcile the diverging metaphysical and physical theories of his time. This approach has been called the "irenic" model. Here we seek to examine the underlying systematic principles to which he appealed, paying attention to the unificational methodology that was characteristic of his philosophical beginnings. Although Kant's first attempt to reconcile two apparently incompatible positions—his treatment of the *vis viva* debate in the *True Estimation of Living Forces*—was unsatisfactory and eclectic, in some crucial respects it set the course for his later work. Compared to Kant's debut, the pre-critical writings of the years 1755/1756 are a milestone in systematicity: as Schönfeld (2000) has shown in detail, they contribute to the project of an all-encompassing philosophy of nature, including theories of the interaction of body and soul, human freedom, and God as the ultimate cause of the world. However, Schönfeld neglects the systematic aspects of Kant's pre-critical project. The young Kant wanted to establish a system of metaphysics in Wolff's style, including Leibniz's principle of sufficient reason, which was compatible with Newton's physics. The present investigation focuses on the methodological problems of this ambitious and demanding task. Given that Wolff's metaphysics and Newton's physics were incompatible at several levels of theory formation, Kant faced a complex unification problem. Wolff's (or Leibniz's) metaphysical principles and the laws of Newton's physics conflicted with regard to the concepts of space and time (relative vs. absolute), force (internal vs. external), the constitution of matter (continuum theory vs. atomism), and the underlying general principles (Newton's belief in divine intervention vs. Leibniz's system of pre-established harmony). The 1755/1756 writings tackled different parts of this multi-level unification problem.

B. Falkenburg, *Kant's Cosmology: From the Pre-Critical System to the Antinomy of Pure Reason*, European Studies in Philosophy of Science 12,
https://doi.org/10.1007/978-3-030-52290-2_1

In order to establish the principles of a coherent system of metaphysics, Kant could not just resolve the above conflicts one by one. He had to employ criteria of systematicity, adequacy conditions, and a well-defined philosophical method. The philosophically relevant cornerstones of the project were the *Theory of the Heavens*, the *New Elucidation*, and the *Physical Monadology*.

1.1 The Reconciliation Principle and First Attempt

The supreme methodological principle of Kant's pre-critical philosophy is the requirement to *reconcile* opposite positions (Adickes 1897; Erdmann 1884; Riehl 1924; Hinske 1970; Friedman 1992). In the light of Kant's later pre-critical project, it is tempting to understand this as a rudimentary *principle of unification*. *True Estimation*, Kant's very first work, finished in 1747 but published only in 1749, sought to exhibit the compatibility of the Cartesian and Leibnizean positions as regards the *vis viva* controversy. Following Bilfinger, a Wolffian, Kant endorsed a principle of locating truth on the middle ground between opposing views, by conceding certain points to both parties:

> If men of sound understanding put forward entirely opposed opinions, and if neither of both of the parties may be presumed to have ulterior motives, then the logic of probability requires that we should look above all for a certain intermediate position which concedes that both parties are to some extent right. (1:32)

Kant's pre-critical approach of mediating between opposite views has been called his *irenic model* of dealing with the metaphysical debates of his time (Hinske 1970, 123). From the above quotation, however, it seems the methodological principle of reconciling opposite points of view does not really go beyond the commonplace that the truth lies somewhere in between them. Indeed, Kant's way of reconciling the opposing positions based on this principle was not convincing at all, as his contemporaries noted—in particular his academic teacher Martin Knutzen, as well as Leonard Euler to whom he sent the draft of the *True Estimation*—and as Kant scholars today still emphasize (Calinger 1979; Schönfeld 2000).

1.1.1 The Vis Viva Controversy

The *vis viva* controversy had been ongoing in the Academy of St. Petersburg since 1725, with the participation of Christian Wolff and others (Calinger 1968, 1969; Schönfeld 2000, 19–35). Its subject was the "true" measure of the force of mechanical motions. The dispute stemmed from a passage of Descartes's *Principia philosophiae* (1644) and Leibniz's criticism of it (Leibniz 1686).[1] Descartes himself

[1]See Mach (1883, 285–289), Dugas (1950); Iltis (1971), Wolff (1987, 304–312), Friedman (1992, 3–4).

did not yet have a concept of mass, but the Cartesians reinterpreted his measure of force in (Newtonian) terms of momentum mv. Wolff and the Wolffians proposed Leibniz's "living force" mv^2, which is proportional to the quantity known as kinetic energy today, as the correct measure of force. Leibniz himself made the distinction between the Cartesian measure mv, which he called *conatus*, the "living force" mv^2 or *vis viva*, and the "dead force" or *vis mortua*, an infinitesimal magnitude of statics, from which the "living force" is obtained via integration over the path of a motion. According to him, only the *vis viva* is the correct dynamic quantity (Leibniz 1695). In hindsight, the Cartesian measure of force corresponds to the time integral, and the Leibnizian measure to the path integral of Newton's force. Hence, both measures are dynamic quantities which both derive from the axioms of Newton's mechanics. This resolution was indicated in the first edition of d'Alemberts *Traité de Dynamique* published in 1743, a book that was not known to Kant when he wrote the *True Estimation*. Historians of science emphasize that d'Alembert gave the full explanation only in the second edition of his *Traité de Dynamique*, published in 1758 (Iltis 1970; Schönfeld 2000). Hence, to a certain extent Kant may be absolved from having suggested a solution on his own and having ignored the contemporary state of the art.

In order to concede the respectively correct points to both parties, Kant's approach was to clarify the relation between mathematics, or mathematical physics, and metaphysics. Given that the dispute about the correct measure of force stemmed from diverging metaphysical presuppositions concerning the concept of matter, his approach responded precisely to the origin of the controversy. For Descartes (and Newton), forces are external and bodies on their own have only inertial motions. For Leibniz, on the contrary, forces are internal, due to the internal activities of the monads underlying all the phenomena of the material world. From a modern point of view, d'Alembert's solution to the dispute had the advantage of abstaining from any metaphysical questions and basing itself on mathematical physics alone. For Kant, a young scholar of the German Leibniz–Wolff school of academic philosophy who was not well trained in mathematics, this way out of the controversy was not obvious.

Both Newton and Leibniz independently criticized Descartes's purely geometrical corpuscular theory of matter. Although they did so for quite different reasons, their respective criticisms made both introduce *dynamic concepts* into physics. But Newton's and Leibniz's concepts of—external or internal, respectively—force were incompatible. According to Leibniz, active and passive forces are inherent to a substance or monad. The monads change their inner states only according to these internal forces, and there are no external interactions between the monads. Quite on the contrary, Newton's mechanics is based on the principle that any change of the state of motion of a body is due to an external cause. This principle bears a debt to Descartes's corpuscular philosophy. According to the law of inertia, a body does not change its state of motion as long as no external force acts upon it. Only Newton's concepts of mass, momentum, force, and gravity give rise to a physical dynamics in the modern sense, and it was on the basis of *this* theory that d'Alembert decided the *vis viva* controversy. His solution demonstrated in particular that both measures of force, the Cartesian as well as the Leibnizian, find their place and make sense within Newton's mechanics. In this way, d'Alembert resolved the controversy about the "true" measure of force by completely detaching the concept of a dynamic quantity from Leibniz's concept of force, and by demonstrating that the quantities that were under debate both belong to Newton's mechanics. The quantities mv and mv^2 simply express different dynamic aspects of a single unified and coherent account of mechanical motion.

1.1.2 Kant's Eclectic Reconciliation

Kant, however, tried it the other way round. In the *True Estimation*, he proposed to maintain the contrary dynamic concepts by restricting the scope of both, i.e., by attributing them to different mechanical phenomena. His suggestion was as follows. The Cartesian measure mv only holds for mechanical motions which stem from the impact of external forces, and which without such an external impact come to rest. Today, such motions are associated with the dissipation of energy. In contrast to them, Kant attributed Leibniz's measure of force mv^2 to so-called "free" motions, for which, as we know today, energy and momentum are conserved (*True Estimation*, §§15–17, 1:28–29).

The shortcomings of Kant's arguments are well known (Schönfeld 2000, 36–55) and need not be repeated here. From a modern point of view, it looks ad hoc to restrict the respective scope of both measures of force in such a way. Kant's solution looks like a case of "piecemeal physics" in Nancy Cartwright's sense (Cartwright 1999). Kant finally introduces a concept of "intension" in order to resolve the *vis viva* controversy (*True Estimation* §117, 1:141–142). It is tempting to trace this concept back to the metaphysical concept of *vis insita* or *vis inertiae* (inertial force), which Newton employed in his *Definition III* of the *Principia* in order to justify the law of inertia. However, both concepts should not be confused or identified; Kant's account is rather based on the *vis insita* of Georg Erhard Hamberger, professor of medicine in Jena (see Massimi and De Bianchi 2013, 487). And in the *True Estimation* he took up the contemporary discussion between the Cartesians and the Leibnizeans, but not yet the principles of Newton's mechanics.

In 1747, Kant defines the "intension" such that force is the product of velocity and intension. He considers this quantity to be a measure of the inherent disposition of matter to maintain a motion. This recalls Leibniz's concept of *vis activa primitiva*, as a primitive activity of matter to resist any change of its state of motion. Kant's "intension" is a mass-independent intensive quantity, unlike Newton's *vis insita* or *vis inertiae*, which is an extensive quantity proportional to mass.[2] On these

[2]See the explanation to *Definition III* in the *Principia* (Newton 1726, 404) compared to § 100 and § 117 of the *True Estimation* (1:110, 1:141–142). Kant interprets the "intension" as the cause of the "living force". In this way, the *vis viva* seems to obtain the status of a Leibnizean *vis*

grounds, he employs his concept of "intension" or *vis insita* as some kind of dynamic principle in order to settle the question of to which kinds of phenomena the Cartesian measure of force applies, and to which not. The intension of a body upon which an external force acts is only instantaneously put into effect, because its disposition to maintain its motion is counteracted. Hence, its force is proportional to velocity (Cartesian measure mv). On the contrary, the intension of a body upon which no external force acts is permanently put into effect, because its disposition to maintain its motion is not counteracted. Hence, Leibniz's *vis viva* holds without any restriction for it, and its force is proportional to the square of velocity (Leibnizean measure mv^2).[3]

Interpreted in such terms, Kant's unsuccessful attempt to combine Descartes's and Leibniz's incompatible concepts of force to a certain extent becomes comprehensible. However, this approach has no coherent foundations. It gives rise to a hybrid dynamic conception: to a half-Leibnizean, half-Cartesian concept of internal forces which are dispositions to maintain the state of motion of bodies, on the one hand, and external forces which counteract these internal forces and change the state of motion, on the other. The concept of *vis insita* seems to serve as a metaphysical principle for deriving the application conditions for the Leibnizean measure of force, and that of *vis inertiae* for deriving the Cartesian measure. Based on this, Kant seems to identify Leibniz's *vis viva* with an internal force which is some kind of *vital principle* of matter in motion, and the Cartesian *conatus* with a merely mathematical description of the result of the effect of an external force, in a "genuine pre-Newtonian" approach that even gave rise to "Cartesian echoes" in his later dynamics (Massimi and De Bianchi 2013, 490–491; see also Ferrini 2018).

Although Kant expends great effort to stick to his principle of combining physics and metaphysics, his 1747 approach still lacks what he would much later, in the *Critique of Pure Reason*, call the *systematicity* of scientific knowledge. What is missing is the reduction of both concepts of force to *uniform* theoretical principles, that is, to a consistent dynamics that is embedded in a coherent metaphysical framework. Kant's 1747 work remains eclectic, in accordance with the philosophical tendencies of his time.

---

activa derivativa which derives from the "intension"; see also § 120 (1:143–144). It is questionable whether Kant really has Leibniz's distinction between *vis activa primitiva* and *vis activa derivativa* in mind here; but his terminology in § 117 suggests this interpretation. There, he distinguishes the "external phaenomenon of force", i.e., motion as the phenomenological effect of the derivative force, and the "basis of activity", i.e., the primitive force as the metaphysical cause of this phenomenon (1:141).

[3]See the arguments at 1:141–144, which cannot be analyzed here. Presumably his concept of "intension" has also to be understood against the background of the contemporary discussion on the infinitesimal in the calculus. The way in which this concept is related to the Wolffian doctrine of quantities should also be taken into account. See the *prima matheseos intensorum principia* in §§ 165–190 of Baumgarten's *Metaphysics* (Baumgarten 1757, 17:61–66), to which Kant's later concept of an intensive magnitude makes reference.

1.2 The Legacy of the *True Estimation*

After the *True Estimation*, Kant did not publish anything for almost a decade, before he reappeared with his 1754 articles about the rotation and the ageing of the earth (1:183–191, 1:193–213). In the years from 1749 to 1754 his focus of attention shifted from Cartesian physics to the principles of Newton's mechanics. His ambitious project of reconciling physics and metaphysics in a more comprehensive sense was taking shape, and he must have become aware how complex and difficult it would be to bridge the discrepancies between Newton's physics and Wolff's metaphysics. The systematic implications of this task are explained in more detail in the next Section (1.3). Before we turn to this, let me sketch how the approach of the *True Estimation*, despite being unsuccessful, shaped his further philosophical thinking. Kant's much more successful writings of the years 1755/1756 indeed continue some views already expressed in the *True Estimation*, which were crucial for his life-long project of unifying physics and metaphysics. To them belong general convictions concerning the relation between physics and metaphysics, as well as more specific views.[4]

Kant's most stable assumptions about the relation between physics and metaphysics were shaped by the academic tradition of Wolff's metaphysics, in which he grew up. According to this tradition, cosmology belongs to metaphysics; and physics deals with concepts and laws which relate to cosmology in the manner of lower-level to higher-level principles. Obviously, physics is an empirical science, whereas metaphysics is not. But the question of how the fundamental concepts physics apply to the phenomena was, for Kant, a metaphysical problem, not a question of empirical research. Hence, for him, in the case of rival physical concepts, what called for metaphysical clarification were the conditions under which the concepts apply to the phenomena. And this was what he had already tried in the case of the *vis viva* controversy, by restricting the respective scopes of both controversial concepts. Later, he became aware that his 1747 attempt had lacked systematicity. What remained, however, was his view that the application conditions of physical concepts have to be clarified by appeal to metaphysical considerations.

It is exactly in this point that Kant's pre-critical as well as critical theories of nature differ from empiricism. For Kant and his followers, the question of under which conditions the concepts of physics apply is not a matter of empirical knowledge but rather a matter of the non-empirical, or metaphysical, presuppositions concerning how to use these concepts. And these presuppositions are closely related to the fundamental concepts of physical dynamics, i.e., the concepts of space, time, and force.

Kant had focused on the concept of force from his very beginnings. The concept of space only came to his attention much later (and the concept of time only in the course of his critical turn). For a long time he affirmed a relational conception

[4]Massimi and De Bianchi (2013) and Ferrini (2018) show that they include Cartesian influences on his later dynamics of the *Physical Monadology* and the *MFNS*.

of space, in accordance with Leibniz's criticism of Newton's concept of absolute space. Albeit in a critically refined version, this relational concept is still present in the *CPR*. In the *True Estimation*, Kant defended a relational account of space which was based on a theory of material substances interacting due to their internal forces:

> It is easy to show that there would be no space and no extension if substances had no force to act external to themselves. For without this force there is no connection, no order, and, finally, without order, no space. (§ 9, 1:23)

According to the young Kant, space is "the sum-total of substantial relations" (Schönfeld 2000, 42), just as for Leibniz. But for the young Kant space was real (and not just ideal, as Leibniz thought), though relational (and not absolute, as it was for Newton).[5] He retained this view until the argument from incongruent counterparts occurred to him, in 1768 (see Sect. 3.4). In the *New Elucidation* of 1755, he supported this concept of interacting material substances via a principle of coexistence, which differentiates his view from the Wolffians' system of *physical influence*. The hybrid concept of force of 1747, too, continued to have an effect on the 1755/1756 cosmology and atomistic theory of matter. The *Theory of the Heavens* employs animist metaphors in order to express the motions of celestial bodies and atoms according to the laws of Newton's mechanics. There, Kant denominates the attractive and repulsive forces of matter as a "source of life" or as principles of self-organization, according to which matter develops from an initial state of absolute chaos to stars and galaxies:

> The elements have essential forces to put each other into motion and they are a source of life for themselves. Matter immediately endeavours to form itself. The dispersed elements of the denser type collect all the matter of lesser specific weight from a sphere around themselves by means of attraction [...]. Nature, however, has still other forces in store which are expressed primarily when matter is dissolved into its particles, by which forces they can repel one another and, by their conflict with the attractive force, bring about that motion that is, as it were, a continuous life in nature. (1:264–265)

The earlier concept of intension, however, is eliminated from the 1755/1756 concept of force.[6] Now Newton's concept of the *vis insita* or *vis inertiae* seems to play a role in it, as a principle which in Kant's 1755/1756 writings serves to form the bridge between Leibniz's and Newton's dynamic principles. This is indicated in the preface to the *Physical Monadology* by Kant's remark that

---

[5]Unlike Vuillemin (1955), but in conformity with Adickes (1924), Friedman (1992, 12 n. 9), too, emphasizes that Kant in 1755/1756 combines Newton's view of a *real* (and not just ideal) space and Leibniz's view of a *relational* space.

[6]Hermann Cohen, in his book on the infinitesimal method, claims that this concept of 1747 is still operative in Kant's critical concept of an intensive magnitude (Cohen 1883, Part II, § 77). Given that Kant revised his concept of force in the writings of 1755/1756 and 1758 (see below), this is improbable. The concept of an intensive magnitude is rather an alteration of the 1747 concept, which seems to be rooted in Wolff's doctrine of *matheseos intensorum*. See also n. 3 above and Kant's notes 3571, 3839, or 4050 on metaphysics (17:64, 17:308, 17:398).

> [...] the principle of all internal actions, in other words, the force which is inherent in the elements, must be a moving force, and one, indeed, which operates in an outward direction, since it is present to what is external [...] (1:476)

Here, Kant identifies the "force which is inherent in the elements" (*vim elementorum insitam*) with a *moving* force, which acts *externally*. In the *True Estimation*, on the contrary, he had sharply criticized the concept of a "moving force" (§ 2; 1:18) and he distinguished the motion of a body as an "external phenomenon" from the "intension" as the "basis of activity" (§ 117; 1:246). The *Physical Monadology* dispenses with this distinction, which was presumably related to Leibniz's account of primitive and derivative forces.[7]

In 1758, in his *New Doctrine*, Kant also criticized Newton's concept of *vis inertiae*, claiming "that the force of inertia has been invented unnecessarily" (2:20). Henceforth he would eliminate any metaphors of living forces as sources of life from his theory of nature, and in particular, from the excerpt of the *Theory of the Heavens* which he authorized to publish in 1791 (Herschel and Gensichen 1791; Ferrini 2004). In 1758 Kant no longer takes up the *vis viva* controversy, presumably due to his former lack of success. Perhaps by then he had become aware of d'Alembert's resolution.[8] His new insight was accompanied by criticism of Newton's concept of absolute motion, which he had used in his 1755/1756 theory of matter without questioning it. In 1786, in the "Phenomenology" part of the *Metaphysical Foundations of Natural Science*, he replaces Newton's concept of absolute motion by the weaker concept of "true" motion, which is based on a relational concept of space and the dynamic difference between rectilinear uniform motion and circular motion (*MFSN*, 4:556–558). In the "Dynamics" part of the *MFNS*, no trace remains of his earlier views about the attractive and repulsive forces as vital principles of matter. Quite on the contrary, in his critical theory of nature Kant considers the law of inertia to be a principle of *lifelessness*, which indicates the inability of material bodies to change their state of motion on their own. In the "Mechanics" part of the *MFNS* he explicitly identifies inertia with the "lifelessness" of matter and rejects any conflation of this principle and the concept of life (4:544).

[7]In 1756 Kant no longer maintains this distinction. This is due to his criticism of Leibniz in the *New Elucidation* of 1755, and in particular the *principle of succession* established there. This principle is incompatible with Leibniz's doctrines of monads and of pre-established harmony. According to Kant's new view, a substance may only undergo changes due to the nexus with *another* substance and not on its own. For this principle and the criticism of Leibniz associated with it see Laywine (1993, 32–35) where it is also suggested that Kant's proof of it is related to the possibility of an objective temporal order in the world.

[8]The dispute continued for decades, however. In the history of scientific concepts it continued to have an effect up to the late nineteenth century. In particular, Hegel's natural philosophy of the 1830 *Encyclopedia* uses a concept of "absolutely free" motion (Hegel 1830, § 268) which is based on distinctions similar to those rejected by Kant in 1758. Indeed, this contributed substantially to the odd impression which Hegel's philosophy of nature gave to later readers. Meanwhile, in physics, kinetic energy was commonly called "living force" until the end of the nineteenth century—Mach, for example, uses this expression throughout his 1883 book (Mach 1883), and not just in his historical account of the origins of the *vis viva* controversy in Descartes's and Leibniz's writings.

1.3 Parting from Eclecticism

In 1755/1756, Kant strengthened his unsatisfactory reconciliation principle of 1747 by adding a *systematicity requirement*. With the *Theory of the Heavens*, the *New Elucidation*, and the *Physical Monadology* he wanted to reconcile Wolff's system of metaphysics with the principles of Newton's physics. Wolff's system had four parts: a *metaphysica generalis*, which contained the general principles of metaphysics, such as Leibniz's principle of sufficient reason; and a *metaphysica specialis* with three sub-disciplines, "rational cosmology", "rational theology", and "rational psychology", which took as their subject matter the World, God, and the soul, respectively.

The metaphysical controversies between Wolffians and Newtonians primarily concerned the relation of Newton's physics to the "rational cosmology" of Wolff's metaphysics. But there were additional controversial issues, in particular concerning theology, as the Leibniz–Clarke debate showed. The philosophical attempts of the time to settle the debates were eclectic, arbitrarily combining certain elements of the respective opposing theories, rather than attempting to establish the principles of a unified theory or metaphysical system in Wolff's sense.

Against this eclectic background, Kant was far from sketching the plan of a metaphysical system in 1755/1756. To escape from eclecticism was a difficult task. He set himself the goal of framing a systematic and all-embracing *theory of everything* encompassing God and the World, which linked cosmology (as a theory of the constitution of matter, physical forces, and the universe as a whole) to rational theology (i.e., the doctrine of God), rational psychology (i.e., the doctrine of an immortal soul), and the general principles of Wolff's metaphysics (above all, Leibniz's principle of sufficient reason). Nor did Kant at this point use the expression "system" in the sense of an architectonics of metaphysics. In his pre-critical writing, the term "system" is above all used in the sense of a metaphysical or scientific doctrine, and sometimes also of a classification system. The *Theory of the Heavens* furthermore brings into play the structure of the solar system (see Appendix A.1.3).

In his reaction to the academic controversies of the 1750s, however, he only succeeded step by step to escape from contemporary eclecticism. In addition, he was dealing with the topics of these controversies as an unknown young scholar in Königsberg: worrying about the conflict between his own views and contemporary Pietism, he published the *Theory of the Heavens* anonymously (Schönfeld 2000; Calinger 1979). Even though he did not sketch the plan of a system of metaphysics in Wolff's style in 1755/1756, he must have already had such a system in mind. Important hints to his pretensions are given by the scattered ambitious remarks in his pre-critical writings, comparing himself in particular with Descartes, for example in the preface to the *Theory of the Heavens* (1:228). Indeed, his 1755/1756 writings represent his first moves in a systematic direction. As Schönfeld (2000) observed, the cosmology of the *Theory of the Heavens*, the metaphysical principles of the *New Elucidation*, and the matter theory of the *Physical Monadology* do not simply complement each other in various respects; rather, they represent a threefold systematic attempt to establish crucial links between the parts of a Wolffian system of metaphysics. The *New Elucidation* aims to settle the metaphysical foundations of such a system; the *Theory of the Heavens* outlines a physical cosmology that makes the bridge to rational theology; and the *Physical Monadology* makes the bridge between this cosmology, which is based on an extended Newtonian mechanics, and a Wolffian theory of physical monads.

In 1747, Kant had forced the combination of the rival dynamic concepts of Newton's and Leibniz's physics in an ad hoc way, based on a hybrid concept of force, by simply restricting the applicability domains of the dynamic concepts at play in the debate. In 1755/1756, he attempted to realize his reconciliatory project in a more coherent way. To do so, he needed a uniform concept of force which was thoroughly compatible with the mathematical description of mechanical phenomena. The systematic basis for this more coherent approach to a metaphysics of nature, or the cosmology part of an all-embracing system of metaphysics, was provided by his very first metaphysical writing, the *New Elucidation* (presented in 1755 in a public dispute at the faculty of Königsberg).

Thus, the *Physical Monadology* of 1756 does not simply continue the reconciliatory project of the *True Estimation*. It belongs to a new, systematic approach to reconciling Newton's physics and Wolff's metaphysics, which was much more ambitious. Kant now proposes to supply metaphysics with new principles, which serve to resolve not just a particular philosophical debate, but rather the whole gamut of contemporary metaphysical controversies. To do so, the *New Elucidation* already takes some decisive steps towards dispensing with the principles of Wolff's metaphysics.[9] Indeed, a non-eclectic unification of the incompatible doctrines of Wolff and Newton could not be achieved without substantially modifying *both* positions.

1.3.1 A First Systematicity Requirement

A first *systematicity requirement* appears in the *Theory of the Heavens*, which substantially strengthens the reconciliation principle of 1747. His further elaboration of this principle would henceforth distinguish Kant from contemporary eclecticism. Balancing the reasons for the law-like organization of the celestial bodies according to the principles of Newton's theory of gravitation, on the one hand, and the problem that in empty space gravitation seems to be an action at a distance, on the other, he strengthens his weak reconciliation principle of 1747 as follows:

> An impartial examination shows that the reasons are equally strong on both sides and both are to be regarded as being completely certain. However, it is just as clear that there must be

[9]Walford (1992) emphasizes that the *New Elucidation* attacks Wolff's position. In contrast, the otherwise instructive comparison of the *True Estimation* and the *Physical Monadology* by Hinske (1970, 42) neglects the systematic rupture between the former and the latter.

> a concept in which these apparently mutually conflicting reasons can and should be united and that we may seek the true system in this new concept. (1:262)

Here, he presents himself as convinced of the existence of an integrative concept which gives rise to a "true system", that is, a coherent cosmological theory capable of unifying the opposing theoretical doctrines. And then he proposes his theory of the genesis of the solar system and its extrapolation to a natural history of the universe, which is based on a theory of structure formation in the early universe with the initial state of a matter distribution as a rotating nebula.

This "true system" is a physical theory that embraces the whole universe. It has to be understood against the background of the prominent opposing metaphysical doctrines of the time. In the preface to the *Theory of the Heavens*, Kant discusses the positions of the "free thinker" and the "defender of religion". He associates the former with Epicurus's atomism (1:222), and hence with a materialistic or naturalistic world view, which he characterizes as "blind mechanism". The corresponding contemporary position was French materialism, which explained the law-like structure of the solar system in naturalistic terms. It was represented in particular by La Mettrie, who had become a member of the Berlin Academy in 1748 (Calinger 1969). The opposing view in the dispute is neither Wolffianism nor the teleological view of nature proposed in the preface of the *Theory of the Heavens*, but the Newtonian view that only repeated actions of God explain the stability of the planetary motions. According to Newton and his followers, God created the world, but has still to intervene in the course of the universe in order to counteract the dissipation of the kinetic energy of the celestial bodies and keep the planetary motions stable.[10] They explain the initial conditions of the law of gravitation, i.e., the mass, position, and velocities of the celestial bodies in the solar system at a given time, by interventions of God rather than by the laws of physics.

Hence, the higher-level goal of the *Theory of the Heavens* is to resolve the metaphysical controversy between the French defenders of materialism (free thinkers, like La Mettrie) and the British defenders of divine intervention into the world (apologists of religion, like Clarke, who shared Newton's views). Given that both referred to Newton's mechanics in support of their arguments, Kant seeks an integrative Newtonian conception of the processes in the universe, from which the arguments of both parties derive. The means to do so, in Kant's view, is already provided by Newton's mechanics, or (more precisely) by an extension of it. The unifying concept or principle is Newton's concept of universal gravitation, and the "true system" is the theory of structure formation in the universe according to the laws of Newton's mechanics. In Kant's view, Newton's theory is stronger and has a larger scope than Newton himself and his followers were aware. Accordingly, he complements it by the assumptions of repulsive forces and by adequate initial conditions. His own theory of structure formation in the universe fills the explanatory gaps in Newton's theory of the solar system, which made Newton and his followers resort to divine intervention. This approach combines anti-Newtonian tendencies

[10] See Leibniz and Clarke (1715/1716).

with Newtonian commitments (Watkins 2013). In order to justify this "true system" (which is indeed a landmark theory of physical cosmology), Kant extrapolates the "systematic constitution" of the solar system (that is, the approximate restriction of the planetary motions to the ecliptic) to the large-scale structure of universe, and to the constitution of the universe as a whole (*Theory of the Heavens* 1:290–291; see Sect. 2.2.1 and Appendix A.1.3). On the other hand, Kant insists that his way of extending Newton's theory to larger and larger scales, up to the entire universe, does not commit him to materialism or naturalism.

In this way, he strengthens the conciliatory commonplace of 1747 (i.e., to concede points on both sides, and look for the truth in between) by the requirement to do so in a systematic way, according to uniform principles. The reconciliation principle of 1755 represents a quest for a unifying conception which supports the best reasons *in favour of* the claims of the two opposing doctrines.

This requirement is contrary to Kant's later methodological principle of finding the strongest reasons *against* a position, the *skeptical method* (see Sect. 4.3.2). The skeptical method is the procedure of countering a position in order to either defeat or confirm it by dispelling all conceivable objections. It is a touchstone for the tenability of arguments and reasons which are to be integrated into a coherent theory according to a unifying principle. Kant did not realize before the mid 1760s that he was in need of such a touchstone. When Swedenborg's writings drew his attention to the unwanted spiritualist consequences of his 1755/1756 system, he wrote the *Dreams* of 1766 (see Sect. 3.2), which Reich (1958, XII) and Kaulbach (1982, 100) consider as the first demonstration of the skeptical method. A further, decisive example of it was Kant's use of the "analytic" method in the *Directions of Space* of 1768, which demonstrated that even with regard to the concept of space the system of 1755/1756 could not be the "true system" (see Sect. 3.4 and Chap. 4).

1.3.2 The Project of Unifying Physics and Metaphysics

Prior to this turning point, for a decade he presumably thought that he had managed to reconcile the principles of Wolff's metaphysics with the laws of Newton's physics. In all modesty, he was convinced he had mastered the difficult metaphysical task mentioned at the beginning of this Section (1.3). At the end of the *Preliminary considerations* of the *Physical Monadology*, he noted

> [...] that anyone who is able to deduce these two principles from the very nature and fundamental properties of the elements will have made a substantial contribution towards explaining the inner nature of bodies. (1:476)

The "two principles" in dispute are the opposite doctrines he there wanted to unify, namely the infinite divisibility of a space filled with matter, and the internal forces of simple material substances (see Sect. 1.4.3). The self-assessment indirectly expressed here was not unjustified. The pros and cons of Wolff's metaphysics and Newton's physics conflicted at several levels and came in several theoretical variants.

1.3.2.1 A Multilayered Problem

From a systematic point of view, Kant faced a multilayered problem. The application conditions of the concepts of mathematical physics were at the same time adequacy conditions of the metaphysical principles underlying rational cosmology. As a part of the *metaphysica specialis* of Wolff's system of metaphysics, cosmology ought to be compatible with the principles of general metaphysics. But in view of the success of Newton's mechanics, it also had to be compatible with the principles of physics as an empirical science. The reconciliation of Newtonian physics with a Wolffian system of metaphysics entailed three levels of theory formation *(horizontal unification)*, and, in addition, their deductive connection *(vertical unification)*:

1. *Mathematical description (geometry):*
   At this level, two conflicting features of Newton's mathematical description of the phenomena, on the one hand, and Wolff's geometrical account of material substances, on the other, had to be reconciled.
   (i) *Atoms vs. continuum*: The continuum properties of space filled by matter emphasized by Euler were in conflict with Wolff's assumption of point-like monads. This is the question Kant wanted to settle with the *Physical Monadology*. The conflicting positions were, however, exactly the other way round in the Leibniz–Clarke correspondence. Leibniz had defended the infinite indivisibility of matter (which Euler now defended on Newtonian grounds, in accordance with his approach to continuum mechanics), and Clarke Newton's atomism (which proposed impenetrable corpuscles, in contrast to Wolff's point-like physical monads).
   (ii) *Absolute vs. relational concept of space and time*: To this complex field of debate was added the controversy about the absolute or relational nature of space and time, which was also initiated by the Leibniz–Clarke correspondence. On this point, Wolff shared Leibniz's criticism, and Kant was on the side of Leibniz and Wolff.

2. *Dynamical principles (metaphysica specialis):*
   At this level, the dynamic principles of Newton's physics had to be reconciled with the cosmology part of Wolff's metaphysics.
   (i) *External vs. internal forces*: Newton's concepts of inertia and external forces were incompatible with Leibniz's and Wolff's conception of internal forces. Within physics, d'Alembert had resolved the *vis viva* dispute exactly in the sense of the systematicity requirement of Kant's *Theory of the Heavens* (see Sect. 1.3.1), that is, by reducing both measures of force to different aspects of the principles of Newton's mechanics. But this solution did not resolve the question of how both concepts relate to Wolff's cosmology and his conception of monads.
   (ii) *Action at a distance vs. physical influence*: On the side of Newtonianism, there was the additional dispute of whether forces are actions at a distance or whether gravitation is due to immediate interactions of mechanical ether particles, as proposed in Descartes's vortex theory, a theory of close-range

effects. On the side of the Wolffians, a variant of the latter, the system of physical influence (*influxus physicus*) came into play. All of these approaches were in conflict with Leibniz's and Wolff's doctrine of pre-established harmony. Kant rejected them all. He wanted to settle the question with his nebular theory of structure formation in the universe, in the *Theory of the Heavens*.

3. *General principles (metaphysica generalis):*
For the foundations of a general metaphysics which could cope with the dynamic and cosmological problems of the second level, four rival systems or metaphysical doctrines were under consideration in the middle of the eighteenth century:
(i) *Divine interaction*. This is the doctrine that the world depends on immediate actions of God, and it came (in our context) in two principal versions. Malebranche assumed that God permanently acts on the world in order to coordinate mind and body (occasionalism). Newton as well as his followers Clarke and Cotes (the author of the preface to the second edition of Newton's *Principia*) believed that God created the initial conditions of the celestial bodies and their motions in the solar system, and has to act upon the world from time to time in order to keep the planetary motions stable. In addition, Newton considered absolute space to be the *sensorium dei*, which guarantees the omnipresence of God within the world.
(ii) *Pre-established harmony*. According to this system of Leibniz and his followers Wolff and Baumgarten, there is no real interaction between the substances in the world, but their apparent interactions are pre-determined by the internal properties and forces of the monads.
(iii) *Physical influence (influxus physicus)*, a rival to Leibniz's system of pre-established harmony. This metaphysical doctrine was associated with the theory of close-range effects mentioned above. Around 1750 it had become accepted by several Wolffians, in particular by Kant's academic teacher Martin Knutzen (Erdmann 1876).
(iv) *Materialism*. The French encyclopedists put forth the naturalistic or materialistic interpretation of Newton's mechanics, according to which no God is needed in order to explain the laws of nature. In particular, d'Alembert, Diderot, d'Holbach, La Mettrie, and Voltaire defended materialism on the grounds of Newton's physics.

Levels (1) and (2) are intertwined. Levels (2) and (3) are, too. Therefore, none of Kant's three 1755/1756 writings deals only with questions that are specific to one level. All of them also aim to establish links between the levels. In particular, the nebular theory of the *Theory of the Heavens* suggests a way out of the conflict (2.ii) of actions at a distance vs. Descartes's mechanistic vortex theory, but also out of the conflict (3.i) vs. (iii) of divine intervention vs. physical influence. In addition, the preface of the *Theory of the Heavens* emphasizes that the theory helps to refute the position of materialism (3.iv). In the *New Elucidation*, Kant attempts to reformulate the principles of Wolff's metaphysics at level (3) in such a way that they become compatible with the mathematical and dynamic principles of levels (1) and (2). In the *Physical Monadology*, he directly tackles the project of unifying physics and metaphysics, i.e., the link between levels (1) and (3) via level (2).

Hence, the task of reconciling a system of metaphysics in Wolff's style with the principles of Newton's physics was complex. A synthesis of Newtonianism and Wolffianism was needed in both dimensions, horizontally as well as vertically. Horizontally, the tenable principles of Newton's physics on the one hand and Wolff's cosmology on the other had to be unified into a consistent theory of matter and the universe. Vertically, the mathematical and dynamic principles of Newton's physics and a cosmology in Wolff's style had to be brought into deductive logical relations with each other and with the principles of the *metaphysica generalis*; and in particular with Leibniz's *principle of sufficient reason*.

It should be noted that Kant's pre-critical project of unifying physics and metaphysics employs both disciplines in at least a two-fold sense. On the one hand, 'physics' means *Newton's physics* together with its metaphysical presuppositions; and 'metaphysics' means the whole system of *Wolff's metaphysics* including its cosmological implications. On the other hand, 'physics' means the required physical theory of the universe, which should result from reconciling the tenable assumptions of Newtonianism and Wolffianism and could be embedded into the rational cosmology part of Wolffian metaphysics; and 'metaphysics' means the general metaphysical principles which are the foundations of such a theory.

1.3.2.2 The Adequacy Conditions of Cognition

In both dimensions, horizontally as well as vertically, views about space and time and the concept of force are at the interface of physics and metaphysics. Kant was familiar with the arguments of the Leibniz–Clarke correspondence, which link the controversy about the absolute or relational nature of space and time to theological questions. According to Baumgarten's metaphysics, the concepts of space, time, and force belong to metaphysics. But space was also a mathematical concept for Kant, according to § 1 of his *Prize Essay* published in 1764:[11]

> The mathematician deals with concepts which can often be given a philosophical definition as well. An example is the concept of space in general. (2:277–278)

The concepts of space, time, and force ought to link physical cosmology to the metaphysical system on which they are grounded. Physics deals with the mathematical description of the phenomena in terms of forces, which metaphysics wants to reduce to the dynamic primary causes of phenomena. Hence, Kant's

[11]See also Kant's notes on metaphysics from phases κ and λ (1769–70). The notes on space and time are correlated with the sections *Totale et partiale, Prima matheseos intensorum principia, Simplex et compositum, Finitum et infinitum, Simultanea* and *Successiva*, which belong to the ontology, not the cosmology part of Baumgarten's metaphysics, in particular to the doctrine of the disjunctive internal predicates of things (17:19).

unifying project was also in need of an *adequacy criterion*, in addition to the systematicity requirement.

This criterion, namely the distinction between the *logical* and the *real* use of reason (see Sect. 2.5), emerges stepwise in Kant's pre-critical writings. In the *New Elucidation*, Kant distinguishes ontic and epistemic grounds, as a first indication of the need he sees to distinguish metaphysical causes and effects, from mere logical grounds and consequences (see Sect. 1.4.2.1). It has to be noted that neither this 1755 distinction nor his later analysis of the *real* use of reason reduces merely to a criterion of *empirical* adequacy, as twenty-first-century philosophers of science may be inclined to think. Kant's later distinction between logical and real grounds distinguishes between logic and metaphysics, not theory and experience. Its 1755 predecessor differentiates the ontic and epistemic grounds of metaphysical arguments, on the one hand, and mere logical judgments on the other. Both distinctions concern the respective scope of logical principles such as the principle of contradiction and Leibniz's principle of sufficient reason (see Sects. 1.4.2, 2.2.2, and 2.5).

At this point let me come back to the legacy of the *True Estimation* emphasized above (in Sect. 1.2). All his life, Kant thought that the conditions under which fundamental physical concepts apply are a matter of metaphysics, not of empirical science. For him, such a criterion was not empirical, but metaphysical. This was already the case long before his critical approach to clarifying the transcendental conditions of the possibility of experience. In this regard, Kant's pre-critical as well as his critical theory of nature substantially differs from empiricism.

Indeed it is easily shown that Kant's pre-critical adequacy conditions of cognition were not just empirical. Together with the rival metaphysical views of Wolffianism and Newtonianism, Kant faced two theories of cognition, and with them, two competing adequacy criteria (Appendix A.3.4). In Wolffianism, they were conceptual. Here, Leibniz's essay *Meditationes de Cognitione, Veritate et Ideis* (Leibniz 1684) was most influential. According to it, adequate ideas are completely clear and distinct; that is, they are concepts which specify all attributes of an object, or which give the *complete notion* of an object. An adequate idea gives the real definition (and not only the nominal definition) of an object (Appendix A.3.2). According to Leibniz's epistemology, the complete notion or real definition of a phenomenon of nature is not available to us. Hence, in Leibniz's view, the concepts of physics cannot be adequate. They are mathematical abstractions, which neglect most of the internal dynamic properties of the substances or monads underlying natural phenomena. Leibniz's principle of the identity of indiscernibles claims that there are *infinitely many such properties*, and that no parts of matter have identical internal properties.

Newton's adequacy criteria, however, were not empirical either, as a simplifying empiricist interpretation of Newton's scientific methodology may suggest. Newton's philosophical views had not much in common with seventeenth-century empiricism, nor did Kant shift from rationalism to empiricism by integrating Newtonian views into a Wolffian metaphysics. Newton's criteria for the adequacy of physical knowledge were methodological. They are expressed in Newton's famous *Rules of*

Reasoning in *Book III* of the *Principia* (see Appendix A.2.2.2). Newton's first rule claims:

> No more causes of natural things should be admitted than are both true and sufficient to explain their phenomena. (Newton 1726, 794)

Neglecting the truth condition, this is a principle of ontological parsimony which is in conflict with Leibniz's principles of sufficient reason and the identity of indiscernibles. Obviously, it is not a criterion of empirical adequacy but a methodological rule. I will show in the next Chapter (Sect. 2.2.2) how Kant makes tacit use of it, in support of his criticism of the scope of Leibniz's principles in the *New Elucidation*.

Newton's methodological rules, the adequacy criteria of empirical science associated with them, and the way in which Kant employs them are analyzed below in Section 2.1. Kant's way of establishing the principles of his pre-critical metaphysics is based on *analytic methods*, which are the subject of (Chap. 2). Before entering into these subtle methodological issues, let us have a closer look at the links established between the parts of a Wolffian system of metaphysics by Kant's 1755/1756 writings.

1.4 The Pre-critical System

Kant never worked out in detail the system of metaphysics in Wolff's style to which his 1755/1756 writings provided the foundations. His pre-critical philosophy is systematic in argument, but not in presentation. The *Theory of the Heavens*, the *New Elucidation*, and the *Physical Monadology* establish horizontal and vertical links between the parts of such a system, arguing for metaphysical principles that reconcile the conflicting metaphysical positions of the time on non-arbitrary middle grounds. Kant now arrived at these middle grounds by systematic reasoning, instead of his former commonplace that the truth lies in between. The *Theory of the Heavens* links the cosmology and the theology parts of a Wolffian system horizontally. The *New Elucidation* aims at establishing vertical links, in order to make the *metaphysica generalis* of the system compatible with the Newtonian principles of the *metaphysica specialis*. The *Physical Monadology* on the one hand establishes a horizontal link within the cosmology part of the system, namely between matter theory and the theory of structure formation in the universe as a whole. On the other hand, it aims at vertical integration by demonstrating that his theory of matter is indeed compatible with the metaphysical principles derived in the *New Elucidation*.

1.4.1 The Physico-Theological Argument

In the first of the three 1755/1756 writings, the *Theory of the Heavens*, Kant demonstrates *in concreto* how the principles of Newton's physics may be extended to a Wolffian cosmology. In doing so, Kant already presupposes the atomistic theory of matter and the dynamics which the *Physical Monadology* of 1756 will justify.[12] In the preface of the *Theory of the Heavens*, Kant points to the atomistic foundations of his cosmology by admitting

> [. . .] that Lucretius' theory or that of his predecessors Epicurus, Leucippus, and Democritus has much in common with mine. (1:226)

An atomistic theory of matter belongs to physics, on the one hand, and to the cosmology part of Wolff's system, on the other. In addition, Kant wants to establish a link from cosmology to theology. In order to do so, he emphasizes that he only shares atomism with his ancient predecessors, but not their materialistic or naturalistic metaphysics, i.e., atheism. According to the preface, the goal of the *Theory of the Heavens* is to find a position in between naturalism and natural theology, on the middle grounds between the arguments of the materialist "freethinker" and the "defender of religion". At the beginning of the preface of the *Theory of the Heavens*, Kant opposes both positions as follows:

> [. . .] religion threatens us with a solemn accusation for the audacity with which one might make so bold as to ascribe to nature, which is left to itself, such consequences in which one can rightly become aware of the immediate hand of the highest being, and is concerned to find protection for the atheist in the forwardness of such observations. (1:221)

The free thinker attributes to "nature [. . .] left to itself" the self-organization of matter in the universe according to mere mechanical laws, whereas the defender of religion explains the structure of the universe by immediate action of God. The position in defence of religion addressed here is the interventionism of Newton and his followers, who explain the stability of the planetary motions by repeated actions of God; whereas Kant explains it by his theory of structure formation in the universe (see Sects. 1.3.1 and 2.2.1). In the preface of the *Theory of the Heavens*, Kant argues for a position on the middle ground which is based on Newton's universal law of gravitation, or, more precisely, on the *universality* of gravity and other forces, which makes it possible to

> [. . .] discover the system that connects the great parts of creation in the whole extent of infinity, to derive the formation of the celestial bodies themselves and the origin of their motion out of the first state of nature through mechanical laws [. . .]. (1:221)

Newton's mechanics is the common ground of the conflicting positions. Its principles serve to reconcile them, in accordance with the systematicity principle given at the beginning of Part Two of the *Theory of the Heavens* (1:262; see Sect. 1.3.1). In the field of physics, Kant employs the theory of structure formation in the universe. To make the bridge to Wolff's cosmology, he employs a *physico-theological*

[12] Massimi (2011) shows in detail how Kant's dynamic atomism of the 1755 writings (including *On Fire*) trace back to the tradition of the speculative Newtonian experimentalism, i.e., to the British and Dutch followers of Newton's *Opticks*. This background fits in well with Kant's use of the Newton's analytic method in the *Theory of the Heavens*, which also traces back to the *Queries* of Newtons *Opticks*; see Sect. 2.2.1 and Appendix A.2.2.2.

argument. He reconstructs the history of the solar system and of the universe as a whole in order to demonstrate that the current structure of the universe evolved from a primordial chaotic matter distribution as its uniform cause, obeying Newton's laws. And then he traces this uniform natural cause back to God, who created the primordial state of the universe and the laws of nature. His summary of this physico-theological argument is as follows:

> Matter, which is the original material of all things, is thus bound to certain laws, if it is left freely to these laws, it must necessarily bring forth beautiful combinations. It is not at liberty to deviate from this plan of perfection. Since, therefore, it is subject to a most wise purpose, it must necessarily have been placed into such harmonious connections by a first cause that ruled over it, and *a God exists precisely because nature cannot behave in any other way than in a regular and orderly manner, even in chaos*. (1:228)

This is a teleological argument. In this point, Kant follows the Newtonian defenders of religion. These, however, interpret God's intention in terms of action against nature's tendency to get into a state of disorder:

> This harmony, people say, is foreign to it; left to its own universal laws, nature would bring about nothing but disorder. These harmonies point to a foreign hand that has been able to force a wise plan onto matter devoid of all regularity. (1:223)

In contrast to their interventionism, Kant proposes that God's plan only came into play at the beginning of the universe, by establishing the laws that bring about the purposeful order of nature:

> But I answer: If the universal laws of causation of matter are also a result of the highest plan, then they can presumably have no purpose other than that which strives to fulfil of their own accord that plan which the highest wisdom has set itself [...] (1:223)

In order to justify his argument, he employs his theory of structure formation in the universe, which predicts that according to the laws of nature matter should develop from primordial chaos to the present order of the universe. In addition, he reminds his readers of Descartes, as a most prominent predecessor whose metaphysics shows that a mechanistic corpuscular philosophy is not committed to materialism:

> Furthermore, I will not be deprived of the right that Descartes always enjoyed from fair judges when he dared to explain the formation of the heavenly bodies from purely mechanical laws. (1:228)

Kant's 1755 cosmology suggests that it is possible to explain the structure of the universe in terms of a *uniform lawful principle*, and hence there is no need to claim repeated interventions of God, as the Newtonians did. The uniform laws of nature give rise to structure formation in the universe, and finally to a "systematic constitution of the universe" (1:246) as an all-embracing system of celestial bodies organized in accordance with Newton's laws. The uniform principle behind the laws of nature is God, as the supreme legislator. Hence, Kant's physico-theological argument makes a stronger claim than just demonstrating that his theory of structure formation in the universe is logically compatible with theology. It comes along as an *inference to the best explanation*, which traces the "systematic constitution of the world" back to God as its ultimate cause. Kant's justification of this inference to God as the best explanation of the structure of the universe becomes more comprehensible against the background of Newton's *analytic-synthetic method* and the way in which Kant makes use of it (see Sect. 2.2.1). In his later writings, he will no longer accept this kind of inference to ultimate causes. In the *Only Possible Argument* of 1763 he already emphasizes that it lacks mathematical rigour (see Sect. 2.4), and in the *CPR* he argues that the quest for ultimate conditions gives rise to the cosmological antinomy.

1.4.2 The Scope of Leibniz's Principles

The *New Elucidation* is much more than an unoriginal, academic writing complying with the standards of Wolffianism. Its innovative features have been emphasized in the Kant literature (Walford 1992, l–li). Kant attacks several contemporary philosophical positions, and in particular some principles of Wolff's metaphysics that are in conflict with Newton's physics. In order to provide the foundations of a Wolffian system of metaphysics, Kant had to make a bridge from the principles of Wolff's *metaphysica generalis* to the principles of a *metaphysica specialis*, which is compatible with Newtonian physics. The *New Elucidation* serves to clarify how it is possible to do so. For his systematic task, the following differences between Newtonian principles and Leibniz's or Wolff's metaphysics are crucial. Here I simply bring these differences together in summary form. His way of justifying them is analyzed in the next chapter (in Sect. 2.2.2).

1.4.2.1 Ontic and Epistemic Grounds

Kant redefines Leibniz's principle of sufficient reason, or Wolff's formulation of it, in terms of the "determining ground". He first explains that a ground "determines a subject in respect of any of its predicates" (1:391). Next, he distinguishes antecedently and consequentially determining grounds. Logically, the distinction concerns the predicates in the antecedent and the consequent of a hypothetical judgment "If S is A, then S is C". The "*antecedently* determining ground" is the predicate A of the antecedent, and the "*consequentially* determining ground" is the predicate C of the consequent (1:392). Unlike his rationalist predecessors, Kant then notes that in metaphysics this logical distinction is not sufficient. And then he makes a first important step towards his later emphasis on the *real use* of reasons which is characteristic of metaphysics, as distinguished from logic. He explains the antecedent predicate A in terms of the "reason *why*", or ontic ground (*ratio essendi*), and the consequent predicate C in terms of the "ground *that*" or epistemic ground (*ratio cognoscendi*) (1:392–393). The distinction is closely related to truth conditions:

> [...] it follows that the determining ground is not only the criterion of truth; it is also its source. (1:392)

He illustrates the distinction by giving the example of the Jupiter satellites. Their eclipses are the ground of knowing the finite speed of light. However, if Jupiter had no satellites or if they had no eclipses, than the speed of light would be the same, even though unknown to us. Hence the eclipses are a consequence of the finite speed of light,

> [...] without which these phenomena could not occur in the way in which they occur. It follows, therefore, that they determine this truth only consequentially. (1:392)

In contradistinction to this epistemic ground, the ontic ground of the finite speed of light is subject to a theoretical explanation such as the light propagation in an elastic medium. Kant refers to Descartes's theory of light, according to which light propagation is due to the collisions of corpuscles (1:393), and adds:

> This would be a ground which determines antecedently. In other words, it would be a ground such that, were it not posited, that which was determinate would not occur at all. (1:393)

In this example, ontic and epistemic ground relate to each other as the cause and the effect of the velocity of light. The Cartesian mechanism of light propagation assumed by Kant is (or may be) the cause of the finite speed of light, and the eclipses of the Jupiter satellites are its effects. In the 1760s, Kant will explicitly distinguish causes and effects as real grounds and consequences from mere logical ones (see Sect. 2.5); but in 1755, he does not yet do this. However, he does differentiate the grounds of truth in metaphysics from mere logical arguments. Given that Leibniz's principle of sufficient reason deals with the former, not with the latter, he criticizes it and replaces it by a purely logical principle (see Sect. 2.2.2).

1.4.2.2 The Principle of Indiscernibles

The *New Elucidation* argues that Leibniz's interpretation of the *principle of the identity of indiscernibles* is a "spurious" corollary [...] incorrectly derived" from Leibniz's principle of sufficient reason. According to Kant, Leibniz wrongly concludes that

> [...] *there is no substance in the entire totality of things which is in all respects like any other substance.* This principle is called the principle of indiscernibles. Taken in its widest sense, as it usually is, it could not be further from truth. (1:409)

The usual reading of Leibniz's principle, according to which there exist no indistinguishable substances with identical properties, is in obvious conflict with atomism. Kant wants to escape this conclusion by adding the position of things to their properties:

> The complete identity of two things demands the identity of all their characteristic marks or determinations, both internal and external. Is there anyone who has excluded place from this complete determination? Accordingly, no matter how great the agreement of things in

respect of their internal characteristic marks, things which are distinguished at least in virtue of their place are not one and the same things at all. (1:409)

The crucial difference from Leibniz's view is that Kant admits of internal as well as external properties or determinations of substances. This point is closely related to Kant's third criticism, discussed below (see Sect. 1.4.2.3), according to which the substances do not only have internal properties, but also external relations and interactions. In line with this view, he harshly criticizes Leibniz's argument against Clarke, which employs the principle of sufficient reason and the permutation symmetry of a constellation of three identical bodies or atoms:[13]

> It is constantly being said that if two substances agree completely in all other respects, then there is no reason why God should assign different places to them. What nonsense! It amazes me that grown men of the greatest gravity should take a delight in such frivolous arguments. Let the one substance be called A and the other B. Let A occupy the place of τoυ B. Since A does not differ from B at all in respect of internal characteristic marks, it follows that in occupying its place, it will be identical with it in all respects, and what was previously called A will now have to be called B [...] For this difference of characteristics indicates a difference only of places. (1:409)

Given that the positions and the respective indices A and B do *not* change, a permutation of the substances would not change their constellation. Hence, in Kant's view there is no real change that has to be explained. If there is no real change of the constellation, no ontic determining ground is at work. And thus, neither Leibniz's principle of sufficient reason nor Kant's principle of the determining ground have to be employed. Therefore, Kant considers the principle of indiscernibles to be compatible with the existence of two indistinguishable substances in the world, like atoms, which differ only numerically.

On the basis of the principles of 1755/1756, this conclusion seemed to be convincing. Kant supports it with additional phenomenological arguments which will be discussed in the next chapter (see Sect. 2.2.2). Nevertheless, when applied to the whole universe, the relational theory of space underlying his 1755/1756 writings was in conflict with the argument carried forward here—as Kant's 1768 article on incongruent counterparts shows (see Sect. 3.4). But in 1755/1756, Kant was not yet discussing the consequences of a Leibnizean relational account of space.

1.4.2.3 The Interactions of Substances

In Kant's view, there are also some "true" consequences which "derive from the principles of the determining ground" (1:410). These consequences are the principles of succession and of coexistence, the *principium successionis* and the *principium coexistentiae* (the pre-critical, "dogmatic" predecessors of the laws of causality and interaction of the *CPR*). Both principles are crucial for the foundations

[13]The first German translation of the Leibniz–Clarke correspondence appeared in 1720, with a preface written by Wolff (Leibniz and Clarke 1720).

of Kant's cosmology of 1755/1756. They serve to reject Leibniz's theory of pre-established harmony and to justify a modified version of the theory of physical influence.

According to the *principle of succession*, interaction or "reciprocal dependence on each other" is *necessary* for the change of state of the substances (1:410). According to the *principle of coexistence*, however, the mere existence of the substances alone is *not sufficient* for the interactions of substances. Kant argues that in addition an ontic ground of being, or coming into being, is needed. To be able to interact with each other, the substances need a "common principle of their existence", which according to Kant is God, or rather the divine idea of conceiving the correlations of the substances in the world:

> [. . .] it does not follow from the fact that God simply established the existence of things that there is also a reciprocal relation between those things, unless the self-same scheme of the divine understanding, which gives existence, has established the relations of things to each other, by conceiving their existences as correlated with each other. (1:413)

In this way, Kant justifies his version of a metaphysical system of *influxus physicus*, which differs from the respective contemporary theories. According to Kant, the physical influences between the substances, and in particular their interactions as described by the universal law of gravitation, are only sufficiently justified due to the existence of God (Friedman 1992, 6–7; Laywine 1993, 37–42). The view that the connections between the substances in the world are ultimately mediated by the "scheme of the divine understanding" (1:413) reconciles the systems of the Newtonians and the Wolffians in such a way that Kant again escapes materialism or naturalism. Indeed, the "scheme of the divine understanding" connects the theory of physical influence to a metaphysical interpretation of the principle of sufficient reason, which supports the physico-theological argument from design of the *Theory of the Heavens*. And in this way he even seems to be capable of reconciling Newton's view of absolute space as the *sensorium dei* with a relational account of space.

1.4.3 How to Mate Griffins and Horses

The *Physical Monadology* aims at reconciling the opposing cosmological positions of the Newtonians and the Wolffians by clarifying how the dynamic principles of matter relate to mathematical concepts. The topic was taken from the academic dispute on monadology initiated by Euler, which took place at the Berlin academy under the direction of Maupertius (Friedman 1992, 3–4). According to the full title, the goal of the *Physical Monadology* is to combine metaphysics and geometry for their employment in natural philosophy. Here, 'natural philosophy' is Newton's term for physics, as used in the title of the *Principia*. The following key sentence indicates that this goal is a most difficult task:

> But how [. . .] can metaphysics be married to geometry, when it seems easier to mate griffins with horses than to unite transcendental philosophy with geometry? (1:475)

Kant uses “transcendental philosophy” synonymously with “metaphysics”. The term is, however, ambiguous. It denotes either special metaphysics, i.e., the dynamic principles underlying physics; or general metaphysics, i.e., the conflicting metaphysical doctrines of the time; or both. According to Baumgarten's metaphysics, “transcendental philosophy” is the *metaphysica generalis*. Baumgarten's *Metaphysica* first appeared in 1737 and had several Latin and German editions. Kant used and commented upon the Latin edition of 1757 (Baumgarten 1757). There, Baumgarten calls the universal internal predicates of *unum, verum*, and *perfectum* “transcendentals” (§§ 73, 89–90, 98–99; 17:43–47). The ‘transcendentals’ are primary predicates denoting the essential attributes of all beings. According to Baumgarten, they belong to ontology. In Baumgarten's metaphysics, the fundamental concepts of a matter theory based on monadology, too, are introduced in general metaphysics. The concepts of *substantia et accidens, simplex et compositum, monas*, and *finitum et infinitum* on which the doctrine of monadology rests, belong to the ontology of Baumgarten's metaphysics.[14]

The *Preliminary Considerations* also contain some more general remarks about the relation between physics and metaphysics, including methodological remarks (see Sect. 2.2.3.1). Therefore, I suggest that Kant here underlines the vertical linkages within his whole systematic endeavour, which are intended to establish relations between all three levels of (1.) mathematics, or geometry, (2.) the dynamical principles which belong to the *metaphysica specialis*, and (3.) the general principles of the *metaphysica generalis* (see Sect. 1.3.2):

1. *Mathematical description (geometry):* Euler's continuum mechanics and Wolff's account of physical monads had to be reconciled, with regard to the opposition of *atoms vs. continuum*.
2. *Dynamical principles (metaphysica specialis):* The dynamic principles of Wolffianism and Newtonianism had to be reconciled, in particular with regard to the oppositions of *internal vs. external forces*, and of *physical influence vs. action at a distance*.
3. *General principles (metaphysica generalis):* The conflicting general metaphysical doctrines of the time, in particular the systems of *pre-established harmony vs. physical influence*, had to be reconciled on some middle ground.

The complexity of the task is not only indicated by the key sentence quoted above, but also by the last sentence of the *Preliminary Considerations* (1:476, quoted

[14]Baumgarten ranks them among the disjunctive internal predicates of existence in general: each being has to be substance *or* accidental, simple *or* composed, finite *or* infinite (see *Synopsis*, 17:19, and §§191–264, 17:66–83). In contrast, Hinske suggests that the term “transcendental philosophy” in the above quotation is *not* to be understood in the sense of Baumgarten's doctrine of the transcendentals (which has its root in the scholastic doctrine of the *unum, verum, bonum*), but as synonymous with ‘metaphysics’, with a primary focus on the *metaphysica specialis*, that is, in the sense of rational cosmology. Much later, the architectonics chapter of the *CPR identifies* transcendental philosophy and ontology (A 845/B 873), interpreting the categories and predicables of ontology, however, as conditions of the possibility of experience.

at the beginning of Sect. 1.3.2). In comparison to the commonplace requirement of the *True Estimation*, to look for truth in the middle between two conflicting views (see Sect. 1.1.2 and 1:32), the reconciliatory goal is much more challenging, and it seems much more difficult to achieve than the one of 1747, in Kant's view. According to its title, the *Physical Monadology* predominantly deals with reconciling the opposing views of levels (1.) and (2.), where the latter, however, is intertwined with level (3.). Kant specifies the problem as follows:

> For bodies consist of parts; it is certainly of no little importance that it be clearly established of which parts, and in what way they are combined together, and whether they fill space merely by the co-presence of their primitive parts or by the reciprocal conflict of their forces. (1:475)

Then, he specifies the difficulties of clarifying the relation between "transcendental philosophy" and "geometry" as follows:

> For the former preemptorily denies that space is infinitely divisible, while the latter, with its usual certainty, asserts that it is infinitely divisible. Geometry contends that empty space is necessary for free motions, while metaphysics hisses the idea off the stage. Geometry holds universal attraction or gravity to be hardly explicable by mechanical causes but shows that it derives from the forces which are inherent in bodies at rest and which act at a distance, whereas metaphysics dismisses the notion as an empty delusion of imagination. (1:475–476)

Hence, the specific task of unifying physics and metaphysics is here to reconcile in particular the following opposite doctrines: (i) Euler's geometrical proof for the infinite divisibility of matter vs. Wolff's thesis of physical monads, which are indivisible constituent parts of matter, or atoms; (ii) Leibniz's relational theory of space vs. the existence of the vacuum, proposed by the atomists; and (iii) the doctrine of pre-established harmony (according to which substances do not interact at all), defended by Leibniz, Wolff, and Baumgarten, but not by Kant's academic teacher Knutzen, vs. the theory of action at a distance defended by the Newtonians, in particular Cotes and Clarke.

In the main text of the *Physical Monadology*, Kant develops a theory of point-like physical monads with attractive and repulsive forces which give rise to the spatial extension of matter. The theorems and proofs of the *Physical Monadology* are intended to demonstrate that indivisible physical monads are compatible with continuous space, if the spaced filled by the monads via their forces differs from their point-like shape. From the point of view of modern physics, the dynamic principles of the theory are just as reasonable as those of Kant's 1755 theory of structure formation in the universe. A similar atomistic theory of matter was independently developed by Boscovich (1758). For Kant, the combined use of metaphysics and geometry in the *Physical Monadology* was the third keystone of his pre-critical project of unifying Newtonian physics and Wolffian metaphysics.

1.5 Conclusions: Kant's Theory of Everything

For a decade, Kant believed that he had successfully reconciled the principles of Wolff's metaphysics with those of Newton's physics. The metaphysical foundations of his pre-critical system were provided by a theory of interacting substances, which, in contrast to the Wolffians' system of physical influence, however, presupposed God's omnipresence in space as the ontic ground of interactions. His system, or doctrine, corresponds to a position on well-justified middle grounds between four conflicting doctrines of the time: Newton's view, according to which God has to intervene in the world from time to time; Leibniz's theory of pre-established harmony, according to which the substances do not interact at all; the Cartesian vortex theory adopted by French materialism, according to which the substances interact by mere mechanical impact; and the theory of physical influence, which emerged in Wolffianism as a physical theory of close-range effects. Kant's pre-critical system attempted to preserve several elements of Newton's physics and Wolff's metaphysics. Its foundations embrace:

- a relational theory of space and time, according to which space and time are the real order of coexistence and succession of interacting substances;
- God as the ontic ground or *ratio essendi* of the substances and their physical interactions, and together with them of space;
- an atomistic theory of matter, according to which matter consists of physical monads, conceived as point-like sources of attractive and repulsive forces which make them fill an infinitely divisible space;
- Newton's concept of force, his law of universal gravitation, and a theory of structure formation in the universe;
- a physico-theological argument, which links physical cosmology to the existence of God as the ultimate cause of structure formation in the universe;
- Leibniz's principle of sufficient reason, re-interpreted in terms of ontic and epistemic grounds, and
- an interpretation of Leibniz's principle of indiscernibles, which does not preclude the existence of atoms.

This theory has two remarkable features. On the one hand, it links physics and cosmology closely to theology, and to general metaphysical principles such as Leibniz's principle of sufficient reason. It is a *Theory of Everything*, in the most ambitious sense of embracing God and the World. On the other hand, the physical parts of this theory are surprisingly modern. It contains three fundamental assumptions about the nature of space, time, and matter, which (in substantially modified versions, however) have remained tenable options in the philosophy of physics up to the present day:

(a) There is no absolute space. Space is a relational structure which arises from the totality of material substances and their interactions.
(b) There are indistinguishable, purely numerically different objects in nature, which only differ by their position within the relational structure of space.

(c) The internal structure of matter is described in terms of a dynamic atomism. The atoms are point-like sources of attractive and repulsive forces.

Thus, by the mid 1750s Kant's unconvincing first attempt at reconciling the conflicting positions of the Newtonians and the Wolffians had unfolded into a complex systematic task. To cope with this task and derive the metaphysical principles of 1755/1756 was a most remarkable systematic success. But Kant's systematic results of 1755/1756 were not only due to the unifying principles discussed in this chapter, but also to his employment of a philosophical methodology. Kant's pre-critical method was the traditional analytic method, which he only defended in his *Prize Essay* published in 1764, when his pre-critical system already began to waver.

References

Adickes, Erich. 1897. *Kants Systematik als systembildender Faktor*. Berlin: Mayer und Müller.

Adickes, Erich. 1924. *Kant als Naturforscher*. Berlin: de Gruyter.

Baumgarten, Alexander Gottlieb. 1757. *Metaphysica*. 4th ed. Halle: Hemmerde. Quoted after *Kants gesammelte Schriften*, Vol. 17.

Boscovich, Roger J. 1758. *Theoria philosophiae naturalis*. Engl.: *A Theory of Natural Philosophy*, ed. J.M. Child. Chicago/London: Open Court Publishing Company 1922. https://archive.org/details/theoryofnaturalp00boscrich.

Calinger, Ronald. 1968. The Newtonian–Wolffian Confrontation in the St. Petersburg Academy of Sciences (1725–1746). In *Cahiers d'histoire mondiale*, XI, 3, Henry Schumann Prize Paper in the History of Science, 417–436.

Calinger, Ronald. 1969. The Newtonian–Wolffian controversy (1740–1759). *Journal of the History of Ideas* 30: 319–330.

Calinger, Ronald. 1979. Kant and Newtonian science: The pre-critical period. *Isis* 70: 348–362.

Cartwright, Nancy. 1999. *The Dappled World. A Study of the Boundaries of Science*. Cambridge: Cambridge University Press.

Cohen, Hermann. 1883. *Das Prinip der Infinitesimalmethode und seine Geschichte. Ein Kapitel zur Grundlegung der Erkenntniskritik*. Berlin: Dümmler.

Dugas, René. 1950. *Histoire de la mechanique*. Preface de Louis de Broglie. Neuchatel: Du Griffon. Transl.: *A history of mechanics*. Neuchatel: Du Griffon (1955).

Erdmann, Benno. 1876. *Martin Knutzen und seine Zeit. Ein Beitrag zur Geschichte der Wolfischen Schule und insbesondere zur Entwicklungsgeschichte Kants*. Reprint Hildesheim: Gerstenberg 1973.

Erdmann, Benno. 1884. Die Entwicklungsperioden von Kants theoretischer Philosophie. In *Reflexionen Kants zur Kritik der reinen Vernunft. Aus Kants handschriftlichen Aufzeichnungen*, ed. B. Erdman. Vol. 2. Reprint Stuttgart-Bad Cannstatt: Frommann-Holzboog 1992.

Ferrini, Cinzia. 2004. Heavenly bodies, crystals and organisms. The key role of chemical affinity in Kant's critical cosmogony. In *The Kantian Legacy (1804–2004): Unsolved Problems*, ed. C. Ferrini, 277–317. Napoli: Bibliopolis.

Ferrini, Cinzia. 2018. Descartes's legacy in Kant's notions of physical influx and space-filling: *True estimation* and *physical monadology*. *Kant-Studien* 109: 9–46.

Friedman, Michael. 1992. *Kant and the Exact Sciences*. Cambridge, MA: Harvard University Press.

Hegel, Georg Wilhelm Friedrich. 1830. *Enzyklopädie der philosophischen Wissenschaften Vol. 2: Die Naturphilosophie*. Engl.: *Encyclopedia of the Philosophical Sciences, Part II: Hegel's Philosophy of Nature*. Oxford: Oxford University Press 2004.

Herschel, William, and Gensichen, Johann Friedrich. 1791. *Über den Bau des Himmels. Drei Abhandlungen aus dem Englischen übersetzt von Michael Sommer. Nebst einem authentischen Auszug aus Kants allgemeiner Naturgeschichte und Theorie des Himmels*. Königsberg: F. Nicolovius.

Hinske, Norbert. 1970. *Kants Weg zur Transzendentalphilosophie. Der dreißigjährige Kant*. Stuttgart: Kohlhammer.

Iltis, Carolyn. 1970. D'Alembert and the *Vis Viva* Controversy. *Studies in History and Philosophy of Science A* 70: 135–144.

Iltis, Carolyn. 1971. Leibniz and the *Vis Viva* Controversy. *Isis* 62: 21–35.

Kaulbach, Friedrich. 1982. *Immanuel Kant. Zweite, durchgesehene Auflage*. Berlin: de Gruyter.

Laywine, Alison. 1993. *Kant's early metaphysics and the origins of the critical philosophy*. Atascadero: Ridgeview Publ.

Leibniz, Gottfried Wilhelm. 1684. Meditationes de Cognitione, Veritate et Ideis. In Acta Eruditorum. Repr. In *Die philosophischen Schriften*, ed. C.I. Gerhardt. Berlin, 1875–90, repr.: Hildesheim: Olms 1965, Vol. 4, 422–426. Engl: Meditations on knowledge, truth, and ideas. In *Philosophical Essays*, ed. G. W. Leibniz, 23–27. Edited and Translated by Roger Ariew and Daniel Garber. Indianapolis & Cambridge: Hackett Publishing Company.

Leibniz, Gottfried Wilhelm. 1686. Brevis demonstratio erroris memorabilis Cartesii et aliorum circa legem naturalem. In *Mathematische Schriften*, ed. C.I. Gerhardt. Berlin, 1849–60, repr.: Hildesheim: Olms 1971, Vol. 6, 117–119. Engl: A brief demonstration of a notable error of Descartes and others concerning a natural law. In: *Philosophical Essays*, ed. G. W. Leibniz, 23–27. Edited and Translated by Roger Ariew and Daniel Garber. Indianapolis & Cambridge: Hackett Publishing Company.

Leibniz, Gottfried Wilhelm. 1695. Specimen dynamicum. In: *Mathematische Schriften*, ed. C.I. Gerhardt, 234–254. Berlin, 1849–60, repr.: Hildesheim: Olms 1971, Vol. 6. Engl.: A specimen of dynamics. In *Philosophical Essays*, ed. G.W. Leibniz, 117–138. Edited and Translated by Roger Ariew and Daniel Garber. Indianapolis & Cambridge: Hackett Publishing Co.

Leibniz, Gottfried Wilhelm, and Clarke, Samuel. 1715/1716. Correspondence. In *Die Philosophischen Schriften*, ed. C.I. Gerhardt, 345–440. Berlin, 1875–90, repr.: Hildesheim: Olms 1965, Vol. 7. Quoted after the Engl. transl. and repr.: Edited, with Introduction, by Roger Ariew. Indianapolis & Cambridge: Hackett Publishing Co. 2000.

Leibniz, Gottfried Wilhelm, and Clarke, Samuel. 1720. *Merckwürdige Schriften [. . .] Zwischen dem Baron von Leibnitz und dem Herrn D. Clarke über besondere Materialien der natürlichen Religion [. . .] mit einer Vorrede Herrn Christian Wolfens*. Jena: Joh. Meyers. (See: *Der Leibniz–Clarke–Briefwechsel*, ed. and transl. by V. Schüller, Berlin: Akademie Verlag 1991, 556–567.)

Mach, Ernst. 1883. *Die Mechanik in ihrer Entwicklung*. Leipzig: Brockhaus.

Massimi, Michela. 2011. Kant's dynamical theory of matter in 1755, and its debt to speculative Newtonian experimentalism. *Studies in History and Philosophy of Science* 42: 525–543.

Massimi, Michela, and De Bianchi, Silvia. 2013. Cartesian echoes in Kant's philosophy of nature. *Studies in History and Philosophy of Science* 44: 481–492.

Newton, Isaac. 1726. *Principia Mathematica Philosophiae Naturae*. 3rd ed. Quoted after: *The Principia: A New Translation* by I.B. Cohen and A. Whitman, 1999. Berkeley: University of California Press.

Newton, Isaac. 1730. *Opticks*. With a preface by I.B. Cohen, based on the 4th edition, New York. 4th ed., London 1930. With a Preface by I.B. Cohen, Dover 1979.

Riehl, Alois. 1924. *Der philosophische Kritizismus. Geschichte und System. Vol. 1: Geschichte des philosophischen Kritizismus*. 3rd ed. Leipzig: Kröner.

Reich, Klaus. 1958. *Einleitung zu: Kant, Immanuel: Über die Form und die Prinzipien der Sinnenund Geisteswelt*. Lat.-dt. Hamburg: Meiner, VII–XVI.

Schönfeld, Martin. 2000. *The Young Kant*. Oxford: Oxford University Press.

Vuillemin, Jules. 1955. *Physique et métaphysique Kantiennes*. Paris: Presses Universitaires de France.

Walford, David. 1992. *Introduction to: Immanuel Kant: Theoretical Philosophy 1755–1770*, xxxv–lxxxi. Translated and edited by D. Walford in collab. with R. Meerbote. New York: Cambridge University Press.
Watkins, Eric. 1992. The early Kant's (anti-) Newtonianism. *Studies in History and Philosophy of Science* 44: 429–437.
Wolff, Michael. 1987. *Geschichte der Impetus–Theorie*. Frankfurt am Main: Suhrkamp.

Chapter 2
Kant's Analytic Method

> *The true method of metaphysics is basically the same as that introduced by* Newton *into natural science and which has been of such benefit to it.* (2:286)

To establish the foundations of a coherent system of metaphysics, Kant relied on the "analytic" method, a quasi-inductive or regressive procedure, following Crusius (Friedman 1992, 20–24). A closer look at Kant's 1755/1756 arguments, however, reveals that his methodology was incoherent. There was no unique "analytic" method in Kant's time, but several variants of an analytic-synthetic methodology which traced back to ancient geometry and had been reinterpreted by the founders of early modern science and philosophy. In post-Cartesian philosophy it was further transformed, giving rise to the methodological debates which Kant's *Prize Essay* addressed. There, Kant claims that Newton's analytic method of physics is more appropriate for philosophy than the synthetic method of mathematics. In his 1755/1756 writings, however, he employed two different variants of the method: on the one hand, conceptual analysis in a Cartesian (or Leibnizean) sense; on the other, analysis of the phenomena in a Newtonian sense. It is by no means obvious how both variants of the analytic method were related in his view, nor why he was convinced that the analytic method(s), in contrast to the "synthetic" procedure *more geometrico*, should be able to bridge the discrepancies between Wolffianism and Newtonianism.

To clarify these questions, this chapter examines Kant's use of analytic and synthetic methods against their traditional methodological background of ancient geometry and their transformations in early modern science and philosophy. The *Theory of the Heavens* employs analysis of the phenomena in Newton's sense, but Kant combines it with a hypothetico-deductive approach which overcomes some of the limitations of Newton's methodology. The proofs of the *New Elucidation* and the *Physical Monadology* are based on conceptual analysis, but the general structure of both writings looks *more geometrico*. The *Prize Essay* seems to merge conceptual analysis and Newton's method. But Kant's 1764 defence of the analytic method

B. Falkenburg, *Kant's Cosmology: From the Pre-Critical System to the Antinomy of Pure Reason*, European Studies in Philosophy of Science 12,
https://doi.org/10.1007/978-3-030-52290-2_2

in philosophy may well be understood in the weaker sense of a methodological analogy. The *Only Possible Argument* of 1763, too, defends the analytic method and rejects the synthetic method for metaphysics, in accordance with the *Prize Essay*. In the end, however, Kant's use of analytic methods was not powerful enough to distinguish logic and metaphysics from each other. In parallel, he had to introduce the distinction between the logical and the real use of reason.

2.1 Analysis and Synthesis in Early Modern Science and Philosophy

In the 1764 *Prize Essay* Kant refers to Newton's "analytic" method. Quite often this method is considered to be identical with the inductive procedures of an empirical science, and the opposite "synthetic" method with the deductive derivation of theorems from the axioms or principles of a theory. This common view traces back to the philosophy of the Enlightenment and nineteenth-century empiricism. Kant himself contributed to this view in his response to the methodological struggles of his time, as set out in the *Prize Essay* (see Engfer 1982, 102). The empiricist tradition of twentieth-century philosophy of science, too, considers Newton's method to be analytic or inductive in this sense, putting it in sharp contrast to the synthetic method of rationalist metaphysics, and in particular of the Leibniz–Wolff school. Kant's way of referring to Newton's analytic method is also usually understood in this way. In particular, Friedman (1992, 20–24) interprets Kant's methodology defended in the *Prize Essay* along this line, suggesting that

> [...] the data for Kant's Newtonian method of metaphysics consists in uncontroversial metaphysical propositions [...] acceptable to all competing schools, on the one hand, and established results of the mathematical exact sciences, on the other. (Friedman 1992, 24)

Although this view is to a certain extent correct, it needs to be qualified in several regards. First, to identify Newton's method with analysis in the sense of induction, and the Wolffian procedures of introducing and justifying philosophical concepts with synthesis in an axiomatic and deductive sense, cannot do justice to the diversity of the seventeenth/eighteenth-century traditions of analysis and synthesis. Second, to use a one-sided analytic method in the sense sketched above would have been of no advantage for Kant's "irenic" unification project. A fundamental methodological opposition between Newton's physics and Wolff's metaphysics would have posed a serious problem for him. Indeed he finally established a methodological analogy between both, which was justified by their common grounds in Pappu analytic-synthetic method (see Sect. 2.3). If he had faced an additional *methodological* discrepancy between Newtonianism and Wolffianism, how would he have been able to reconcile the diverging principles of both doctrines in his 1755/1756 writings? Finally, Kant's pre-critical metaphysics does not rely on "uncontroversial metaphysical propositions [...] acceptable to all competing schools". We have already seen that in the *Theory of the Heavens* and the *New Elucidation*, he argued in favour of metaphysical principles that differed from the views of *all* competing schools and were located on some middle ground between them (see Sects. 1.4.1 and 1.4.2.3). It is only in the *Prize Essay* that he first appeals to uncontroversial demonstrations and principles (see Sect. 2.2.3.2).

Actually, seventeenth-century science and philosophy shared common methodological grounds, namely the method of analysis and synthesis, which was adopted from mathematics in order to establish reliable scientific knowledge. As we have noted, it traced back to ancient geometry, and specifically to Pappus's commentary on Euclid's geometry (Pappus 1589). According to Pappus, analysis and synthesis are the complementary parts of a joint regressive-progressive method. The analytic part is regressive: it proceeds from something given to the underlying principles. The synthetic part is progressive or deductive: it aims at confirming the principles by deriving what was originally given as the starting point of analysis (see Appendix A.2.1). The method was very popular in early modern science, and in seventeenth-century philosophy, too. Galileo's resolutive-compositive method and Newton's account of analysis and synthesis are similar variants of it, while Descartes's methodology of the *Discourse* is a generalized version: the method is obviously taken up in the second and third rule of the *Discourse* (Descartes 1637; see Appendix A.2.2.1). The method of the *Meditations* (Descartes 1641) consists in applying the second rule, by analyzing the contents of our consciousness and doubting all of it in order to find the unquestionable truth of the *cogito*. Even Descartes's dualism may be considered to result from decomposing the human being into an extended and a thinking substance. Hobbes, too, in *De corpore* (Hobbes 1655), identified scientific reasoning with the method of decomposition and recomposition (see Appendix A.2.2). Descartes's methodology and Newton's scientific method became very influential in science and metaphysics, with the philosophical systems being modeled after mathematical methods and based on the method of analysis and synthesis. This may be easily demonstrated for philosophers as disparate as Descartes, Hobbes, Spinoza, or Leibniz. After Descartes, however, different variants emerged, which finally gave rise to the methodological struggles of eighteenth-century metaphysics.

2.1.1 Newton's Analytic-Synthetic Method

Newton's methodology is Pappus's combined method of analysis and synthesis (Pappus 1589; see Appendix A.2.1), which he transforms on the lines of Galileo's resolutive-compositive method in order to apply mathematics to the phenomena of physics (see Appendices A.2.2 and A.2.2.2). In natural science, the method combines inductive and deductive arguments in two complementary steps, analysis and synthesis (Newton 1730, 404). Analysis is the regress from some given phenomena to the underlying principles. Newton expresses the underlying principles in his famous *Rules of Reasoning* (Newton 1726, 794–796), which result in a kind of inference to the best explanation. The *Rules of Reasoning* combine causal inferences with inductive generalizations (see Appendix A.2.2.2); they do not reduce to induction in the usual sense of empirical generalization. Newton follows Bacon in employing the term "induction" for his method, in Rule IV, but he also emphasizes that well-established principles have to be "deduced from the phenomena", in the famous passage where he criticizes "hypotheses" concerning the cause of gravity (Newton 1726, 943). His method has therefore been termed "deduction from the phenomena" (see Achinstein 1991, 32–50). The complementary step of synthesis is the progress from the principles to the phenomena, resulting in the explanation of known phenomena and the prediction of new ones.

Galileo and Newton elaborated the analytic-synthetic method in physics. Both combined experimental analysis with mathematical deductions, with Galileo's resolutive-compositive method tracing back to Zabarella (Engfer 1982, 90–99). For Galileo, resolution included the experimental and mathematical decomposition of phenomena, and composition included the derivation of the observed phenomena from mathematical principles. Newton followed Galileo and Pappus (Engfer 1982, 100–102). For him, the inductive part of the method (analysis or *resolutio*) was the regress from the phenomena, as given, to their components and causes, as the principles sought. In order to support the regress from the given phenomena to the principles of a physical theory, Newton performed experiments and analyzed the observations. For him, the regress from the phenomena to the principles includes mathematical idealization and conceptual analysis.

Just as Galileo's resolutive-compositive method does, Newton's analytic-synthetic method has two predominant aspects. The analysis consists in the decomposition of phenomena, and in causal analysis. Both aspects further combine experimental and mathematical features with conceptual analysis. The experimental and mathematical decomposition of the phenomena gives rise to causal analysis, and the causes are expressed in terms of idealized concepts of mathematical physics. That is, the experiments aim at the decomposition of the phenomena into causal components, such as forces, which are described in idealized mathematical terms in order to establish theoretical principles, such as Newton's laws of force and gravitation. Synthesis, in turn, is the recomposition of the phenomena, on the one hand, and the derivation of effects from the causes, on the other. Both aspects may be realized in experiments which aim at recovering a phenomenon from its causal components found in the analysis, and in mathematical descriptions which derive phenomenological laws from the theoretical principles.

In the *Principia* (Newton 1726), Newton explains the motions of the celestial bodies and other mechanical phenomena in terms of gravitation as a universal force, in accordance with his method of analysis and synthesis. Here, the analysis of the phenomena concerns the motions of the planets around the sun according to Kepler's laws and the free fall of bodies on earth according to Galileo's law. The analysis finally gives rise to the assumption of gravitation as the cause of all mechanical motions. In turn, the law of gravitation makes it possible to explain these phenomena in the synthetic part of the method. The synthesis is carried out by means of mathematical deductions, which enable Newton to explain the planetary motions, the tides, and much more. In the *Opticks* (Newton 1730), he explains the constitution of light and other optical phenomena in terms of colours and hypothetical light atoms. Here, the analysis consists in detailed experiments on the decomposition of light into colours, the experimental recomposition of white light from the colours, and many other optical phenomena. In particular, he shows how white light is decomposed into the colours by a prism and recomposed by superposing the spectra from two parallel prisms (Newton 1730, 147). However, these experiments do not give rise to mathematical laws which precisely explain the nature of light, but only to qualitative speculations about the atoms of light and matter, as set out the *Queries* of the *Opticks*.

In both fields of physics, the phenomena are the starting point of causal analysis. The crucial difference between the *Principia* and the *Opticks* is that Newton is able to give a full-fledged mathematical theory of gravitation in order to explain the phenomena of mechanics, whereas he is not able to do so for the atoms of light or matter in terms of forces and mathematical laws. Without establishing well-defined, empirically confirmed theoretical principles that explain the phenomena under investigation, the second step of the analytic-synthetic method, the synthetic part, is missing. Hence, in the *Opticks* the method remains incomplete. The very fact that Newton here only writes *Queries* about the atoms indicates that in his view the analysis part of the method on its own does not give a sufficient scientific explanation. The complementary synthesis part is necessary, too, and it has to be carried out by mathematical deduction of the phenomena from the theoretical principles.

2.1.2 Analysis and Synthesis in Post-Cartesian Philosophy

In early modern science, analysis and synthesis (or resolution and composition) were considered to be two complementary tools of scientific method, both of which have to be employed one after the other in order to demonstrate truths or to prove principles. Descartes and Newton considered both complementary steps of the method to belong together. In post-Cartesian philosophy, however, the two parts of the method split off. They were considered to be two independent rival methods of metaphysics and this gave rise to the methodological struggles of eighteenth-century philosophy.

The splitting of analysis and synthesis had already begun Descartes's day. According to the *Discourse*, both parts of the method form two out of four rules of the scientific method (Descartes 1637), but in the *Meditations*, Descartes followed the analytic method in order to find his supposedly unquestionable principles of philosophy which were beyond any doubt for him (Descartes 1641). Mersenne, in his objections to the *Meditations*, asked Descartes to present the principles of his philosophy also *more geometrico*, that is, according to the synthetic method (Descartes 1934, Vol. II, 29). Descartes replied that he considered the analytic method to be appropriate for demonstrating the way in which something has been discovered (ibid., 48–50). In addition, though, he presented the contents of his philosophy *more geometrico*, in terms of definitions, postulates, and axioms (ibid., 52–59), giving an impression of metaphysics as an axiomatic, quasi-mathematical discipline. It was Spinoza, in his ethics *more geometrico* (Spinoza 1677), who primarily followed the synthetic method of giving definitions and axioms and deducing theorems from the axioms.

Post-Cartesian philosophical understanding of analysis and synthesis was based on the highly influential Port-Royal Logic (Arnauld and Nicole 1685). This combined several methodological models, some of which had even been explicitly rejected by Descartes, in his aversion to traditional Aristotelian logic (Engfer 1982, 131–132). The Port-Royal Logic distinguished analysis and synthesis as follows:

> Hence there are two kinds of method, one for discovering truth, which is known as *analysis*, or the *method of resolution*, and which can also be called the *method of discovery*. The other for making the truth understood by others once it is found. This is known as *synthesis*, or the *method of composition*, and can be also called the *method of instruction*. (Arnauld and Nicole 1685, 233)

In the Port-Royal Logic, analysis, as an inductive method, is still interpreted in the traditional double sense of decomposition and causal analysis. But the examples of decomposition given there do not only concern the analysis of natural phenomena into their parts, but also into their properties. In this way they make the bridge to the analysis of concepts (e.g., in mathematics).

Leibniz's account of analysis and synthesis is closely connected to his logic and to his projects of an *ars characteristica* and an *ars combinatoria*. He considers analysis and synthesis to be separate logical methods. The synthesis proceeds from the principles to their consequences, while the analysis traces from a given problem back to the principles, as in the tradition. In contradistinction to Descartes or Newton, however, he emphasizes the primacy of synthesis (Leibniz 1679; Engfer 1982, 196). A more detailed examination shows that Leibniz no longer unambiguously relates the analytic method to an inductive, heuristic *ars inveniendi*, and the synthetic method to a deductive *ars demonstrandi*, as the Port-Royal Logic did (Engfer 1982, 199–208). Wolff is a particular case. He predominantly follows Proklos' account of the synthetic method, i.e., Euclid's axiomatic organization of geometry, rather than Pappus's analytic-synthetic method (Engfer 1982, 231–245). His *German Logic* of 1713 neglects the concepts of analysis and synthesis, but he takes up their versions of the Port-Royal Logic in the *Latin Logic* of 1728 (Engfer 1982, 227–228). Hence, in eighteenth-century philosophy the earlier assimilation of analysis and synthesis to the well-defined parts of Pappus's method had become quite unclear.

In the tradition of Leibniz's logic, analysis was henceforth interpreted as *conceptual analysis* (see Appendix A.2.2.3). According to the essay *Meditationes de Cognitione, Veritate et Ideis* (Leibniz 1684), conceptual analysis aims at distinct and adequate concepts, where adequacy means distinctness of all the attributes of a substance (see Appendix A.3.2). The essay is an elaborated version of Descartes's theory of clear and distinct ideas, adapted to Leibniz's logic. For Leibniz, the distinctness of all the attributes of an idea is an ideal of perfect cognition which in general cannot be accomplished. In his view, conceptual analysis most often only provides nominal definitions, except in logic and mathematics. Wolff follows Leibniz in calling for conceptual analysis to make confused concepts distinct. His doctrine of concepts and definitions relies on Leibniz (1684).

Wolff combines conceptual analysis with synthetic derivations of propositions (see Engfer 1982). In addition, he considers the analytic method as an appropriate procedure for deriving distinct metaphysical concepts. According to him, this conceptual analysis precedes the synthetic formation of a doctrine in terms of definitions, axiomatic principles (such as the principle of sufficient reason), and the demonstration of theorems (Engfer 1982, 245–255). In Wolffianism it became usual to present a metaphysical doctrine in this way, strictly separating its systematic presentation *more geometrico* from its justification.

In Newtonianism, on the other hand, conceptual analysis was criticized as being too speculative. In the preface to the second edition of the *Principia*, Cotes accuses Cartesian physics of not taking the phenomena sufficiently into account. Without explicitly referring to Descartes, Cotes marks a strict methodological contrast between Cartesian physics and Newton's approach. In contrast to the former, the latter is the combined analytic-synthetic method of

> those whose natural philosophy is based on experiment. Although they too hold that the causes of all things are to be derived from the simplest possible principles, they assume nothing as a principle that has not yet been thoroughly proved from phenomena. They do not contrive hypotheses, nor do they admit them into natural science otherwise than as questions whose truth may be discussed. Therefore they proceed by a two-fold method, analytic and synthetic. From certain selected phenomena they deduce by analysis the forces of nature and the simpler laws of those forces, from which they then give the constitution of the rest of the phenomena by synthesis. This is that incomparably best way of philosophizing which our celebrated author thought should be justly embraced in preference to all others. This alone he judged worthy of being cultivated and enriched by the expenditure of his labor. (Cotes 1713, 386).

According to Cotes, Descartes neglects the *analytic* aspect of the only correct method of knowledge of nature in favor of a one-sided synthetic or deductive approach, which with all due precision does not get beyond ingenious speculation. His criticism aims at the physics part of the *Principia philosophiae* (Descartes 1644), a work that indeed lacks methodological coherence (Engfer 1982, 150)

In the Leibniz–Clarke correspondence, Clarke's Fourth Reply may be understood in a similar way. There, Clarke accuses Leibniz of not giving a proof of the principle of indiscernibles "from the nature of things" (Leibniz and Clarke 1715/1716, 29). Such criticism gave rise to several authors interpreting Newton's analytic method in empiricist terms; and, indeed, a one-sided empiricist interpretation of Newton's method has remained influential from eighteenth-century methodological struggles up to current research (see Appendix A.2.2.2).

Hence, Pappus's analytic-synthetic method, which in the progression from medieval science to Descartes and Newton had already become many-faceted, differentiated into confusingly many variants in post-Cartesian philosophy. In addition, the philosophers split the synthetic part off from the analytic part of the method, replacing the *combined* analytic-synthetic method of Descartes's philosophy and Newton's physics by usage of *either* the analytic *or* the synthetic method, claiming that only one of them can be "the" adequate method of metaphysics.

Descartes and Newton had both defended the analytic part of the method as the appropriate procedure for deriving their respective philosophical principles and systems, even though Descartes's *Principles of Philosophy* may have given another impression. After Descartes, a one-sided use of the synthetic method in metaphysics became widely accepted. Spinoza's *Ethics more geometrico* (Spinoza 1677) provided the most prominent example. The metaphysical system of Wolff, on which eighteenth-century German school philosophy rested (Baumgarten 1757), had indeed strong scholastic features. Based on Leibniz's principle of sufficient reason and the metaphysical concept of a monad, it combined the synthetic method of starting from definitions and axioms with conceptual analysis, proceeding *more geometrico*. Nevertheless, Wolff was also strongly impressed by Newton's physics, and wanted to take the empiricist objections of his contemporaries into account, too; surprisingly many references to Locke's philosophy and Newton's physics are to be found in his *German Logic* and his *Preliminary Discourse* (Hettche and Dyck 2019, Sect. 7, n. 65, 66). Against this complex methodological background, then, Kant's pre-critical reconciliation project took shape.

To shed light on the way in which Kant invoked Newton's analytic method, let me summarize the most important of the methodological developments sketched above. In post-Cartesian philosophy, as we have noted, the parts of Pappus's combined analytic-synthetic method split into separate methods. According to the Port-Royal Logic, the analytic method is an inductive, heuristic *ars inveniendi*, whereas the synthetic method is a deductive *ars demonstrandi*. Leibniz ties the method to his logic, identifying synthesis with the progress from principles to their consequences and analysis with the regress from given problems to their principles. In the Leibniz–Wolff school, the analytic method is identified with conceptual analysis. Here, conceptual analysis aims at analyzing confused concepts into simple and clear attributes, which make it possible to define distinct and adequate metaphysical concepts; and the synthetic method is identified with the systematic presentation *more geometrico* of a doctrine in terms of definitions, axioms, and the demonstration of theorems. In Newtonianism, on the contrary, the analytic method continues to be Pappus's combined analytic-synthetic method in a version which is similar to Galileo's resolutive-compositive method.

In Newtonian science, analysis does not just mean conceptual analysis; rather, it means in addition the experimental and mathematical analysis of the phenomena. Newton's analytic method (or "induction") aims at decomposing the phenomena and tracing the components back to their causes. The Newtonians agree with the Wolffians on what the synthetic method is, namely the presentation of a theory *more geometrico*. However, they have completely different accounts of the analytic method. For the Newtonians, the analytic method is primarily the experimental and mathematical analysis of the phenomena for their components and causes. For the Wolffians, it is only conceptual analysis. But, in addition, the physicists and the philosophers disagree about the relationship between analysis and synthesis. In

Newtonian physics, the full-fledged combined analytic-synthetic method counts as the only correct method; it is in post-Cartesian philosophy that we see the beginning of the methodological struggles over which is the correct method, *either* the analytic, *or* the synthetic one.

2.2 Kant's Methodology in the 1755/1756 Writings

Before writing his *Prize Essay* published in 1764 and the *Only Possible Argument* of 1763, Kant was not concerned with the methodological struggles of his time. In his 1755/1756 writings he uses the Newtonian method of analysis of the phenomena and the Leibniz–Wolffian method of conceptual analysis in parallel, depending on the topic he was dealing with. In addition, he combines the latter with the synthetic method *more geometrico* of giving definitions and demonstrating theorems. Kant employs these parts and variants of the analytic-synthetic method without questioning whether they are coherently used. His goal is to combine physics and metaphysics, in order to lay the foundations for a philosophical system, or doctrine, that resolves the metaphysical debates of the Wolffians and the Newtonians (see Sects. 1.3 and 1.4). Let me now re-examine the methodology of his 1755/1756 writings against the background of the analytic-synthetic method, or respectively the variants of the analytic and synthetic methods, discussed in the previous Section (2.1).

As I show in the following, the *Theory of the Heavens* uses the Newtonian method of analysis of the phenomena and in addition a hypothetico-deductive approach that goes beyond Newton's inductive methodology. In contrast, the *New Elucidation* and the *Physical Monadology* combine conceptual analysis with the deductive demonstration of theorems as well as additional arguments in order to reveal certain limitations of such deductions. This methodological difference is not really surprising, given that the *Theory of the Heavens* lays the grounds for a physical cosmology in Newton's tradition, whereas the *New Elucidation* and the *Physical Monadology* belong to a metaphysics in Leibniz's and Wolff's tradition. The most intriguing question is then: What are Kant's methodological grounds for mediating between both traditions and combining various approaches, in order to establish the foundations for the unified system of metaphysics described in the first chapter?

2.2.1 Cosmology: Model-Based Analysis of the Phenomena

The *Theory of the Heavens* of 1755 has two explanatory goals. One is to explain structure formation in the universe, the other is to make the bridge from cosmology to theology and to explain the structure of the universe by reference to God as the ultimate cause (see Sect. 1.4.1). Kant establishes the link between both explanatory goals by demonstrating that the structures observable in the universe trace back to a unique cause. The argument combines a model-based hypothetico-deductive approach with inferences from the observable phenomena to their causes. The latter are inductive inferences drawn in accordance with Newton's methodological Rules I–III (see Appendix A.2.2.2). However, Kant goes beyond Newton's methodology, extending the analysis of the phenomena in four crucial regards.

First, he *explains the initial conditions* of the celestial bodies in the solar system, which Newton considered to be given and maintained by God. In particular, Kant sketches the evolution of the solar system from an initial matter vortex to its present shape, including the elliptic planetary orbits, and the approximate restriction of all motions of celestial bodies in the solar system to a common plane, the ecliptic. The model explains the form of Kepler's motions without assuming the shape of the solar system. Kant even calculates the duration of the axial rotation of Saturn, and later Herschel's observations indeed confirmed this calculation.

Second, in order to give these explanations, Kants develops a *theoretical model of structure formation in the universe*, the model which today is known as the Kant-Laplace hypothesis. According to it, the sun and the other celestial bodies of the solar system emerged by a clustering process from a rotating nebular matter distribution. The model is based on Newton's mechanics including the law of gravitation, and the atomistic theory of matter which is later justified in the *Physical Monadology*. Kant's model assumes the following initial conditions. At the beginning of the universe, matter was continuously but inhomogeneously distributed around its centre of gravity and formed a rotating nebula of decreasing density towards the exterior. From this model, Kant predicts the formation of vortexes and the clustering of matter into planets, stars and other celestial bodies, planetary systems such as the solar system, and star systems such as the Milky Way. His model takes up the Cartesian vortex theory and gives it an anti-Cartesian interpretation by combining it with Newton's theory of gravitation. The model is speculative, or purely hypothetical, in two regards. On the one hand, it does not derive from the decomposition and causal analysis of the observed phenomena according to Newton's *Rules of Reasoning*. On the other hand, it is mainly qualitative and gives no detailed deduction *more geometrico*. Kant downplays the deviations from Newton's methodology and emphasizes that he leaves the deductive derivation to the reader:

> One might, if one wished to be expansive, ultimately arrive at the framework I propose to present of the origin of the universe by pursuing on one's own a series of conclusions following from one another in the way of a mathematical method with all the splendour this involves [...]; however I would prefer to present my opinions in the form of a hypothesis and leave it to the insight of the reader to examine their worthiness [...]. (1:263)

Third, Kant *extrapolates* his theory of structure formation to the *large-scale structure of the universe*: from the solar system to the Milky way, to other galaxies, and finally, to the whole universe. This extrapolation is already made in the initial condition of his above model, which is indeed a cosmological model of the initial matter distribution of the universe. Nonetheless, Kant can give some empirical support to this extrapolation by explaining important astronomical phenomena in terms of his model. In particular, he interprets the Milky Way as a rotating system of stars and the fixed-star nebulae as galaxies like the Milky Way, in an account which is more precise than the one given in Lambert's *Cosmological Letters* (Lambert 1761). This extrapolation is indeed based on an analysis of the astronomical phenomena that follows Newton's methodological Rules I–II of causal analysis. In order to support it, Kant defines the concept of the *systematic constitution of the universe* (see also Appendix A.1.3). The solar system is organized in such a way that the planetary motions are approximately restricted to a common plane, the ecliptic. This phenomenon agrees with the prediction from Kant's theory of structure formation in the solar system. Therefore, Kant specifies the general concept of a system of celestial bodies to that of a rotating dynamically bound system, in which the motions around the center of gravity are approximately restricted to a common plane, in accordance with the phenomenological properties of the solar system:

> In this treatise, I shall frequently use the expression of a *systematic constitution of the universe*. [...] Actually, all the planets and comets that belong to our universe constitute a system simply because they orbit around a common central body. But I take this term in a narrower meaning [...]. The orbits of the planets relate as closely as possible to a common plane, namely to the extended equatorial plane of the Sun; the deviation of this rule occurs only at the outermost border of the system, where all motions gradually cease. If, therefore, a certain number of heavenly bodies that are arranged around a common central point and move around this, are simultaneously restricted to a certain plane in such a way that they have the freedom to deviate from it to either side only as little as possible; if such deviation occurs gradually only in those that are most remote from the centre point and thus participate less in the relationships than the others: then, I say, that these bodies are related to each other in a *systematic constitution*. (1:246)

The "systematic constitution of the universe" consists in disk-like rotating systems of celestial bodies, underpinned by the constraint that the vertical deviations of their motions may only gradually increase towards the outer boundaries of the system. This constraint restricts the motions of the celestial bodies in the system to Kepler orbits with approximately the same center of gravity and angular momentum. Hence, the "systematic constitution" of celestial bodies is a specific phenomenon explained by the evolution of matter according to the laws of Newton's mechanics, with the initial conditions of Kant's model. Based on his definition of the "systematic constitution", Kant makes two inferences: first, the Milky Way is a rotating disk-like system of fixed stars in this sense, observed from a lateral standpoint; second, the nebulae observable by a telescope in the night sky are disk-like rotating systems of fixed stars, just like the Milky Way. According to Newton's Rules I–II of causal analysis, all these structures are traced back to the same initial conditions and the same laws of Newton's physics; in particular, to the law of universal gravitation (with an implicit assumption of the conservation of angular momentum, which neither Newton nor Kant explicitly expressed). In a further inductive step, Kant generalizes the observation that the celestial bodies form disk-like rotating systems like the solar system from the nebulae to the large-scale structure of the universe.

So far, Kant's argument combines hypothetico-deductive reasoning based on a theoretical model and analysis of the phenomena in Newton's sense within this model. Schönfeld (2000, 88–93) claims that the underlying methodology is mere analogical reasoning; this is correct, but it has to be noted in addition that it follows Newton's methodology of *Query 31* of the *Opticks* (see Sect. 2.1.1). Given that Newton had no mathematical theory of atoms and light, his analysis of the optical phenomena could only rely on analogical reasoning, that is, on his famous principle that

> Nature will be very conformable to her self and very simple, performing all the great Motions of the heavenly Bodies by the Attraction of Gravity with intercedes those Bodies, and almost all the small ones of their Particles by some other attractive and repelling Powers which intercede the Particles. (Newton 1730, 397)

In the *Theory of the Heavens*, Kant can only follow the qualitative methodology of Newton's *Queries*, given that he, too, had no mathematical theory of the atoms, to say nothing of a probabilistic theory of the way in which atoms cluster to vortex structures and rotating systems of celestial bodies. Therefore, in contrast to Newton's full-fledged analytic-synthetic method of the *Principia*, the synthetic part of the argument, i.e., the hypothetico-deductive theory of structure formation, is not spelled out in mathematical terms. It is not based on the results of a mathematical analysis of the astronomical phenomena, but rather on a qualitative cosmological model of the initial matter distribution of the universe and an atomistic theory of matter. Kant adds the latter to Newton's mechanics on the lines of *Query 31* of Newton's *Opticks*. Remarkably enough, in this regard Kant's theory is closer to modern physical cosmology than to Newton's mechanics. What is more, even though it is basically qualitative, it also has some quantitative predictive power, as Kant's calculation of the axial rotation of Saturn shows. As a precursor of modern physical cosmology, Kant's theoretical model of the universe has additional interesting aspects which cannot be discussed here (Schönfeld 2010). Its most remarkable feature, however, is perhaps the way in which it explains the dark night sky in an infinite universe with infinitely many stars (the so-called Olbers paradox, which traces back to Kepler), by assuming an expanding star distribution and a finite lifetime of the stars (De Bianchi 2013).

The *fourth* crucial regard in which Kant extends Newton's scientific methodology is his physico-theological proof of the existence of God. In the preface of the *Theory of the Heavens*, Kant makes an inference from the universal laws of nature, which give rise to the same kind of structure formation at small and large scales of the universe, to God as their unique cause (1:228; see Sect. 1.4.1). Methodologically, this inference looks like an *inference to the best explanation*, but it is speculative and transcends the limitations of empirical science. Newton's own scientific methodology did not support such an inference, as his famous dictum "I do not feign hypotheses" concerning the unknown cause of gravitation shows (Newton 1726, 943). At most, Rules I and II of the *Principia* support the assumption that the systematic organization of the celestial bodies in the universe traces back to a common physical cause, that is, a primordial rotating matter vortex made up of atoms, which is the initial condition of structure formation in the universe in Kant's cosmological model. In a further regressive step, Kant traces this initial condition back to God as the creator of the universe. By doing so, he provides his physical cosmology with a theological foundation, following Newton's theism rather than the scientific methodology of the *Principia* and the *Opticks*.

Eight years later, in the *Only Possible Argument*, Kant would present an extended version of this physico-theological argument, together with an abridged version of the reasoning of his 1755 cosmology. However, from now on he would no longer consider this kind of metaphysical inference to be tenable, emphasizing that

> [...] in spite of all its excellence, this mode of proof will never be capable of mathematical certainty or precision. (2:160)

2.2.2 *Metaphysics: An Approach Purely* More Geometrico*?*

In Kant's pre-critical unification project, the *New Elucidation* aims at making the principles of Leibniz's metaphysics, and in particular the principle of sufficient reason, compatible with the principles of Newton's physics. In order to do so, Kant proceeds *more geometrico*, in combination with Wolff's method of justifying metaphysical definitions and principles by means of conceptual analysis (see Sect. 2.1.2). Therefore, to a certain degree Hinske was right to emphasize that the *New Elucidation* is "the most conservative and orthodox scholarly piece ever written by Kant" (Hinske 1970, 88; my translation). But with regard to Kant's goal of reconciling a *metaphysica specialis* in Wolff's sense with Newton's physics, the *New Elucidation* is much more than just an orthodox contribution to Wolffianism (see Sect. 1.4.2).

The crucial methodological question of the *New Elucidation* is: How does Kant, by means of conceptual analysis and an approach *more geometrico*, manage to make Leibniz's principle of sufficient reason conformable with Newtonian views? One might suspect that he has to impose certain constraints on Leibniz's principle of sufficient reason in order to achieve his goal; the question, then, would be where they come in. However, Kant's strategy is simply to show that Leibniz's principle is logically weaker than Leibniz and Wolff assumed, and that therefore its logical consequences do not contradict the Newtonian principles of the *Theory of the Heavens*. He delimits the scope of Leibniz's principle of sufficient reason in order to demonstrate that it does not preclude the possibility of atoms and of real interactions between the atoms (see Sect. 1.4.2). In addition, he employs arguments which are supported by Newton's methodology.

Accordingly, the *New Elucidation* proceeds *more geometrico* with proofs based on conceptual analysis, which are complemented by phenomenological arguments based on Newton's methodological rules. His line of reasoning is conclusive in so far as he shows that the metaphysical claims of the Wolffians do not derive from Leibniz's principle (or the logical principle of the determining ground, which replaces it in the *New Elucidation*; see Sect. 2.2.2.1). His way of appealing to the phenomena and to Newton's methodological rules concerning the principle of indiscernibles is also convincing (see Sect. 2.2.2.2). His proofs of the principles of coexistence and succession are more debatable (Sect. 2.2.2.3), yet they serve to rebut Leibniz's system of pre-established harmony in order to establish his own, specific account of interacting substances in the world.

2.2.2.1 The Determining Ground: Conceptual Analysis

The *New Elucidation* proceeds as follows. First, Kant clarifies the principle of sufficient reason by means of conceptual analysis, redefining it in terms of the "determining ground", which determines a subject with respect to a predicate (1:391). On this basis, he distinguishes ontic and epistemic grounds and discusses the logical consequences of the redefined principle (see Sect. 1.4.2). Kant's version of Leibniz's principle is the "principle of the determining ground":

> *Proposition V*: Nothing is true without a determining ground. (1:393)

The principle is an analytic judgment and Kant demonstrates it by conceptual analysis. It is based on the following definitions:

> To determine is to posit a predicate while excluding its opposite. That which determines a subject in respect of any of its predicates, is called the ground. (1:391)

The proof starts with the traditional nominal definition of truth, according to which a judgment is true if its subject has the properties attributed to it by its predicates. Here, truth means for Kant that "the subject is determinate in respect of a predicate" in such a way that "the opposite of the predicate in question should be excluded" (1:393). Then, the subject of a true judgment is nothing but the "determining ground" of its truth, and in fact the *ratio essendi* (see Sect. 1.4.2.1). To put it in other words, the "determining ground" is the truthmaker of the judgment, and the "principle of the determining ground" is an analytic truth. Kant's proof is lengthy and, in addition, he gives a further analytical proof.

The aim of demonstrating the principle of sufficient reason based on conceptual analysis belongs to Wolff's tradition (see §22 of Baumgarten's *Metaphysics*, 17:31); Kant, however, shows that the principle demonstrated in such a way is logical and not metaphysical. Based on it, he then criticizes Spinoza's conception of God as the *causa sui*:

> *Proposition V*: To say that something has the ground of its existence within itself is absurd. (1:394)

Kant proves this proposition by conceptual analysis, analyzing the concept of the ground of existence. He first argues that the ground of existence (or *ratio essendi*, see Sect. 1.4.2.1) of something is its cause. Then, he concludes from an analysis of the concept of cause that something which has its cause in itself "would be simultaneously both earlier and later than itself, which is absurd" (1:394). This argument may pass for a convincing example of the use of conceptual analysis.

Next, Kant adds his own proof of the existence of God which is much more dubious. He claims that it derives from his logical principle of the determining ground as well, whereas in fact it amalgamates logical and metaphysical reasoning. The proof is based on inferences from existence to logical possibility, from logical to real possibility, and from real possibility to the existence of an absolutely necessary being, as the ultimate ground of possibility (1:395). Kant takes this proof up again in the *Prize Essay* and the *Only Possible Argument*. In 1762/1763, however, he was already less convinced of the certainty of his proof than in 1755 (see Sect. 2.4 below).

2.2.2.2 Indiscernibles: Analysis of the Phenomena

Kant also criticizes Leibniz's principle of indiscernibles, claiming it to be a conclusion incorrectly drawn from Leibniz's principle of sufficient reason. He argues that it does not contradict the assumption of atomism, provided the principle is interpreted in such a way that position is admitted as a sufficient distinguishing mark of two substances. His argument is based on an analysis of the concept of the permutation of two substances (1:409, quoted above in Sect. 1.4.2.2). Kant argues that to permute indistinguishable objects does not give rise to any consequences. If there are two indistinguishable individuals A and B, they merely differ in name. Hence, their only distinguishing marks are the positions denoted by 'A' and 'B'. This, however, amounts to the following nominalistic claim: if object A and object B interchange their positions, then object A becomes object B, given that it is only individuated by its position, and vice versa. In support of his claim, he appeals to an argument by Leibniz against Clarke (which is found in Leibniz's fourth letter to Clarke); turning it back, however, against Leibniz:

> To suppose two things indiscernible is to suppose the same thing under two names. (Leibniz and Clarke 1715/1716, 22)

The difference between two solely numerically different objects A and B is just the difference in their positions, or the "characters" or indexes 'A' and 'B' that denote them. So Kant supports Clarke in his insistence in his fourth letter that:

> Two things by being exactly alike do not cease to be two. (Leibniz and Clarke 1715/1716, 29)

Since, of course, the permutation of things changes neither the positions nor the indices corresponding to them, there is no real change in the exchange of places. Therefore, there is no need to ask for a sufficient reason. The principle of sufficient reason is thus consistent with the assumption that two indistinguishable, merely numerically different individual things may exist without contradiction.

The argument is intended to defeat Leibniz's assumption that the existence of indiscernibles would indeed require a sufficient reason. If the permutation of two individual objects has no real consequences, there is no quest for a sufficient reason. Here, Kant already makes tacit use of the distinction between the logical and the real use of reason, which he would make explicit in 1762/1763 (see Sect. 3.1). The crucial point of his 1755 argument against Leibniz's principle of indiscernibles is that logical principles such as the principle of the determining ground (which is a logical rule connecting antecedent and consequent) are metaphysically neutral. These principles are consistent with various alternative principles of a *metaphysica specialis* and the corresponding assumptions about space, time, and the substances in the world.

In order to select adequate metaphysical principles from the various logical possibilities, phenomenological justifications are needed. Kant indeed supports his permutation argument against Leibniz's principle of indiscernibles by appealing to phenomenological considerations which rely on Newton's methodological Rules I–III (explained in Appendix A.2.2.2). His very approach of considering *position* to be among the distinguishing properties of a substance is based on an appeal to the phenomena. In addition, he points out that the structure of the phenomena in many areas of natural science does *not* support Leibniz's principle of indiscernibles:

> For that bodies which are to be said similar, such as water, mercury, gold, the simplest salts, and so forth, should agree completely in their primitive parts in respect of their homogeneous and internal characteristic marks, corresponds to the identity of the use and function which they are defined to fulfill. This is to be seen from their effects, which we observe by issuing from those same things, always the same and never with any discernible difference. (1:409–410)

Similar bodies or substances found in nature are thus identical in use and function. Now Newton's Rule III comes into play, claiming the universality of the extensive properties "that belong to all bodies on which experiments can be made" (Newton 1726, 395). According to this, such bodies or substances are therefore similar in their "primitive parts". In addition, Newton's Rule II is employed, according to which the same effects in nature should be attributed to the same causes. This supports the claim that the same observable effects of the parts of such homogeneous empirical substances are due to the same causes. Kant completes his argument with a kind of principle of ontological parsimony for internal properties:

> Nor it is proper here to suppose that there is some hidden difference which escapes the senses [. . .], for that would be to search for knots in a bullrush. (1:409–410)

His mordant comment that to assume hidden different causes would be "to search for knots in a bullrush" follows Newton's Rule I, which recommends ontological parsimony in order refute occult qualities. Kant, however, concedes that Leibniz, too, can rely on phenomenological arguments to justify his interpretation of the principle of indiscernibles, namely on the internal diversity of organic and other complex structures in nature:

> We admit that Leibniz, the originator of this principle, always detected a discernible difference in the structure of organic bodies or in the organisation of other bodies of extreme complexity, and we admit that one may with justification assume that there is a discernible difference in all cases of this kind. (1:410)

But an inversion of Newton's Rule II, i.e., the inference from different causes to different effects, immediately explains why this assumption makes sense for organic structures:

> For, in cases where it is necessary that a number of different factors have to harmonise together in a very high degree before something can be produced, it is obvious that they cannot always yield the same determinations. Thus among the leaves of the same tree, you will scarcely find two which are completely alike. (1:410)

The last sentence obviously refers to the famous passage from Leibniz's Fourth Letter to Clarke:

> An ingenious gentleman of my acquaintance [...] thought he could find two leaves perfectly alike. The princess defied him to do it, and he ran all over the garden a long time to look for some; but it was to no purpose. Two drops of water or milk, viewed with a microscope, will appear distinguishable from each other. This is an argument against atoms, which are confuted, as well as a vacuum, by the principles of true metaphysics. (Leibniz and Clarke 1715/1716, 22)

Finally, Kant points out that by his argument

> only the metaphysical generality of this principle should be rejected. (1:410)

Leibniz's interpretation of the principle of indiscernibibles is just one of two (or more) logically possible options, and none of them is a logically necessary consequence of the principle of the determining ground. In order to select among them, phenomenological criteria are required. At this point of the *New Elucidation*, Kant does not want to prove any more or any less. In this way, here he takes a first decisive step towards distinguishing between the logical use of predicates in logic or mathematics and the real use of predicates in metaphysics or natural science, which first appears in the *Only Possible Argument* and the *Negative Magnitudes* of 1763 (see Sect. 3.1).

2.2.2.3 The Principles of Succession and Coexistence

The *New Elucidation*, however, does not yet clearly distinguish between logical and real grounds. Here, Kant demonstrates several principles of a *metaphysica specialis* which he considers to be metaphysical consequences of his "principle of the determining ground". In particular, he gives proofs of the principles of succession and coexistence. They are the dogmatic predecessors of the critical principles of causality and interaction (or community) explained in the *CPR*. The pre-critical principles of succession and coexistence are non-logical, *metaphysical* principles concerning the relation of grounds and consequences. It is hard to see how they may derive from a merely logical principle such as the "principle of the determining ground", as a principle of logical grounds and consequences. However, both principles are crucial for Kant's pre-critical system of 1755/1756. They serve to refute Leibniz's theory of pre-established harmony and justify his modified version of the theory of physical influence.

According to the principle of succession, the interaction or "reciprocal dependency" is a necessary condition for changes of the states of substances (1:410). Kant gives three similar proofs in terms of determining grounds in order to show that Leibniz's concept of a substance is logically too weak to explain the change of the inner state of a substance. So far, so good. In order to describe the change of the attributes of a substance, a *temporal* concept of a substance is needed, which cannot without additional assumptions be derived from the *logical* principle of the determining ground.

But Kant makes further claims. In the second proof, he argues that "whatever is posited by a determining ground" must be "posited simultaneously" with that ground (1:411). From a modern logical point of view, it is confusing that he does not differentiate the merely logical structure of predication from a temporal theory of substances, and thus seems to be merging the logical and the semantic aspects of his theory of substances—where the latter, in the context of Kant's pre-critical project, are metaphysical. To a certain extent, this missing distinction may be justified by the existential presuppositions of traditional logic, which bring tacit semantic assumptions into play (see the analysis of traditional vs. modern logic in Strawson 1952, 164–194, and Sect. 5.3.4.2). However, all three proofs look awkward. Laywine discusses them in detail and concludes that they

> are rather strange arguments. At best, they seem contrived; question-begging at worst. [...] Nonetheless, I have come to think that something quite interesting is going on in these proofs, especially in the second of the three. [...] In a nutshell, Kant's argument against pre-established harmony is that the favored system of Wolff and Leibniz cannot explain how the order of succession is possible. (Laywine 1993, 33)

According to the principle of coexistence, the mere existence of substances is not a sufficient condition of their interactions. Therefore, Kant introduces an additional ontic ground or *ratio essendi* for their interactions, a "common principle of their existence", which he identifies with the "divine understanding" or respectively its "schema" (1:413–414). This principle serves to justify his modified version of a metaphysical system of *influxus physicus*, according to which the physical influences between the substances only exist due to the existence of God (Laywine 1993, 37; Friedman 1992, 7). His theory that the interactions of the substances in the world are due to the "schema of the divine understanding" is indeed another cornerstone of the pre-critical reconciliation project. It not only gives additional support to the physico-theological proof for the existence of God given in the *Theory of the Heavens*, but moreover mediates between Newtonianism and Wolffianism in such a way that it strikes a balance between Newton's interpretation of absolute space as the *sensorium dei* and Leibniz's relational account of space.

The proof of the principle of coexistence is again based on merging the logical structure of predication and the semantic (or metaphysical) assumptions about the substances in the world. Here, the core of the argument is that substances in Leibniz's sense have only non-relational, monadic properties which rule out any relations between them. The proof not only employs conceptual analysis, but also appeals to the premise that "it certainly does not fall to finite beings to be the causes of other substances" and "all the things in the universe are found to be reciprocally connected with each other" (1:413). It is hardly possible not to consider this proof as question-begging.

Hence, Kant's proofs of the principles of succession and coexistence are not sound. They make unjustified transitions from the logical principle of the determining ground to its metaphysical use, i.e., the principle of sufficient reason, and to the principles of a specific metaphysical doctrine. The weakness of the proofs has long been noted by Kant scholars (Riehl 1924, 277; Hinske 1970, 88; Laywine 1993, 33), who have demonstrated that Kant's system of 1755/1756 lacks a strict demarcation between logic and metaphysics as well as a clear separation between the principles of the *metaphysica generalis* and the *metaphysica specialis*. Even though Kant is aware of Crusius's distinction between real and ideal grounds (see (Walford 1992, xl)), and even though he brings some aspects of it into play, he does not yet distinguish between the logical and the real use of reason. We have to conclude that the *New Elucidation* employs mixed methods which are incoherent and unclear.

2.2.3 *Physical Monadology: A Case for Conceptual Analysis*

Methodologically, the *Physical Monadology* at first sight looks similar. It proceeds *more geometrico* and employs conceptual analysis in order to demonstrate theorems, like the *New Elucidation*. However, the approach is different. The *Physical Monadology* contains only two definitions, and no principles. Hence, the approach *more geometrico* only concerns the structure of the writing, but not the deduction of theorems from principles. From Kant's point of view, this is a clear case of conceptual analysis, as his examples in the *Prize Essay* of 1764 demonstrate. In addition to conceptual analysis, the *Physical Monadology* makes use of geometrical constructions and arguments, and it starts with preliminary considerations concerning the relation between physics and metaphysics. Here we find Kant's first explicit methodological remarks on the relation between natural science and philosophy.

2.2.3.1 The *Preliminary Considerations*

Kant's preface to the *Physical Monadology* serves to make the main goals of the writing clear. The first is to combine metaphysics and geometry, the second is to reconcile two rival metaphysical doctrines concerning the constitution of matter. In the *Preliminary Considerations*, Kant argues that metaphysics is indispensable for natural science, even though the investigation of nature depends on experience and the use of mathematics. Criticizing a one-sided empiricist or sensualist attitude, he emphasizes that to rely on experience in search of truth should not be carried to excess:

> [...] there have been some who have observed this law to such a degree that, in searching out the truth, they have not ventured to commit themselves to the deep sea but have considered better to hug the coast, only admitting what is immediately revealed by the testimony of the senses. (1:475)

Then, he emphasizes that the restriction of natural science to immediate sensory experience misses out knowledge of the first causes of the phenomena, and the nature of bodies. Hence, physics is in need of metaphysics:

> And, certainly, if we follow this sound path, we can exhibit the laws of nature though not the origin and causes of these laws. For those who only hunt out the phenomena of nature are always that far removed from the deeper understanding of the first causes. Nor will they ever attain knowledge of the nature itself of bodies [...] Metaphysics, therefore, which many say may be properly absent from physics is, in fact, its only support [...]. (1:475)

This sharp criticism of an empiricist attitude towards natural science might be addressed to Newton and his followers, who restrict the analytic-synthetic methodology to inferences based on the analysis of the phenomena, consistent with a long-standing stereotype of Newton's empiricism. Kant might be thinking of Newton's famous demarcation between metaphysics and empirical science based on the *General Scholium* of the *Principia*:

> [...] I do not feign hypotheses. For whatever is not deduced from the phenomena must be called an hypothesis; and hypotheses, whether metaphysical or physical, or based on occult qualities, or mechanical, have no place in experimental philosophy. In this experimental philosophy, propositions are deduced from the phenomena and are made general by induction. (Newton 1726, 943)

Kant does not accept this demarcation, as his pre-critical system shows. The *Theory of the Heavens* goes beyond Newton's methodology, employing a hypothetico-deductive approach which extrapolates the analysis of the phenomena on the grounds of atomism to a cosmological model, and moreover to the physico-theological proof of God's existence (see Sect. 2.2.1). The *New Elucidation* defends the atomism of the *Theory of the Heavens* on phenomenological grounds and gives a metaphysical explanation of the interactions of substances in terms of the "schema of the divine understanding" (see Sect. 2.2.2). The *Physical Monadology* finally provides the atomism of the *Theory of the Heavens* and the *New Elucidation* with a metaphysical justification.

2.2.3.2 Conceptual Analysis Supported by Geometrical Arguments

The *Physical Monadology* is structured into two sections, each beginning with a definition followed by theorems and their proofs. The underlying concept of force is not defined. The *Physical Monadology* only gives definitions for the concepts of a simple substance and the contact of bodies. In the proofs Kant does not employ the results of the *New Elucidation*. He makes use only of conceptual analysis and geometrical constructions and arguments.

Section 2.1 begins with the definition of a simple substance or monad, as a substance which does not consist of separable parts. Logically, a simple substance is a mereological atom. On the basis of his definition, Kant proves the theorem that bodies consist of simple parts. The proof starts with the phenomenological premise that bodies consist of parts which independently of each other have enduring existence. Then, it proceeds by conceptual analysis of the part–whole relation and the definition of a simple substance. The crucial argument is that the composition of bodies is a contingent relation of the parts. Hence, this relation does not change the independent or separate existence of the parts; if the composition was removed, the simple parts of which a body consists would remain (1:477). The proof is analytical. It claims only that the simple parts of a compound are mereological atoms. But the proof *per se* does not us tell anything about whether there are finitely or infinitely many parts. Both options are still logically possible. In order to exclude the latter, i.e., the assumption of infinitely many infinitely small monads, Kant first employs a geometrical construction in order to demonstrate that the space which bodies fill is infinitely divisible (1:478), relying on contemporary proofs of the infinite divisibility of matter (Walford 1992, 422, n. 6). Then he demonstrates that an infinitely divisible compound cannot consist of simple parts. The proof makes use of the geometrical properties of any extended part of a continuum, specifically that any finite part of it is again extended. This property gives rise to the conclusion that no extended part of a body can be assumed to consist of points (a position which from a pre-Cantorian point of view is plausible); hence, an extended body cannot consist of (infinitely many) simple parts (1:479). (In the *CPR*, Kant will take this proof up again as a typical dogmatic argument that supports the thesis of the second antinomy; see Sect. 5.4.2.1.)

Subsequently, Kant develops his theory according to which the physical monads are simple or point-like, whereas due to their forces they nevertheless fill an extended spatial region. Section 2.2 proceeds in the same way as Section 2.1. It begins with the definition of the contact of bodies in terms of impenetrability, understood as a reciprocal repulsive force, which is explained in contradistinction to the Newtonian concept of action-at-a-distance (1:483). Then, Kant demonstrates several theorems concerning the repulsive and attractive forces of the physical monads, which explain the constitution of matter.

The demonstrations of the *Physical Monadology* are methodologically important because Kant takes them up as examples of the correct, analytic method of metaphysics in his *Prize Essay* of 1764. The analytic proof of the theorem that all bodies consist of simple substances or parts is indeed repeated twice in the *Prize Essay* as an example of conceptual analysis (2:279, 2:286), followed by an abridged demonstration of the other features of physical monads proven in 1756 (2:287–288) (see Sect. 2.3). In the *Physical Monadology*, Kant adds a methodological remark concerning his analytic method to the proof of the theorem that bodies consist of simple substances, emphasizing that he did not deduce this theorem from Leibniz's principle of sufficient reason (that is, according to the synthetic method):

> I have deliberately omitted the celebrated principle of sufficient ground from the present demonstration. In omitting it, I have accomplished my purpose by means of the ordinary combination of concepts to which all philosophers subscribe, for I was apprehensive that those who would not accept the principle of the sufficient ground would be less convinced by an argument which was based on it. (1:477)

Walford mentions that Crusius in particular did not accept the principle (Walford 1992, 422, n. 5). Nor, however, did the Newtonians. In his Fourth Reply to Leibniz, Clarke criticizes the way in which Leibniz employs the principle to derive the principle of indiscernibles and to reject atomism (Leibniz and Clarke 1715/1716, 29–32). Kant takes up this criticism in the *New Elucidation* (1:409–410; see Sects. 1.4.2.2 and 2.2.2). His methodological remark quoted above indicates that concerning his method he is also attempting to follow his "irenic" strategy of reconciling the positions of the Wolffians and Newtonians. That is, he makes use of a method which both parties should accept. In addition, he emphasizes that the uncontroversial method he employs is the traditional analytic-synthetic method: the analytic part of it is conceptual analysis, which aims at clarifying confused concepts in order to determine their distinct marks; the corresponding synthetic part is the combination of these distinct conceptual marks within his conception of physical monads, according to a method which in his logic lectures he calls the "synthesis of coordination" (see 9:59, 24:81, and Sect. 2.3.2.3).

Thus, Kant must have thought that his demonstrations in the *Physical Monadology* were uncontroversial and compelling, given that they invoked the common methodological grounds of Wolffianism and Newtonianism and that they were only based on conceptual analysis and geometrical arguments. The strongest evidence for this conviction is that Kant employed the arguments of the *Physical Monadology* thirty years later in a practically unchanged version, for the critical purposes of his cosmological antinomy. The proofs of the thesis and the antithesis of the second antinomy take up the reasoning of 1756 as an uncontroversial (though uncritical) and fallacious argument of pure reason (see Sect. 5.4.2 and *CPR*, A 434/B 462).

2.3 The Defence of the Analytic Method in the *Prize Essay*

For the year 1763 the Royal Prussian Academy of Sciences raised the *Prize Question*, "Are the metaphysical sciences capable of the same evidence as the mathematical sciences?" The question was addressed to the analytic and synthetic methods, which were at the time the subject of intense metaphysical debate. In response to this "either-or" question, Kant's *Prize Essay* is a prominent defence of the analytic method. He wrote the *Prize Essay* in 1762, but it was only published in 1764. As is well known, Mendelssohn was awarded the prize for his positive answer to the question: he attributed the method of conceptual analysis to philosophy as well as to mathematics, and argued that the certainty of philosophical analysis is comparable to that of mathematics, but has a lower degree of comprehensibility (Mendelssohn 1764; Schönfeld 2000, 211). Kant also attributed the analytic method to philosophy, but he associated it with the analytic method of Newtonian science, contrasted it with the synthetic method of mathematics, and rejected the latter for philosophy.

As shown above, in Kant's time there was no unambiguous analytic method (Sect. 2.1), nor did Kant employ such a method (Sect. 2.2). In his 1755/1756 writings he had used several unconnected methodological strategies, and at first glance it seems he associates them with "the" analytic method defended in the *Prize Essay*. There, he seems to put conceptual analysis in a Wolffian sense on a par with the analysis of the phenomena in Newton's sense, downplaying their substantial differences. Given that both kinds of analysis emerged from common methodological grounds, i.e., the analytic-synthetic tradition of early modern science and philosophy, he may even have had a certain justification for doing so.

The *Prize Essay* supports the defence of the analytic method with examples taken from the *Physical Monadology* of 1756. However, now Kant emphasizes an opposition between the analytic and the synthetic method that he had *not* stressed in his earlier writings. Hence, careful analysis is needed in order to understand his defence of the analytic method in the *Prize Essay*. In the following, I want to show that Kant in fact does not identify the conceptual analysis of metaphysics with Newton's method. His goal in the *Prize Essay* is rather to illustrate a structural affinity between both variants of the method, which gives rise to a methodological analogy between philosophy and Newtonian science.

The *Prize Essay* is Kant's first work to deal explicitly with methodological questions. It distinguishes analytic and synthetic procedures of deriving concepts, discusses them, and finally compares the correct method of metaphysics to the analytic method of Newtonian science. Kant claims that metaphysical definitions can only be obtained by analysis, in contrast to the synthetic method of mathematics which is built on concepts arbitrarily composed from simpler concepts. He argues that metaphysics is a discipline of conceptual analysis, and that its goal is to clarify the distinct marks of given concepts, as far as it is possible to do so with certainty.

The *Prize Essay* does not directly criticize the synthetic method of a metaphysics *more geometrico* for generating a dogmatic doctrine in terms of definitions, axiomatic principles, and the demonstration of theorems. Kant primarily attacks the foundations of such a doctrine, namely the way of obtaining the definitions of metaphysical concepts as the starting point of the synthetic method. He attacks

> [...] the shallowness of the proofs offered by the metaphysicians when, in accordance with their custom, they confidently establish their conclusions on the basis of definitions which have been laid down once and for all as the foundations of their argument. The conclusions instantly collapse if the definitions are defective. (2:288)

The context is Kant's conceptual analysis of the nature of bodies (2:288), his major example in the *Prize Essay*, which takes up the proofs given in the *Physical Monadology* of 1756. The above quotation attacks the definitions of metaphysics, i.e., the results of conceptual analysis in Wolff's sense. In his view it is not the method *more geometrico* as such which is defective, but rather its starting point, the metaphysical definitions. To rule the synthetic method out for metaphysics, in contradistinction to mathematics, is for Kant merely a consequence of his criticism of arbitrary, barely justified metaphysical concepts.

2.3.1 The Background: Wolff, Leibniz, Descartes

Before looking in more detail at Kant's argument, it is useful to sketch the relevant rationalist background. The *Prize Essay* refers to Leibniz and Wolff (in §1, 2:277), whereas it does not explicitly mention Descartes. However, there are several indirect allusions to Descartes, and in fact the argument of the *Prize Essay* is related to all three rationalist thinkers' views on analysis and synthesis, and in particular on conceptual analysis. The textbook which Kant used all his life for his Logic lectures, Meier's *Auszug aus der Vernunftlehre* (1752), defines conceptual analysis as follows:

> The action, which brings about a certain degree of distinctness in our cognitio, is called the *dissection of cognition*[1] (resolutio, analysis, anatomia cognitionis). (Meier 1752, § 139, 16:340)

Kant's emphasis on conceptual analysis is far from original. The philosophical tradition of conceptual analysis traces back to the Cartesian theory of clear and distinct ideas (Appendix A.3.1). To clarify the characteristic marks of confused concepts and to make them distinct was indeed part of the philosophical program of the Enlightenment (Aufklärung). "Aufklärung" remained the key concept of eighteenth-century German philosophy until Kant's transcendental idealism was received (Engfer 1982, 26–27). Therefore, the main achievement of the *Prize Essay* cannot just be its demand for conceptual analysis. Lambert (whose prize essay remained a draft) defends the analytic method, but Kant's approach to the 1763 *Prize Question* has two distinguishing marks. On the one hand, it demarcates the philosophical method of conceptual analysis as against the synthetic procedure of constructing concepts in geometry; on the other, it connects the conceptual analysis of philosophy to Newton's methodology.

Newton's analytic-synthetic method, as well as the rationalist account of analysis and synthesis, derive from Pappus's model. Newton, like the authors of the influential Port-Royal Logic, follows Pappus's explanation that the analytic method is the regress from given consequences to the principles, and the synthetic method is the progress from the principles to their consequences. Even though Newton and the authors of the Port-Royal Logic share this general understanding of analysis and synthesis, they realign it in different ways. On their common ground, however, Kant is able to make the bridge from conceptual analysis to Newton's methodology (see Sect. 2.3.2.3). In accordance with the Port-Royal Logic, Meier's Logic textbook

[1] "Die Handlung, wodurch ein gewisser Grad an Deutlichkeit in unserer Erkenntniss hervorgebracht wird, heisst *die Zergliederung der Erkenntniss*"; my translation.

expresses Pappus's distinction between analysis and synthesis in terms of logical grounds and consequences:

> Thus, the grounds are either put prior to the consequences, or subsequently. The former is *the synthetic method of instruction* (methodus synthetica), the latter *the analytic* (methodus analytica).[2] (Meier 1752, § 422, 16:786)

On a closer inspection, Kant's argument for the use of the analytic method in philosophy comes close to the Cartesian account of analysis and synthesis. Descartes assigned analytic demonstrations to philosophy and synthetic demonstrations to geometry. To Mersenne's objection against the employment of the analytic method of reasoning in the *Meditations* he replied:

> It was this synthesis alone that the ancient Geometers employed in their writings [...] But I have used in my Meditations only analysis, which is the best and truest method of teaching. On the other hand synthesis, [...] though it very suitably finds a place after analysis in the domain of geometry, nevertheless cannot so conveniently be applied to these metaphysical matters we are discussing.
>
> For there is this difference between the two cases, viz. that the primary notions that are the presuppositions of geometrical proofs harmonize with the use of our senses, and are readily granted by all. Hence, no difficulty is involved in this case [...]. On the contrary, nothing in metaphysics causes more trouble than the making the perception of its primary notions clear and distinct. (Descartes 1934, Vol. II, 49)

However, Kant's methodological stance is far from identical with Descartes's, who favors the analytic rather than the synthetic method for geometry (Hintikka and Remes 1974, 112). In addition, Kant distances himself from the Cartesian ideal of a *mathesis universalis*, which was taken up by Leibniz and received in Wolffianism. Concerning the distinction between the methods of philosophy and mathematics, he emphasizes that

> [...] nothing has been more damaging to philosophy than mathematics, and in particular the *imitation* of its method in contexts where it cannot possibly be employed. (2:283)

Even though he does not explicitly mention Descartes in the *Prize Essay*, Kant leaves no doubt that the Cartesian as well as all other metaphysics (including his own of 1755/1756!) failed to arrive at certainty. In contradistinction to mathematical knowledge, metaphysical theories share the "fate of opinions" to be untenable:

> Claims to philosophical cognition generally enjoy the fate of opinions and are like the meteors, the brilliance of which is no guarantee of their endurance. Claims to philosophical cognitions vanish, but mathematics endures. Metaphysics is without doubt the most difficult of all the things into which man has insight. But so far no metaphysics has ever been written. (2:283)

Nevertheless, his own methodological approach, which stands up to the metaphysical struggles of the time, is still indebted to the Cartesian tradition of renewing the foundations of metaphysics in epistemological terms. The *Second reflection* of the

[2]"Also werden die Gründe entweder den Folgen vorgesetzt, oder nachgesetzt. Jene ist *die synthetische* (methodus synthetica), diese aber *die analytische Lehrart* (methodus analytica)"; my translation.

Prize Essay begins with the Cartesian definition of metaphysics as the doctrine of the first principles of our cognition, just as Baumgarten's *Metaphysics* does:

> § 1. *Metaphysica* est scientia primorum in humana cognitione principiorum. (Baumgarten 1757, 17:23)

> Metaphysics is nothing other than the philosophy of the fundamental principles [über die ersten Gründe] of our cognition. (2:283)

> [...] the true philosophy, the first part of which is metaphysics, which contains the principles of knowledge [...] Thus philosophy as a whole is like a tree whose roots are metaphysics, whose trunk is physics, and whose branches, which issue from this trunk, are all the other sciences. (Descartes 1934, Vol. I, 211)

The first part of Descartes's *Principia philosophiae*, the metaphysics, has the title: *Of the Principles of Human Knowledge*. Descartes's definition of metaphysics, which is still to be found at the beginning of Baumgarten's *Metaphysics*, belonged to the common Cartesian inheritance of Wolffianism. The main distinction between Kant's and Baumgarten's definitions is that Kant differentiates philosophy in general, on the one hand, and metaphysics as the particular philosophical doctrine of the first principles of cognition, on the other, as did Descartes in his metaphor of the tree structure of philosophy.

2.3.2 *The Argument of the* Prize Essay

In line with the Leibniz–Wolffian tradition, Kant considers analysis and synthesis to be radically opposed ways of arriving at concepts. According to § 1 of the *First reflection* of the *Prize Essay*, synthesis consists in the construction of concepts by an arbitrary combination of characteristic marks, whereas analysis consists in the decomposition of given complex concepts into their distinct marks.

2.3.2.1 Mathematical vs. Philosophical Definitions

In contrast to the Leibniz–Wolffian tradition, however, Kant relates the synthetic method of defining concepts via the combination of marks to mathematics, and the analytic method of defining concepts based on conceptual analysis to philosophy. He justifies this classification as follows: given that mathematics is a discipline which defines its concepts in an arbitrary way, it can only arrive at definitions by combining concepts, i.e., by the synthetic method.

> The concept which I am defining is not given prior to the definition itself; on the contrary, it only comes into existence as a result of that definition. [...] In this and in all other cases, the definition obviously only comes into being as a result of *synthesis*. (2:276)

He illustrates this claim with the examples of a trapezium and a cone. Contrary to mathematics, philosophy is wisdom or cognition of the world (*Weltweisheit*). As such, it has the task of clarifying the confused concepts of given things, by making their characteristic marks distinct via conceptual analysis.

> In philosophy [*Weltweisheit*], the concept of a thing is always given, albeit confusedly or in an insufficiently determined fashion. The concept has to be analysed; the characteristic marks which have been separated out and the concept which has been given have to be compared with each other in all kinds of contexts; and this abstract thought must be rendered complete [*ausführlich*][3] and determinate. (2:276)

As an example, he considers the concept of time. The difficulties of arriving at a philosophical definition of the concept of time are well known from Augustine's famous remarks in the *Confessions*, which Kant also quotes (2:283).

> The idea of time has to be examined in all kinds of relations if its characteristic marks have to be combined together to see whether they yield an adequate concept [*zureichenden Begriff*][4] [. . .]. If, in this case, I had tried to arrive at a definition of time synthetically, it would have had to have been a happy coincidence indeed if the concept, thus reached synthetically, had been exactly the same as that which expresses the idea of time which is given to us. (2:276–277)

The crucial difference between mathematics and philosophy is obvious. Philosophical concepts apply to the world and hence they have to satisfy adequacy conditions, in contradistinction to purely mathematical concepts. Kant adds sharp criticism of Leibniz's arbitrary, synthetic definition of the philosophical concept of a slumbering monad (which for him is an invention rather than a definition), and of Wolff's analytical definition of the concept of similarity (2:277).

Finally, he admits that some mathematical concepts are also susceptible to philosophical explanation or exposition. His detailed example, the concept of space, refers to the 1756 project of unifying geometry and metaphysics in the *Physical Monadology* (see Sects. 1.3.2 and 1.3.2.2). It is exactly this concept, however, which will give rise to a related criticism of Leibniz's project of an *analysis situs* and turn out to be the weak point of his pre-critical cosmology, five years later (see Sect. 3.4).

2.3.2.2 Two Methodological Rules

In the *Second reflection*, Kant proposes two methodological rules which in his view guarantee as much metaphysical certainty as possible:

---

[3]The literal translation of 'ausführlich' is 'in extenso'. Here, however, Kant refers to a concept contained in his lectures on logic. The Cambridge translation (Walford and Meerbote 1992, 444, n. 14) refers to §16 of the *Logic*: "A *completely* distinct concept can be so [. . .] either in regard to the totality of its *coordinate* marks or with respect to the totality of its *subordinate* marks. *Extensively* complete or sufficient distinctness of a concept consists in the total clarity of its coordinate marks, which is also called *exhaustiveness*" (9:99). See also the notes 2406–2408 on logic, which refer to §147 of G. F. Meier's logic (14:349–350). The background is Leibniz's account of a "complete concept" (see Appendix A.3.2), which Kant's *Prize Essay* no longer maintained for metaphysical concepts.

[4]The literal translation of 'zureichend' is 'sufficient'. The term 'adequate' suggests an implicit reference to Leibniz's "adequate idea", which is the complete concept in Leibniz's sense (see Appendix A.3.2). Again, Kant refers to §147 of Meier's logic and the subjects of his own logic (9:99), dealing with the adequacy conditions of philosophical concepts.

> The *first* and the most important [*vornehmste*][5] rule is this: one ought not to start with definitions, unless that is, one is merely seeking a nominal definition, such as, for example, the definition: that of which the opposite is impossible is necessary. (2:285)

Here, he claims that a metaphysical theory does not admit of real definitions as its starting point, but only of nominal definitions. This "most important" rule precludes a synthetic approach *more geometrico* which is based on arbitrary, unjustified definitions that go beyond the logical truth of analytic judgements. Kant's claim has to be understood against the background of Leibniz's theory of definitions. According to Leibniz, a real definition must satisfy the condition that the possibility of the object so defined follows from it (Leibniz 1684, 25, 26). In the context of Leibniz's theory of the complete concept and the compossibility of all substances, this is a very strong condition. According to Leibniz, a non-contradictory definition obtained by conceptual analysis and based on finitely many conceptual marks in general does not give rise to an adequate, complete, concept (Leibniz 1684, 24) (see Appendix A.3.2). Detached from Leibniz's logic and metaphysics, only the very weak condition of logical possibility remains, i.e., the logical consistency of a definition.

A nominal definition in Leibniz's sense suffices to distinguish objects simply by some of their characteristic marks. It does not give rise to a well-justified metaphysical definition in Kant's sense, but at most to the *exposition* of a concept (*Logic* §102, 9:141). The exposition of a metaphysical concept must not be mistaken for a nominal nor for a real definition. According to the *Prize Essay*, it just may serve as the starting point for further conceptual analysis that lays the ground for the synthetic approach *more geometrico*:

> The *second* rule is: one ought particularly to distinguish those judgements which have been immediately made about the object and relate to what one initially encountered in that object with certainty. Having established for certain that none of these judgements is contained in another, these judgments are to be placed at the beginning of one's inquiry, as the foundation of all one's inferences, like the axioms of geometry. (2:285)

Hence, the immediate judgements which can be made with certainty have to be fixed. In so far as they are logically independent of each other, they may serve as a starting point for the approach *more geometrico*. These immediate judgements give rise to the exposition of metaphysical concepts. The second rule recalls the following remark in Leibniz's *Meditationes de Cognitione, Veritate et Ideis*:

> Furthermore, the rules of common logic, which even the geometers use, are not to be despised as criteria for the truth of assertions, as, for example, the rule that nothing is to be admitted as certain, unless it is shown by careful testing or sound demonstration. (Leibniz 1684, 27)

Kant, however, reinterprets these Leibnizian criteria of truth in a Newtonian sense. In his view, to apply Newton's method in metaphysics requires to "seek out the rules" for the exposition of metaphysical concepts "on the basis of certain

[5]The literal translation of 'vornehmste' is 'most distinguished'.

experience and, if need be, with the help of geometry" (2:286) (see below). In this way, his views about conceptual analysis in metaphysics come close to an analysis of the phenomena, which starts with confused concepts and attempts to find their distinct characteristic marks, in order to make reliable judgments about the distinct marks. Such judgments then may serve as a quasi-axiomatic basis for the synthetic method.

A substantial problem with Kant's suggestion however remains. How might the exposition of metaphysical concepts give rise to *logically independent* definitions? How might Kant guarantee the independence of the quasi-axiomatic principles of a theory which is built on an incomplete exposition of its basic concepts, given that the complete nature of the things in question is unknown? Kant's methodological rules amount to squaring the circle. The *Prize Essay* suggests a method according to which metaphysics differs from mathematics in achieving expositions, but not in providing definitions of its fundamental concepts. Nonetheless, this method is considered to preserve the certainty of the synthetic method *more geometrico* as far as possible.

2.3.2.3 The Relation to Newton's Method

The passage in which Kant compares Newton's analytic method to his own metaphysical method following his two methodological rules, begins with the famous claim:

> The true method of metaphysics is basically the same as that introduced by *Newton* into natural science and which has been of such benefit to it. (2:286)

However, just as for Newton, for Kant the synthetic method *more geometrico* is only the second step of a *combined* analytic-synthetic approach (see also Grier 2001, 31–32). First comes the analysis of the phenomena, which aims at finding the laws of nature according to which the phenomena occur:

> *Newton*'s method maintains that one ought, on the basis of certain experience and, if need be, with the help of geometry, to seek out the rules in accordance with which certain phenomena of nature occur. (2:286)

Kant admits that the laws established in such a way may be not fundamental:

> Even if one does not discover the fundamental principles of these occurrences in the bodies themselves, it is nonetheless certain that they operate in accordance with this law. (2:286)

Nevertheless, the phenomena are sufficiently explained once the synthetic step of the analytic-synthetic method has been carried out:

> Complex natural phenomena are explained once it has been clearly shown how they are governed by these well-established laws. (2:286)

The parallel between conceptual analysis in philosophy and the analysis of natural phenomena in physics is that both methods seek to perform a regress, from the thing that is given, to certain underlying principles, in accordance with Pappus's account of analysis. The distinct marks of metaphysical concepts have to be obtained from an analysis of the phenomena found in our mind, just as the laws of physics are obtained from an analysis of the phenomena of nature:

> Likewise in metaphysics: by means of certain inner experience, that is to say, by means of an immediate and self-evident inner consciousness, seek out these characteristic marks which are certainly to be found in the concept of any general property. And even if you are not acquainted with the complete essence of the thing, you can still safely employ those characteristic marks to infer a great deal from them about the thing in question. (2:286)

Here, too, Kant specifies both the analytic and the synthetic step of the method. The analysis consists in looking for the distinct marks which a concept certainly has, according to inner experience. The synthesis, in turn, consists in deriving several crucial features of the nature of a thing. Against the Leibniz–Wolffian background, however, it is clear that the analysis of mental phenomena and the corresponding empirical concepts do not attain to grasp "the complete essence of the thing", i.e., they do not give rise to a complete or adequate concept in Leibniz's sense. But Kant's proviso, cited above, "Even if one does not discover the fundamental principles of these occurrences in the bodies themselves" (2:286), indicates the view that the analysis of natural phenomena in physics does neither.

Kant scholars differ on the significance of Kant's methodological comparison. According to Schönfeld (2000, 224), Kant *identifies* the inner experience of certain marks of metaphysical concepts with Newton's outer experience. However, Kant's way of equating both kinds of experience may also be understood in a weaker sense, that is, by analogy. Friedman's observation that according to Kant metaphysics has to "adopt a quasi-inductive or regressive method" (Friedman 1992, 22) is also compatible with such a weaker, analogical sense.

We have already seen that in the methodology of the *Theory of the Heavens* (see Sect. 2.2.1), Kant relies on analogies in the inductive sense, that is, of the analogical reasoning which belongs to Newton's analytic method as described in the *Opticks* (Newton 1730). In addition, the *Prize Essay* brings into play the concept of an analogy in the sense of a quasi-mathematical proportion. Kant still uses this conception in the *CPR* (Callanan 2008; Pieper 1996) and the *CPJ* (Breitenbach 2009, 70–75). According to it, an analogy is a double ratio

$$A : B = C : D,$$

just as in the traditional metaphysical analogy "God relates to the World as the watchmaker to the clock". Kant indeed employs this analogy in the *Only Possible Argument*, in the context of the physico-theological proof of the existence of God (2:153). Here it is a causal analogy, whereas the analogy of the *Prize Essay* is merely methodological. But this methodological analogy, too, has the structure of a double ratio. According to the above quotations, Newton's analytic-synthetic method relates to natural phenomena and their explanation in terms of laws, just as conceptual analysis and the subsequent deductions *more geometrico* relate to the phenomena of inner experience and the inferences drawn from its characteristic marks. This methodological analogy fits in well with the reconciliatory strategy of Kant's 1755/1756 project of incorporating the principles of Newton's physics into a Wolffian system of metaphysics. A related analogy between the methods of Newtonian science and of metaphysics will recur in Kant's experiment of pure reason, in the second edition of the *CPR* (see below, Sect. 6.2.2).

Kant's comparison of the analytic-synthetic methods of metaphysics and physics is remarkable in several respects. It refers to the common methodological grounds of the positions involved in the methodological debates of the time, that is, the tradition of analysis and synthesis which was still effective in Wolffianism as well as in Newtonianism (even though in different ways). The bridge between the methods of metaphysics and physics has two pillars. One is Kant's claim that both should rely on the *complete* method of analysis and synthesis in Pappus's sense, according to which analysis precedes synthesis. The other is the criterion of certainty, or empirical evidence, which according to Kant applies to Newton's analysis of the phenomena of physics as well as to clear and distinct concepts *à la* Descartes or Leibniz. Kant implicitly recalls the analysis and synthesis of problems according to the second and third rules of Descartes's *Discourse*, but also Descartes's way of assuring himself of certain knowledge via the *cogito*. The reference to "immediate and self-evident inner consciousness" seems to invoke the first rule of Descartes's *Discourse*, i.e., Cartesian evidence.

In this way, Kant's comparison establishes an analogy between the regress from confused concepts to their distinct marks and the regress from phenomena to their principles in physics. Kant merely had to combine two variants of the analytic method, which were familiar to him from Wolffian logic: the analysis of complex concepts into distinct partial concepts, and the regress from given consequences to their principles.[6] In his Logic Lectures, he combines both variants of analysis, calling them the analysis of *coordination* and of *subordination*, respectively (see 9:59 and 24:81). Analysis and synthesis of coordination consist in the decomposition and recomposition of a given whole and its parts; analysis and synthesis of subordination consist in the inference from consequences to their principles, and the deduction of consequences from the principles, respectively.

Hence, it is not true that Kant had already forgotten the Wolffian tradition of justifying metaphysical concepts (as claimed by Engfer 1982, 150–151, n. 56). Quite on the contrary, Kant invokes it as part of an impressive rhetoric. The way in which he combines coordinative and subordinate analysis and synthesis is a logical justification for his methodological analogy between Newton's analysis of the phenomena and the Leibniz–Wolffian analysis of confused concepts.

---

[6]See Meier's logic, §139 and §422 (Meier 1752, 16:340,16:786). In his lectures on logic, Kant repeatedly refers to both paragraphs of Meier's logic textbook. In his comments to § 139, he emphasizes that concepts may not only be clarified by analysis, but also by synthesis, i.e., by arbitrarily combining clear conceptual marks. He distinguishes the procedures of making a concept clear and distinct (analysis) or making a clear and distinct concept (synthesis). See the early Blomberg logic (presumably 1771) (24:130–131), or the late Vienna logic (1796) (24:843 pp).

2.4 Comparison with the *Only Possible Argument*

The *Only Possible Argument* was probably completed shortly before the *Prize Essay* (Walford 1992, lix, lxiii). Schönfeld (2000, 214) notes an important tension between defending the analytic method of metaphysics in the *Prize Essay* and playing down the certainty of all demonstrations of the existence of God in the *Only Possible Argument*. I cannot see such a tension, given that neither of these writings claims absolute certainty for the use of the analytic method in metaphysics, as compared to the certainty of mathematical proofs. The *Prize Essay* applies the analytic method to the principles of natural theology (2:296–297), taking up the dubious 1755 proof of the existence of God (1:395; see Sect. 2.2.2.1). But I cannot see that the *Prize Essay* "portrayed the ontological proof as the paradigm of certainty", as Schönfeld (2000, 214) claims. Concerning this proof, Kant now emphasized two issues. First, the distinctness and certainty of these principles is restricted to the concept of an absolutely necessary being and the results of its conceptual analysis. Second, these principles are not absolutely certain, but have only a "high degree of certainty":

> Metaphysical cognition of God is thus capable of a high degree of certainty in all those areas where no analogon of contingency is encountered. (2:297)

Methodologically, the approach in the *Only Possible Argument* is completely compatible with the defence of the analytic-synthetic method in the *Prize Essay*, as will now be sketched. The *Only Possible Argument* discusses the traditional arguments in favour of the existence of God and proposes its own distinct demonstration, which takes up the 1755 proof. In order to clarify the crucial concepts of existence and of God, Kant rebuts synthetic definitions of the concepts, in accordance with the first methodological rule of the *Prize Essay* (see Sect. 2.3.2.2). With regard to the concept of existence ("Dasein"), he explains his approach as follows.

> It is not to be expected that I shall begin by offering a formal definition of existence. Such a procedure is always undesirable when the correctness of the suggested definition is so uncertain. [. . .] My procedure will be like that of someone who is searching for a definition and who first of all assures himself of what can be said with certainty, either affirmatively or negatively, about the object of the definition, even though he has not yet established the concept of the object in detail. (2:71)

These remarks, which describe the method of conceptual analysis explained in the *Prize Essay*, are followed by an explicit rejection of the synthetic method of mathematics:

> To aspire to a definition is to venture upon unnecessary difficulties. The mania for method and the imitation of the mathematician, who advances with a sure step along a well-surfaced road, have occasioned a large number of such mishaps on the slippery grounds of metaphysics. (2:71)

Kant then proceeds to his famous argument that existence is not a predicate, which lays the grounds for his criticism of the traditional ontological proof of the existence of God. Even though "existence" is sometimes grammatically used as a predicate, this use should not be confused with the attribution of characteristic marks to a thing, and it should not be misused to derive existence from mere possibility (2:72–73). Subsequently, he criticizes Wolff's definitions of existence as "completion of possibility" and Baumgarten's related "concept of thoroughgoing internal determination" (2:76), both of which confuse the predicates of a real and a possible thing. Hence, the crucial point is that existence is no *real* predicate, but a modal predicate which is opposed to mere possibility:

> But we have already seen that the difference between a real thing and a mere possible thing never lies in the connection of this thing with all the predicates which can be thought in it. (2:76)

With regard to the concept of God, Kant also rejects the synthetic method in favor of an analytic approach. In particular, he emphasizes that he restricts himself to "an analysis" which does not aim at a "formal doctrine", but may just contribute to finding the foundations of such a theory. Here, he associates the method *more geometrico* with a *systematic* approach which starts with definitions, using the term "systematic" for the first time in the sense of a systematic theory, i.e., a cognitive system (see Appendix A.1.3):

> I am not here offering a determinate definition of the concept of God. If it were my purpose to treat the matter systematically, I should have to provide such a definition. But what I am here setting forth is intended to be an analysis which may serve as a foundation for the formal doctrine proper. (2:89)

The analysis is the criticism of the traditional proofs of the existence of God, and the suggestion of his own distinct proof, which is based on an analysis of the concepts of logical and real possibility as well as absolute necessity, as in the *Prize Essay* (2:297) and the *New Elucidation* (1:395). To go into more detail is beyond the scope of the present work. One more feature of the *Only Possible Argument*, however, matters here. In the last section, Kant classifies the arguments for the existence of God as follows:

1. Proofs from the concept of the *possible*:
    - 1.1 *Synthetic* argument, which proceeds "from the possible as a *ground* to the existence of God as a *consequence*". (2:156)
    - 1.2 *Analytic* argument, which proceeds "from the possible as a *consequence* to the divine existence as a *ground*". (2:156)
2. Proofs from the concept of the *existent*:
    - 2.1 *Synthetic* argument, which proceeds from "the existence of what we experience to the existence merely of a first and *independent cause*" (first analytic step) "and then, by subjecting this concept to analysis" (second analytic step) proceeds "to the derivation of its divine characteristics" (synthetic step). (2:156)
    - 2.2 *Analytic* argument, which proceeds "directly from that which experience teaches us to both the existence and the properties of the Divine Being". (2:156)

Schönfeld (2000, 196) correctly emphasizes that this classification does not deal with specific proofs or arguments, but just with their foundations, as indicated by the German expression *Beweisgrund.* Kant classifies the *types* of possible proofs or arguments in favour of the existence of God, and he discusses some *examples* of specific proofs belonging to the four classes. The classification is based on the concepts of "possibility" (1.) and "existence" (2.), on the one hand, and the logical distinction between analysis (regress from consequences to grounds: 1.2 and 2.2) and synthesis (progress from grounds to consequences: 1.1 and 2.1), on the other.

It is not obvious that the proofs under the heading 2.1 are synthetic, but Kant argues that they are. He discusses the Cartesian ontological proof as a case of a synthetic argument from the concept of the possible (1.1) and rejects it by referring to his own conceptual analysis, according to which 'existence' does not belong to the real predicates of a thing (2:156). Proofs of the kind 2.1 in his view just pretend to follow a combined analytic-synthetic method, whereas in fact they do not, but rather employ the principle of sufficient reason. Kant attributes them to the Wolffian school (2:157) and explains that all of their variants are defective. He claims that

> [...] no matter how they are revised, these proofs can never be anything but arguments from concepts of possible things, not inferences from experience. At best, therefore, they are to be counted among the proofs of the first kind. (2:159)

Why do they reduce to proofs of the kind 1.1, in his view? Probably, he considers them to be based on arbitrary definitions rather than on conceptual analysis, violating the methodological rules of the *Prize Essay*.

Unfortunately, his own proof is not much better off. It makes an inference from the internal possibility of the things in the world to God as the real ground of this possibility, and it is ranked among those headed 1.2 (2:157). As in 1755, Kant attempts to derive God's existence from the impossibility that nothing exists and the necessity that the real possibility of things must have a ground of being, or *ratio essendi* (see Sect. 2.2.2.1). He considers this to be an improved version of the traditional ontological proof. Schönfeld (2000, 197–205) discusses Kant's ontological proof (and its flaws) in detail against the background of the *New Elucidation*.

However, Schönfeld's very instructive analysis of Kant's arguments in the *Only Possible Argument*, their background and their flaws, misses a crucial issue concerning the relation of the proofs 1.2 and 2.2. The remaining proof under heading 2.2 is the "cosmological proof" to which the physico-theological proofs discussed in Section 2.2 of the *Only Possible Argument* belong, and in particular Kant's physico-theological proof of the *Theory of the Heavens* of 1755 and its elaboration in the *Only Possible Argument*. Kant still avers that it is "so natural, so persuasive", but emphasizes that it nevertheless lacks mathematical rigour (2:160). Thus, the outcome of Kant's discussion of the four kinds of proofs seems to be that, according to Kant, even his own refined version of the physico-theological proof (2.2) is untenable, and only the ontological proofs of type 1.2 remain.

In addition, it remains unclear why in Sect. 2.3 of the *Only Possible Argument* Kant presents *two* possible proofs (among them the physico-theological proof), on the one hand (2:159–161), and at the same time claims that there is only one demonstration, on the other (2:162–163). According to Schönfeld (2000, 192–196), this is an unresolved puzzle in Kant research. In my view, Kant's account of the analytic-synthetic method sheds at least some light on this puzzle. Once Kant's ontological proof (1.2) and his physico-theological proof (2.2) are understood as two parts of a combined analytic-synthetic argument, it becomes clear that in his view they complement each other and only together give a complete demonstration.

Indeed, the two proofs do not stand apart. In accordance with Newton's methodological rules, the tenable part of the physico-theological proof (2.2) is the inference to a common physical cause of the observable "systematic constitution" of the universe, in approximately planar rotating systems of celestial bodies such as the solar system (see Sect. 2.2.1). Only in a further regressive step does Kant trace the uniform initial condition of his cosmological model back to God as the creator of the universe, in 1755 as well as in 1763. At the end of Section 2.1 of the *Only Possible Argument*, Kant establishes the following link between this proof result and his ontological proof (1.2):

> Our mature judgement of the essential properties of the things known to us through experience enables us, even in the necessary determinations of their internal possibilities, to perceive unity in what is manifold and harmoniousness in what is separated. It follows that the *a posteriori* mode of cognition will enable us to argue regressively to a single principle of all possibility. We shall thus finally arrive at the self-same fundamental concept of absolutely necessary existence, from which the *a priori* mode of cognition initially started out. (2:92).

Hence, even though the proofs 1.2 and 2.2 are analytic with regard to the logical regress from consequences to their grounds, in another regard, namely concerning their epistemic grounds or *ratio cognoscendi*, they relate as the two complementary steps of a combined analytic-synthetic argument. The epistemic grounds of the physico-theological argument (2.2) are *a posteriori*, while those of the ontological proof (1.2) are *a priori*. In accordance with Newton's methodology and its metaphysical analogue of the *Prize Essay*, they combine into an analytic-synthetic argument as follows. Its *analytic step* is the physico-theological proof (2.2). It consists in an analytic regress from the *a posteriori* observed phenomena to God as their common cause. (Here, Newton's methodology supports the inference to a common physical cause of the systematic constitution of the universe, and by extension to metaphysics: to God as the ultimate cause of the observable order of the universe.) The physico-theological proof, however, lacks mathematical certainty. To make the argument complete, it has to be complemented by the synthetic part of the analytic-synthetic method. This *synthetic step* is the ontological argument (1.2), "from which the *a priori* mode of cognition initially started out". The argument seems to be synthetic with regard to Kant's claim that it is *a priori*, starting from the *a priori* concept of the possible.

According to Schönfeld (2000, 201–205), the crucial flaw in Kant's argument (1.2) is that here the conceptual construction of a material condition enters, which refers to Leibniz's concept of the (real) possibility of substances. Hence, in the last analysis, Kant's ontological argument cannot escape the kind of flaws that he himself criticized in the arguments of types 1.1 or 2.1. Further questions remain. Even though it makes sense to assume that the proofs 2.2 and 1.2 combine to one analytic-synthetic proof, why does Kant in the *Only Possible Argument* first present the synthetic or *a priori* part (1.2) of the argument, which starts "from [...] the *a priori* mode of cognition", and only then the analytic or *a posteriori* part (2.2) of a regress from the phenomena to God as their cause? Why does synthesis precede analysis here, as against Newton's claim in the *Opticks* that it should be the other way round? (Newton 1730, 404; Appendix A.2.2.2). And how do synthesis and analysis as the partial steps of a combined analytic-synthetic argument relate to classifying their epistemic proof grounds as *a priori* and *a posteriori*? One cannot escape the impression that in order to focus on other metaphysical problems, Kant did not sufficiently work out his ontological proof (2.2) and its relation to the physico-theological proof (1.2).

2.5 Conclusions: The Scope of the Analytic Method

In the 1755/1756 writings, Kant made use of the analytical method (or its Newtonian and Wolffian variants, respectively) in order to make bridges between the parts of a system of metaphysics in Wolff's sense and Newton's physics. The *Theory of the Heavens* employs Newtonian analysis of the phenomena, extending Newton's method to a hypothetico-deductive approach based on a cosmological model of structure formation in the universe. The *New Elucidation* and the *Physical Monadology* make use of conceptual analysis in Wolff's sense in order to support metaphysical definitions. The *Physical Monadology* combines conceptual analysis with geometrical demonstrations, whereas the *New Elucidation* combines it with deductions of theorems from logical and metaphysical principles, on the one hand, and analysis of the phenomena in the sense of Newton's methodology, on the other. These deductions bring the principle of sufficient reason into play, which Kant reformulates as a merely logical principle in terms of the determining ground. However, the *New Elucidation* does not yet clearly demarcate the logical use of this principle against its use in metaphysics, given that a clear distinction between logical and real determining grounds is still lacking in it. On the grounds of the logical principle, Kant criticizes the concept of a *causa sui* which Spinoza's metaphysics employs as a self-contradictory concept. In addition, he criticizes Leibniz's interpretation of the principle of indiscernibles, arguing that it is not a mere logical consequence of Leibniz's principle of sufficient reason. But his own proofs of the existence of God and the principles of succession and coexistence combine logical conclusions and undisputed metaphysical premises in an unclear way. Hence, the *New Elucidation* does not make the scope of conceptual analysis in metaphysics clear.

Only the *Prize Essay* finally attempts to do so. Here, Kant defends the analytic method of conceptual analysis by analogy with Newton's method. His convincing examples stem from the *Physical Monadology*. His use of the expression

Weltweisheit (wisdom of the world) for philosophy marks the difference between mathematical and metaphysical concepts. It indicates that metaphysical concepts refer to objects in the world, in contradistinction to logical principles and the arbitrary definition of mathematical concepts. Therefore, the meaning of metaphysical concepts depends on experience. Kant's comparison with Newton's method demonstrates that in order to find adequate definitions, conceptual analysis should rely on "certain experience" or "certain phenomena", which in the case of metaphysics is "inner experience" (2:286). But the resulting metaphysical concepts have no absolute certainty. Kant concedes this restriction of metaphysical cognition indeed for the only example taken from the *New Elucidation*, the principles of natural theology. In the *Prize Essay* as well as in the *Only Possible Argument*, he attributes to them a lower degree of certainty than in 1755.

The analogy between the analysis of the phenomena in Newton's sense and the analysis of metaphysical concepts marked the difference between mathematical and metaphysical concepts. However, it did not help to mark the differences between logic, metaphysics, and empirical science, in which Kant was increasingly interested from the early 1760s. The distinction between logical and real determining grounds did not yet enter the *Prize Essay*; it emerged in the *Only Possible Argument* and is made precise in the *Negative Magnitudes*. In the early 1760s, Kant still assumed that the foundations of his 1755/1756 doctrine were non-contradictory, adequate, and sufficiently justified by his use of the analytic method, with the only qualification being that his physico-theological proof as well as his ontological proof of 1755 lacked mathematical rigour and absolute certainty. His distinction between the logical and the real use of reason concerned the adequacy of metaphysical concepts. His attempts at further clarifying their adequacy conditions, however, resulted in the collapse of his pre-critical metaphysics.

References

Achinstein, Peter. 1991. *Particles and Waves. Historical Essays in the Philosophy of Science*. New York: Oxford University Press.

Arnauld, Antoine, and Nicole, Pierre. 1685. *La logique, ou l'art de penser*. Quoted after the Engl. transl.: *Logic or the Art of Thinking*. Transl. and ed. by J.V. Buroker. Cambridge University Press 1996.

Baumgarten, Alexander Gottlieb. 1757. *Metaphysica*. Halle: Hemmerde. Quoted after *Kants gesammelte Schriften*, Vol. 17.

Breitenbach, Aangela. 2009. *Die Analogie von Vernunft und Natur. Kantstudien-Ergänzungshefte 159*. Berlin: Walter de Gruyter.

Callanan, John J. 2011. Kant on analogy. *British Journal for the History of Philosophy* 16: 747–772.

Cotes, Roger. 1713. Editor's preface to the second edition. In Newton (1726), 385–399.

De Bianchi, Silvia. 2013. The evolution of the sphere. Kant's conception of matter and the expanding universe. In *The Harmony of the Spheres. Kant and Herschel on the Universe and the Astronomical Phenomena*, ed. S. De Bianchi, 17–45. Newcastle upon Tyne: Cambridge Scholars Publishing.

Descartes, René. 1637. *Discours de la méthode*. Quoted after the Engl. transl.: *Discourse on the method of rightly conducting the reason*. In Descartes (1934), Vol. I, 79–130.

Descartes, René. 1641. *Meditationes*. Quoted after the Engl. transl.: *Meditations on first philosophy*. In Descartes (1934), Vol. I, 131–199.

Descartes, René. 1644. *Principia philosophiae*. Engl.: (Selections from) *The principles of philosophy*. In: Descartes (1934), Vol. I, 201–302.

Descartes, René. 1934. *The Philosophical Works of Descartes in Two Volumes*, transl. by Elizabeth E. Haldane and G. R. T. Ross. Cambridge University Press.

Engfer, Hans-Jürgen. 1982. *Philosophie als Analysis. Studien zur Entwicklung philosophischer Analysiskonzeptionen unter dem Einfluß mathematischer Methodenmodelle im 17. und frühen 18. Jahrhundert*. Stuttgart-Bad Cannstadt: Frommann-Holzboog.

Friedman, Michael. 1992. *Kant and the Exact Sciences*. Cambridge, MA: Harvard University Press.

Grier, Michelle. 2001. *Kant's Doctrine of Transcendental Illusion*. New York: Cambridge University Press.

Hettche, Matt and Dyck, Corey. 2019. Christian Wolff. In *The Stanford Encyclopedia of Philosophy*, ed. Edward N. Zalta (Winter 2019 Edition). https://plato.stanford.edu/archives/win2019/entries/wolff-christian.

Hinske, Norbert. 1970. *Kants Weg zur Transzendentalphilosophie. Der dreißigjährige Kant*. Stuttgart: Kohlhammer.

Hintikka, Jaako and Unto Remes. 1974. *The Method of Analysis: Its Geometrical Origin and Its General Significance*. Boston Studies XXV. Dordrecht: Reidel.

Hobbes, Thomas. 1655. *De corpore*, Ch. IV. In *Thomas Hobbes: The Elements of Law, Natural and Politic*, ed. J.C. Gaskin, 193–212. New York: Oxford University Press.

Lambert, Johann Heinrich. 1761. *Kosmologische Briefe über die Einrichtung des Weltbaues*. Augsburg 1761, Repr.: *Philosophische Schriften*, Vol. 5. Hildesheim: Olm 2006. *Cosmological Letters on the Arrangement of the World Edifice*. Scottish Academic Press Ltd. 1976.

Laywine, Alison. 1993. *Kant's Early Metaphysics and the Origins of the Critical Philosophy*. Atascadero: Ridgeview Publ.

Leibniz, Gottfried Wilhelm. 1679. De synthesi et analysi universali seu arte invendi et judicandi. In *Die Philosophischen Schriften*, ed. C.I. Gerhardt, 292–298. Berlin, 1875–90, repr.: Hildesheim: Olms 1965, Vol. 7. Engl.: On universal synthesis and analysis, or the art of discovery and judgment. In *Gottfried Wilhelm Leibniz, Philosophical Papers and Letters*, ed. LĖ. Loemker, 129–134. Dordrecht: Reidel 1969.

Leibniz, Gottfried Wilhelm. 1684. Meditationes de Cognitione, Veritate et Ideis. In: Acta Eruditorum. Repr. In *Die Philosophischen Schriften*, ed. C.I. Gerhardt, 422–426. Berlin, 1875–90, repr.: Hildesheim: Olms 1965, Vol. 4. Quoted after the Engl. transl.: Meditations on Knowledge, Truth, and Ideas. In *Philosophical Essays*, ed. G.W. Leibniz, 23–27. Edited and Translated by Roger Ariew and Daniel Garber. Indianapolis & Cambridge: Hackett Publishing Company.

Leibniz, Gottfried Wilhelm, and Clarke, Samuel. 1715/1716. Correspondence. In *Die Philosophischen Schriften*, ed. C.I. Gerhardt, 345–440. Berlin, 1875–90, repr.: Hildesheim: Olms 1965, Vol. 7. Quoted after the Engl. transl. and repr.: Edited, with Introduction, by Roger Ariew. Indianapolis & Cambridge: Hackett Publishing Co. 2000.

Meier, Georg Friedrich. 1752. *Auszug aus der Vernunftlehre*. Halle: Gebauer. Quoted after: *Kants gesammelte Schriften*, vol. 16.

Mendelssohn, Moses. 1764. *Abhandlung über die Evidenz in Metaphysischen Wissenschaften*. Berlin: Haude und Spener.

Newton, Isaac. 1726. *Principia Mathematica Philosophiae Naturae*. 3rd ed. Quoted after: *The Principia: A New Translation by I.B. Cohen and A. Whitman*, 1999. Berkeley: University of California Press.

Newton, Isaac. 1730. *Opticks*. With a preface by I.B. Cohen, based on the 4th edition, New York. 4th ed., London 1930. With a Preface by I.B. Cohen, Dover 1979.

Pappus of Alexandria. 1589. *Pappi Alexandrini Mathematicae Collectiones*. Venedig. Engl. Transl.: Book 7 of the Collection. Part 1. Edited with Translation and Commentary by Alexander Jones. New York 1986.

Pieper, Annemarie. 1996. Kant und die Methode der Analogie. In *Kant in der Diskussion der Moderne*, ed. G. Schönrich and Z. Katu, 95–112. Frankfurt a. M.

Riehl, Alois. 1924. *Der Philosophische Kritizismus. Geschichte und System. Vol. 1: Geschichte des Philosophischen Kritizismus*. 3rd ed. Leipzig: Kröner.

Schönfeld, Martin. 2000. *The Young Kant*. Oxford: Oxford University Press.

Schönfeld, Martin. 2010. Kant's early cosmology. In *A Companion to Kant*, ed. G. Bird. Chichester: Wiley-Blackwell.

Spinoza, Baruch. 1677. *Ethica ordine geometrico demonstrata*. Engl.: *The Ethics*. Translated from the Latin by R. H. M. Elwes. http://www.gutenberg.org/files/3800/3800-h/3800-h.htm. Accessed 12 Feb 2020.

Strawson, Peter F. 1952. *Introduction to Logical Theory*. London: Methuen.

Walford, David. 1992. *Introduction to: Immanuel Kant: Theoretical Philosophy 1755–1770*. Translated and edited by D. Waldford in collab. with R. Meerbote. Cambridge University Press, xxxv–lxxxi.

Walford, David, and Meerbote, Ralf. 1992. *Immanuel Kant: Theoretical Philosophy 1755–1770*. Translated and edited by D. Waldford in collab. with R. Meerbote. Cambridge University Press.

Part II
The Critical Turn

The second part of this book investigates the critical turn in detail. Kant's pre-critical systematic efforts could not escape the fact that his methodology gave rise to incoherent results. Perhaps this was no wonder, given the incoherency in the ways he employed different variants of the analytic and synthetic methods in his 1755/1756 writings.

Chapter 3 studies the collapse of the pre-critical system. In the 1760s Kant increasingly questioned the principles of his 1755/1756 metaphysics, beginning with the distinction between logical and real grounds introduced in the *Negative Magnitudes* and the *Only Possible Argument* of 1763, culminating in the *Dreams of a Spirit-Seer* of 1766 and the *Directions of Space* of 1768. According to the *Dreams*, the pre-critical system failed with regard to the interaction between body and soul, or, in terms of a system of metaphysics in Wolff's style, the relation between rational cosmology and rational psychology. The argument from incongruent counterparts of 1768 made him think that things were even worse, resulting in the refution of Leibniz's relational theory of space which Kant had adopted from his very first writings. Yet in view of Leibniz's invariance arguments set out in the correspondence with Clarke, Kant could not adopt Newton's theory either. Now he realized that the pre-critical project of giving metaphysical foundations to physics by means of the analytic method had definitely gone astray. Even the cosmology part of the system remained without viable foundations, and only the epistemic shift of metaphysics first articulated in the *Dreams* offered a solution.

Chapter 4 reconstructs the road from the "great light" of 1769 to the 1770 *Dissertation*, and then through Kant's silent decade towards the antinomy of pure reason. Our detailed reconstruction disproves a long-established view of the critical turn, according to which the *Dissertation* should have provided a resolution of the cosmological antinomy. In fact, at that time, there was no cosmological antinomy. In 1770, Kant was still attempting to reconcile his new theory of space and time as pure forms of intuition with a rational cosmology in Wolff's style. His 1770 arguments concerning the limitations of metaphysical cognition supported a strategy of avoiding any conflict between the claims about the sensible and the intelligible worlds. Only later, when he developed a new theory of objective cognition, did

Kant reverse his 1770 strategy. His distinction between logical and real grounds was decisive, and required the real use of the mathematical concept of infinity. It was this insight that made him claim the antinomy. After the failure of the analytic-synthetic method, Kant now tried the skeptical method—without, however, adopting Hume's skepticism. The first clear indications of the cosmological antinomy are to be found in his notes on metaphysics from 1773–1775.

Chapter 3
The Collapse of the Pre-critical Cosmology

Metaphysics, with which [. . .] I have fallen in love [. . .], offers two kinds of advantages. The first is: it can solve the problems thrown up by the enquiring mind, when it uses reason to spy after the more hidden properties of things. But hope is here all too often disappointed by the outcome. And, on this occasion, too, satisfaction has escaped our eager grasp. (2:367)

In the 1760s Kant started to acknowledge the weak points of his pre-critical system, and as a consequence he began to notice that the use of analytic methods in metaphysics had led him astray. Schönfeld (2000, 216–217) argues that Kant had already abandoned his pre-critical project in the early 1760s, not only because of the weakness of the physico-theological and the ontological proofs of the existence of God, but also in the face of the problem of convincingly incorporating human freedom in the pre-critical system. Schönfeld shows that, starting with the *Only Possible Argument*, Kant came to see increasing problems concerning the foundations of metaphysics, culminating in the *Dreams of a Spirit-Seer* of 1766. In contrast, I propose that in 1762/1763 Kant was not yet aware that his pre-critical project had failed. Otherwise, neither his defence of the analytic method in the *Prize Essay*, with its recourse to the proofs of the *Physical Monadology*, nor the application of the method in the *Only Possible Argument* makes sense. In the *Negative Magnitudes* and the *Only Possible Argument*, he tightened and extended his criticism of the unjustified real use of logical principles, still trusting in the principles of his pre-critical system, even though admittedly he now considered them as less certain than in 1755/1756.

The decisive collapse of his pre-critical system came in two steps. In 1766, Kant took notice of Swedenborg's writings and faced up to the problem that his own pre-critical theory of the interactions between body and soul lay the grounds for spiritism. It was now that he saw the arguments of the *New Elucidation* as defeated, and so wrote the *Dreams*. Schönfeld's book ends with a discussion of the *Dreams*, neglecting the crucial role of the 1768 paper *Directions in Space* for Kant's critical turn. So far, however, only the relation between rational psychology and rational

B. Falkenburg, *Kant's Cosmology: From the Pre-Critical System to the Antinomy of Pure Reason*, European Studies in Philosophy of Science 12,
https://doi.org/10.1007/978-3-030-52290-2_3

cosmology was affected, whereas the foundations of the pre-critical cosmology still seemed unaffected. With the *Directions in Space*, it became obvious that neither the metaphysical foundations of physics nor the way of establishing them by the analytic method had been left untouched. The 1768 argument from incongruent counterparts convinced him that the analytic method had also gone astray as regards the concept of space, and hence, as regards the foundations of cosmology. The puzzle posed by it was clearly decisive, given that Kant's search for a solution resulted in the 1770 theory of space and time as pure forms of intuition.

3.1 The Logical vs. Real Use of Reason

The first crucial step towards Kant's critical philosophy was to differentiate between the logical and the real use of reason, in order to distinguish logic (and mathematics) from metaphysics. The *New Elucidation* of 1755 had merged both uses, employing rather strange arguments to prove the metaphysical principles of succession and coexistence by appeal to the logical principle of the determining ground (see Sect. 2.2.2.3). The 1762/1763 writings distinguish the logical and the real use of reason based on the difference between the principle of contradiction and Leibniz's principle of sufficient reason. Kant elaborates the distinction in terms of logical vs. real grounds, non-real vs. real predicates, and logical vs. real oppositions. The distinction is explained the *Only Possible Argument* and in the *Negative Magnitudes*; in the former, Kant explicitly associates the logical ground with the principle of contradiction, whereas he seems to associate the real ground with the principle of sufficient reason:

> [...] the actuality, by means of which, as by means of a ground, the internal possibility of other realities is given, I shall call the first real ground of this absolute possibility, the law of contradiction being in like manner its first logical ground [...] (2:79)

Kant's criticism of the traditional ontological proof of God's existence in the *Only Possible Argument* is based on the distinction between logical and real predicates. Kant argues that existence is not a real predicate which would add any further property to the properties such as perfection, absolute justice, etc., traditionally attributed to God (2:76) (see Sect. 2.4). In this way, he finally makes a clear distinction between the logical and the real use of his principle of the determining ground, which according to the *New Elucidation* of 1755 was a logical principle that nevertheless ought to have metaphysical consequences. In 1763, Kant argues that the *ratio essendi* for the existence of a thing, i.e., its ontic ground, cannot be found in the predicates attributed to this thing. Therefore, the logical and real grounds of being have to be distinguished.

Kant further elaborates the distinction in the *Negative Magnitudes*. Here, he makes the distinction between the *logical* opposition of predicates, which gives rise to contradiction, and the *real* opposition of predicates, which gives rise to the mutual cancelling of respective quantities such as opposite motions, attractive and repulsive forces, having money and having debts, etc. According to the *Negative Magnitudes*, logical grounds are associated with consequences that can be found by means of conceptual analysis, whereas real grounds cannot be found in such a way. Logical grounds contain their consequences analytically; real grounds do not.[1]

> I fully understand how a consequence is posited by a ground in accordance with the rule of identity: analysis of the concepts shows that the consequence is contained in the ground. [...] But what I should dearly like to have distinctly explained to me, however, is how one thing issues from another thing, though not by means of the law of identity. The first kind of ground I call the logical ground, for the relation of the ground to its consequence can be understood logically. In other words, it can be clearly understood by appeal to the law of identity. The second kind of ground, however, I call the real ground, for this relation belongs, presumably, to my true concepts, but the manner of relating can in no wise be judged. (2:202)

The distinction between logical and real opposition in the *Negative Magnitudes* is parallel to the distinction between logical and real grounds. It applies to predicates. Logical opposition occurs between contradictory predicates, while real opposition occurs between predicates denoting properties which may compensate each other, such as the motions of a body within a larger body moving in the opposite direction (e.g., a ship, to take Galileo's famous example from the *Dialogue*):

> The former distinction, that between logical opposition and real opposition, is clearly understood by means of the law of contradiction. And I understand how, if I posit the infinity of God, the predicate of mortality is cancelled by it, and it is cancelled because mortality contradicts infinity. But how the motion of one body is cancelled by the motion of another body – that is another question, for the motion of the second body does not stand in contradiction to the motion of first body. (2:203)

Kant's distinction between the logical and the real, as applied to grounds, predicates, and their opposition, was decisive for his development towards the critical turn. Several authors have noted that in making this distinction, Kant was influenced by Crusius (Allison 2012, 140; Guyer 1987, 11; Walford 1992, xl). According to Friedman (1992, 20–24), Kant's pre-critical analytic method followed Crusius in being a quasi-inductive or regressive procedure. The quasi-inductive or regressive aspects of Kant's method, however, were also in agreement with the analytic method(s) of the Cartesian and Wolffian tradition. Even though Kant in a certain sense followed Crusius in his introduction of the distinction between logical and real grounds, he was not as close to empiricism as some Kant scholars are inclined to think. Kanzian (1993) gives a concise account of Kant's distinctions and the way in which they differ from Crusius's empirical realism.

We have seen that Kant's system of 1755/1756 lacks a clear separation of the *metaphysica generalis* and the *metaphysica specialis*, as well as a strict demarcation between logic and metaphysics (see Sect. 1.4.2.1). In the *New Elucidation*, Kant

[1]Here, 'analytically' is meant in the sense of Kant's later, critical account of analytic judgments. The bridge from Kant's pre-critical analytic-synthetic method, as defended in the *Prize Essay*, to his later account of analytic and synthetic judgments, is conceptual analysis (see Appendix A.2.2.3).

differentiates between ontic and epistemic grounds, but he does not yet explicitly distinguish the logical and the real use of reason. His 1755 criticism of Leibniz's principle of indiscernibles is at least a first crucial step towards this distinction. He argues that Leibniz's principle of indiscernibles is not a logical consequence of the principle of sufficient reason and rejects its metaphysical generality (1:410). Nevertheless, he tries to derive metaphysical consequences from it (see Sect. 2.2.2.3).

In 1755, Kant follows Crusius. In 1763, in contrast, he stresses the way in which his new distinction between logical and real grounds differs from Crusius's distinction between ideal (or epistemic) and real (or ontic) grounds (2:203; see also Longuenesse 2001, 75). Now Kant criticizes Crusius's distinction, observing that on his account the ideal and real grounds may coincide, given that the real or ontic grounds may coincide with our ideal or epistemic grounds for knowing that a phenomenon will happen; whereas real and logical grounds are always strictly different (ibid.). Kant's distinction between logical and real grounds amounts to that of logical and causal relations, or logical grounds and consequences, on the one hand, and causes and their effects, on the other.

Kanzian (1993) in addition notes that Kant's 1762/1763 account of real grounds underlies the condition that all objects or substances in the world must be compossible:

> Whether an object is "real", depends on the question of whether its concept holds as an integral part of the system of all possible concepts and their combinations. Thereby, Kant adopts an essential rationalist assumption.[2] (Kanzian 1993, 403)

Kanzian relates this assumption to Wolff and Baumgarten. Indeed, the background here is Leibniz's principle of the compossibility of all substances in the world, which according to Schönfeld (2000, 201–205) is also responsible for the crucial flaw in Kant's 1763 argument for the existence of God (see Sect. 2.4).

I will argue below that this background is in addition crucial for understanding Kant's 1768 argument from incongruent counterparts (see Sect. 3.4), which has provoked so many misunderstandings. Another closely related point is decisive for understanding Kant's 1768 argument and its consequences for his critical turn: the distinction between logical and real grounds in the *Only Possible Argument* and the *Negative Magnitudes* substantially strengthens Kant's critique of the unjustified consequences of Leibniz's principle of sufficient reason, first expressed in the *New Elucidation* (see Sects. 1.4.2 and 2.2.2). However, the distinction adds the principles of succession and coexistence to these unjustified consequences, that is, two keystones of the pre-critical system. The 1762/1763 writings do not yet indicate that Kant realized this problem. The *Dreams* of 1766 does.

After all, the distinctions between logical and real grounds, and between logical and real opposition, are crucial for Kant's critical turn and his later views on Leibniz. He carries these distinctions on to the *CPR*, where they enter the chapter entitled

[2]"Ob ein Gegenstand 'real' ist, hängt davon ab, ob sein Begriff als Bestandteil des Systems aller möglichen Begriffe und Begriffsverbindungen gilt. Damit übernimmt Kant eine grundlegende rationalistische Voraussetzung." My translation.

the *Amphiboly of concepts of reflection*. In the first amphiboly, Kant criticizes Leibniz's version of the principle of indiscernibles as a confusion of *noumena* and *phaenomena*, that is, as mistaking the logical and empirical use of reason (B 316–317). In the second amphiboly, he criticizes the confusion between the logical and the real opposition of predicates: the latter is logically possible, the former is not (B 320–321).

3.2 Against Spiritism

The *Dreams* have an exceptional status in Kant's intellectual development. The writing contains wild polemics rather than careful arguments. The rhetoric fluctuates between sarcastic denigration of Swedenborg's approach, and self-mockery; and he shifts by turns between criticizing and defending metaphysics. Methodological considerations remain marginal, and they only concern Kant's caustic criticism of Swedenborg's way of arguing. Above all, the *Dreams* bear witness to Kant's despair about his unfortunate affection for metaphysics. Methodologically, the writing may best be understood as demonstrating his tremendous skepticism concerning the adequacy of his pre-critical system, and may indeed represent his first probings of the skeptical method itself (Reich 1958, XII; Kaulbach 1982, 100).

The *Dreams* express Kant's insight that his 1755/1756 views about the interactions between body and soul had failed, and with them the horizontal links between the specific parts of the pre-critical project, that is, his account of the three disciplines of the *metaphysica specialis*. The rhetoric of the *Dreams* is testimony to Kant's mixed feelings about his 1755/1756 theory of physical influence, given that it did not preclude the possibility of physical actions by immaterial souls upon the physical world. As already observed by Riehl (1924, 289) and investigated in detail by Laywine (1993, 72–100), he now realized that his metaphysical system of 1755/1756 opened the floodgates to phantoms and spirit seeing. In the face of this undesired consequence, the relation between rational cosmology and rational psychology he wanted to establish with the *New Elucidation* had turned out to be fallacious.

Hence, the systematic endeavour of 1755/1756 had gone astray in substantial respects. Kant now realizes that his attempt to reconcile Leibniz's system of pre-established harmony and Newton's metaphysical theory of the omnipresence of God in the world, though in a modified system of physical influence, had fallen apart. The *New Elucidation* gave the proof for a principle of coexistence, according to which only the scheme of God's understanding mediates the interactions of substances (1:412–413). This proof was indeed flawed. From a modern point of view, it merges logical and semantic assumptions and gives the impression of being circular (see Sect. 2.2.2.3). Against the background of the 1762/1763 distinction between logical and real grounds, Kant now sees the principle of coexistence as untenable. Whether he now also realizes the flaw in his ontological proof for the existence of God given in the *Only Possible Argument* is hard to say (see Schönfeld

2000, 197–205). However, as a consequence of these painful insights, he begins to sketch his first ideas concerning the program of a critique of reason which aims at determining the limitations of the real use of reason:[3]

> Metaphysics, with which, as fate would have it, I have fallen in love, but from which I can boast of only a few favours, offers two kinds of advantages. The first is: it can solve the problems thrown up by the enquiring mind, when it uses reason to spy after the more hidden properties of things. But hope is here all too often disappointed by the outcome. And, on this occasion, too, satisfaction has escaped our eager grasp. [...]
>
> The second advantage of metaphysics is more consonant with the nature of the human understanding. It consists both in knowing whether the task has been determined by reference to what one can know, and in knowing what relation the question has to empirical concepts, upon which all our judgments must at all times be based. To that extent metaphysics is a science of the *limits of human reason*. (3:367–368)

The claim that "metaphysics is a science of the *limits of human reason*" indicates a radical turn in Kant's approach to metaphysics. His notes on metaphysics after the *Dreams* demonstrate that he now began to consider metaphysical concepts as subjective and epistemic (see Sect. 4.2.1). In the course of elaborating this new, epistemological approach to metaphysics, he reassessed his concept of space, which after the failure of his 1755 theory of physical influence was in need of revision too.

3.3 The Concept of Space

From his very first writings, Kant presupposed a relational theory of space. §9 of the *True Estimation* relates it to the forces of substances, anticipating his theory of physical influence and his dynamic atomism of 1755/1756:

> It is easy to show that there would be no space and no extension if substances had no force to act external to themselves. For without this force there is no connection, without connection, no order, and, finally, without order, no space. (1:23)

In his writings until 1756, this relationalism was based on a *realistic interpretation of Leibniz's view of space and time* as the orders of the coexistence and succession of the monads. Kant tacitly assumed this relationalism to be compatible with a Newtonian account of objects in space (Walford 1999). Yet it was exactly this compatibility that was now at stake.

[3]Kant employs the notion of the *focus imaginarious* to explain why the spirit seer locates the objects of his imaginations outside the mind (2:346). Grier (2001, 37) emphasizes that this notion "will become central to Kant's account of the illusory metaphysical ideas of reason in the *Critique*" and she observes a "striking semblance to the account of the optical illusion related to mirror vision in Newton's *Opticks*", that is, another analogy between metaphysics and Newton's methods. This analogy, however, was now in the service of criticizing metaphysics.

3.3.1 In Favour of Relationalism

It is only in 1758 that Kant gives an explicit argument for his relationalism, by claiming that the existence of an empty space or a vacuum within the world is absurd:

> Even though I might imagine a mathematical space empty of all creatures as a container for the bodies, this would not help me. For how might I distinguish its parts and the various places that are not occupied by anything corporeal? (*New Doctrine*, 2:17)

Here, Kant indirectly appeals to Leibniz's invariance arguments against Clarke's defence of Newton's absolute space. The argument is based on the principle of indiscernibles, which according to Leibniz derives from the principle of sufficient reason (Leibniz and Clarke 1715/1716, 14–15, 22–24). Indeed, Kant upheld Leibniz's arguments against absolute space and time all of his life.[4] In the *New Elucidation* of 1755, he supported a Leibnizian theory of space and time, too, but here reinterpreted in terms of really interacting entities. This reinterpretation stood in manifest opposition to Leibniz's genuine relational theory, according to which space and time are not real, but *ideal*. They are relational *well-founded phenomena (phaenomena bene fundata)* which are founded by the monads, i.e., relationless substances.

In contradistinction to Leibniz, for whom absolute space was only an abstract mathematical concept (Leibniz 1715, 669; see below Sect. 3.4.1), the young Kant considered space as a real entity. The *New Elucidation* explained space and time in terms of the relations of coexistence and succession of physically interacting monads. The background was Kant's version of a theory of physical influence, according to which the connections between the substances are mediated by the "scheme of the divine understanding" (1:413; see Sects. 1.4.2.3 and 2.2.2.3).

Against this background, it perhaps becomes clearer why Kant considered his relational theory of space to be compatible with a Newtonian account of objects in space. From 1746 on, he thought that absolute space in Newton's sense was something like the "sum-total of substantial relations" (Schönfeld 2000, 42) of the physical monads, which are mediated by God. In 1758, he assumed that space, as a system of substantial relations, is substantive, even though not absolute in Newton's sense of an absolute inertial frame (Schönfeld 2000, 187).

At this point, however, a methodological problem arises. Is Kant's use of Leibniz's principle of indiscernibles incoherent? Leibniz's principle claims that two objects are identical if they have identical *internal* properties. According to the Leibniz–Clarke correspondence, it precludes the existence of two indistinguishable entities in the world, such as atoms, as well as the parts of an empty space inside or outside the world. Kant accepts Leibniz's refutation of absolute space and time, but he rejects Leibniz's refutation of atomism. In order to defend atomism,

[4]The argument reappears in the antinomy of pure reason, in the proof of the antithesis of the first antinomy, which Kant did not consider to be fallacious from a critical point of view (*CPR*, A 427–431/B 455–459; see Sect. 5.4.1.2).

the *New Elucidation* defends a *Newtonian theory of individuation*. Numerically identical objects such as crystals are observed in nature. To ask for unobservable distinguishing marks behind the phenomena would mean "to search for knots in a bullrush" (*nodos in scirpo quaerere*) (1:410; see Sects. 1.4.2.2 and 2.2.2).

In 1755, Kant argues that Leibniz's interpretation of the principle of indiscernibles does not derive from the principle of sufficient reason. The former exceeds the logical scope of the latter, to which it adds metaphysical premisses. In the light of Kant's later distinctions, its use belongs to the real use of reason. But are Kant's criteria for this use uniform? The use of the principle according to Newton's analytical method leads to the claim that indistinguishable objects such as crystals or atoms may exist in the world. The application of the principle to empty space and time, on the other hand, favours Leibniz's view that space and time are relational. According to this, the world has no well-defined position in space and time. Both applications seem to be associated with conflicting assumptions about the absolute or relational character of space.

They become compatible, however, if we understand Kant's pre-critical account of Leibniz's principle in the sense of a verificationist criterion of meaning. Understood in this way, Kant's 1755 and 1758 arguments are completely analogous: To speak of the position of the world in an absolutely empty space is as meaningless as to speak of the effect of exchanging two only numerically different substances: it makes no real difference. Indeed, in 1758 Kant again draws on the famous argument against absolute space that Leibniz presents in his fourth letter to Clarke, which he had turned back *against* Leibniz in 1755 (see Sect. 2.2.2.2):

> To suppose two things indiscernible is to suppose the same thing under two names. (Leibniz and Clarke 1715/1716, 22)

Here, Leibniz proposes that the concept of the position of the universe is subject to a nominal definition without any corresponding real possibility. The position of the world in space and time is fictitious, given that the positions of an empty space and the moments of an empty time are indistinguishable. Therefore, it is impossible to give a real definition of *where* in space and time the world is located. From a modern point of view, here an invariance argument is linked to a verificationist criterion of meaning. The world as a totality is invariant against spatial and temporal translations, while to apply the concept of location to the whole world is meaningless, because the meaning of the concept cannot be empirically verified (Earman 1989, 119).

Leibniz's criterion of meaning is stronger than a modern verificationist criterion, and stronger than Kant's pre-critical account of it, too. For Leibniz, the concept of the position or location of the universe lacks content in the sense that any proposition which attributes to the universe a concrete position or a specific location is unsatisfiable. In terms of modern formal semantics, this means that the extension of this concept is the empty set. According to Kant's later critical terminology, this corresponds to the interpretation that the position or location of the universe is a *nihil negativum*, an "empty object without concept" or "non-entity" (*CPR*, B 348).

Kant's pre-critical interpretation of 1758 is weaker. It implies only that empty space or time is a non-contradictory but fictitious *ens rationis* or *ens imaginarium*, a mere fiction of thought or perception.[5] In the notes on metaphysics dating from 1766–1768, when he already had substantial doubts about the adequacy of his pre-critical system, he raised a related objection to absolute space and time:

> Whether there is a *spatium absolutum* or *tempus absolutum*, would mean to say as much as whether between two things in space everything [...] in between can be destroyed and yet the specific empty gap would remain, and whether, if [...] for a whole year all motions and changes would stop, the following could not begin, so that **in between an empty year** would have happened. (N. 3892, phase θ; 17:330)[6]

This consideration, too, refers to the Leibniz–Clarke debate. In the postscript to the fourth letter to Clarke, Leibniz again deploys the principle of sufficient reason, here against the existence of the void *within* the world:[7]

> God could have placed some matter in it without derogating in any respect from all other things; therefore he has actually placed some matter in that space; therefore, there is no space wholly empty; therefore all is full. [...] I shall add another argument grounded on the necessity of a sufficient reason. It is impossible that there should be any principle to determine what proportion of matter there ought to be, out of all the possible degrees from a plenum to a vacuum, or from a vacuum to a plenum. (Leibniz and Clarke 1715/1716, 27)

With his arguments against the real possibility of a void in the world and of Newton's absolute space, Leibniz himself wanted to prove that space and time, as the relations of coexistence and succession, are *ideal*. The young Kant, on the contrary, interprets space and time as the *real* relations of real substances in the world. By doing so, he cannot justify the real use of Leibniz's arguments against absolute space or time by means of direct analysis of the phenomena, but only indirectly, by an epistemic criterion of meaning. Adopting Leibniz's principle of indiscernibles for space and time seems to run counter to his earlier effort to refute it for indistinguishable objects. It seems that in 1755 and 1758 Kant justifies both views by epistemic criteria, which we may understand in terms of verificationism; and he even continues to employ both views after his critical turn. In the *CPR*, he criticizes the concept of the position of the universe in the doctrine of the antinomies (A/505–507/B 533–

---

[5]The distinction between an *ens rationis* or "empty concept without object", and an *ens imaginarium* or "empty intuition without concept" (*CPR*, 348) does here not yet apply, given that it presupposes the theory of space and time as forms of pure intuition, which Kant first presented in 1770.

[6]"Ob es ein *spatium absolutum* oder *tempus absolutum* gebe, würde soviel sagen wollen, ob man zwischen zwey Dingen im Raume alles [...] dazwischen liegende vernichten könne und doch die bestimte leere Lücke bleiben würde, und ob, wenn [...] ein gantzes Jahr Bewegungen und veränderungen überhaupt aufhöreten, nicht das folgende Anheben könne, so dass ein **leeres zwischen Jahr** verlaufen wäre." My translation (note not contained in the Cambridge edition). See also Sect. 3.4.3.

[7]Based on a variant of this argument, Kant later proved the antithesis of the first antinomy (see Sect. 5.4.1): "[...] the relation of the world to empty space would be a relation of the world to *no object*. Such a relation, however, and hence also the boundedness of the world by empty space, is nothing [....]" (*CPR*, B 457; see Sect. 5.4.1). Contrary considerations anticipating the proof of the thesis of the first antinomy can also be found in the notes on metaphysics prior to 1769; cf. N. 3840 and N. 3912. See Sect. 4.2.1. However, there is still no mention of a cosmological antinomy. These notes on metaphysics belong at most in a very vague sense to the so-called problem of the antinomy.

535), and the application of the principle of indiscernibles to objects with identical properties in the amphiboly (B 320).

Nevertheless, the question remains to what extent Kant's pre-critical uses of the principle of indiscernibles may give rise to a coherent system of metaphysics. In 1755/1756, he thought that his theory of physical influence made the relational concept of space established by Leibniz's principle compatible with Newton's assumption of indistinguishable atoms. With the *Dreams* of 1766, however, this metaphysical justification of his concept of space can be seen to have gone astray.

3.3.2 The Quest for a Metaphysical Exposition

Within Kant's pre-critical system, the concept of space is at the interface between metaphysics and mathematical physics. In the *Prize Essay*, Kant notes that the mathematical concept of space may also be subject to philosophical explanation or metaphysical exposition:

> The mathematician deals with concepts which can often be given a philosophical definition as well. An example is the concept of space in general. (2:277–278)

He uses this ambiguity of the concept of space as an example in order to clarify the distinction between the synthetic definitions of mathematics and the analytic method needed for metaphysics:

> It is the business of philosophy [Weltweisheit] to analyse concepts which are given in a confused fashion, and to render them complete [ausführlich][8] and determinate. The business of mathematics, however, is that of combining and comparing given concepts of magnitudes, which are clear and certain, with a view to establishing what can be inferred from them. (2:278).

Then, Kant makes the following distinction. The *geometrical* concept of space, which in 1762/1763 for him still belongs to the logical use of reason, gives rise to "synthetic" mathematical definitions. The *metaphysical* (or cosmological) concept of space, however, belongs to the real use of reason.

Since it is not obvious whether our logico-mathematical concepts are adequate for the real use of reason, the cosmological concept of space needs a metaphysical foundation provided by the analytic method. In addition, the metaphysical exposition of the concept of space is prior to its mathematical definition:

> But before I set about the task of defining what space is, I clearly see that, since this concept is given to me, I must first of all, by analysing it, seek out those characteristic marks which are initially and immediately thought in that concept. (2:281).

In the *Theory of the Heavens* and the *Physical Monadology* of 1755/1756, Kant had used the analytic method in order to establish a Newtonian explanation of structure formation in the universe and an atomistic view of matter which is compatible with the infinite indivisibility of space. In the *Prize Essay*, he simply recalled the properties of space employed in the demonstrations of the *Physical*

[8] See n. 3 of Chap. 2.

Monadology; but he did not yet make any effort to establish his concept of space by means of analyzing our experience of space. In 1768 he apparently became aware that the relational concept of space employed in his previous writings lacked an adequate metaphysical exposition. The topic of the 1768 paper is indeed to contribute to the philosophical explanation of the concept of space. According to the methodology defended in the *Prize Essay*, such an explanation had to be based on the methodological rule

> to distinguish those judgements which have been immediately made about the object and relate to what one initially encountered in that object with certainty. (2:285)

Kant's 1764 criterion of certainty is the (Cartesian) evidence of our idea of space, for which our "certain inner experience" (2:286) is identical with the empirical characteristics of space given in our outer experience. Applied to the concept of space, at the interface of physics and metaphysics, Kant's analytic method of metaphysics obviously collapses into a Newtonian analysis of the phenomena. In this regard, its metaphysical exposition is in sharp contrast to other metaphysical principles such as the principles of natural theology, which according to the *Prize Essay* and the *Only Possible Argment* do not have mathematical certainty, and hardly any empirical support either. Kant's comparison of Newton's method and the analytic procedure of metaphysics in the *Prize Essay* culminates in stating:

> And even if you are not acquainted with the complete essence of the thing, you can still safely employ those characteristic marks to infer a great deal from them about the thing in question. (2:286)

This is exactly what Kant finally does in his 1768 paper concerning the relational concept of space, which he defended in 1758 along Leibniz's line of reasoning. He compares it with our inner (and outer) experience of space, or spatial objects. Together with the above remarks from the *Prize Essay*, even the very title of the paper, *Concerning the Ultimate Ground of the Differentiation of Directions in Space*,[9] indicates that Kant is asking for the *real grounds* of the concept of space in view of certain spatial phenomena, from which he wants to draw some certain characteristic marks of space.

3.4 The Puzzle of Incongruent Counterparts

Kant's 1768 analysis of the concept of space takes up the phenomenon of incongruent counterparts, i.e., left-handed and right-handed objects with the evident mark of having opposing orientations or chirality. In nature we find chiral objects,

[9]This is the adequate translation of "Vom ersten Grund des Unterschieds der Gegenden im Raum" given in the Cambridge edition of Kant's works (Walford 1992). For the confusion in the Anglo-American literature generated by inadequate or even absurd translations, see Byrd (2008) and Rusnock and George (1995); and Walford and Meerbote (1992, 456–457, n. 1).

e.g., snail shells, screws, and hands; and in three-dimensional Euclidean space it is impossible to transform such chiral objects by translation or rotation into their opposing counterparts. Kant argues that this phenomenon is at odds with Leibniz's relational account of geometry.

Kant's argument from incongruent counterparts (enantiomorphs) has drawn much attention from philosophers and historians of science (van Cleve and Frederick 1991; De Risi 2007). However, the discussion thus far has neglected the context of Kant's attempt to lay the grounds for a system of metaphysics in the face of the Leibniz–Clarke debate. His 1768 puzzle concerned the question of whether an object is unambiguously determined by means of a relational account of space, if that object is *the world*.

The decisive role of Kant's 1768 paper for his critical turn has also been neglected by Kant scholars. With the exception of a few remarks, Schönfeld (2000, 187–188) excludes it from his analysis. Beiser (1992) mentions it in passing without discussing its role for Kant's development. Several Kant scholars, most notably Buroker (1981), Friedman (1992), and Walford (1992), emphasize the significance of *Directions in Space* for Kant's turn towards his theory of space and time as forms of pure intuition, which he first presented in his 1770 *Inaugural Dissertation*. Other articles focus on Kant's criticism of Leibniz's principle of sufficient reason (Byrd 2008) and the influence of Euler's 1748 *Réflexions sur L'Éspace* (Walford 1999). But a more comprehensive view of Kant's 1768 argument against the background of his pre-critical writings is still missing.

Kant always upheld Leibniz's famous invariance arguments against Newton's absolute space and time. In his 1755/1756 cosmology and metaphysics, he presupposed that space and time are relational. The *New Elucidation* already partially dispensed with Leibniz's metaphysics, that is, with the principle of indiscernibles understood as the claim that infinitely many internal properties are the only distinguishing marks of objects. *Directions in Space* intensifies this criticism, sharply attacking Leibniz's project of an *analysis situs*. Leibniz had proposed but never completed this mathematical project. It aimed at establishing geometry by means of spatial relations alone. From a modern point of view, it aimed at giving geometry axiomatic foundations in terms of a few primitive relations. De Risi (2007, 101–107) shows that Kant and his contemporaries could have had no detailed knowledge of it, since Leibniz's writings on it were unpublished. Nor were Wolff's or Lambert's remarks on it particularly revealing. The meager knowledge that Kant had of the project prompted him to relate it to a Leibnizean metaphysics of space (De Risi 2007, 238) modified in terms of real relations, that is, to a relational concept of physical space which indeed was his own.

Kant's harsh criticism of Leibniz's project of an *analysis situs* wrongly led several interpreters to reduce his 1768 argument and its tenability to a *mathematical* issue.[10] Yet although a mathematical analysis of it is instructive, Kant's argument

[10]In particular Buroker (1981, 1991). The otherwise excellent analyses of Mühlhölzer (1992), George and Rusnock (1994), and Rusnock and George (1995) clarify some misunderstandings

belongs to the context of his pre-critical cosmology, that is, to physical geometry; and its validity for the latter has long been a matter of debate. The more recent discussion mainly deals with the questions of whether Kant's argument supports space-time substantivalism (Earman 1989; Nerlich 2009), and whether the phenomena of parity and CP violation of current particle physics add something new to it (Hoefer 2000; Lyre 2005). The first question amounts to modern versions of a Newtonian vs. Leibnizean view of spacetime, while the second concerns its significance for the questions of symmetries and symmetry violations in twentieth-century physics. In the following, I will not take up this debate but seek only to reconstruct Kant's argument in its historical context and discuss its validity in epistemological respects.

The 1768 paper becomes much more telling when it is related to Kant's project of establishing a system of metaphysics by means of the analytic method. It is a crucial turning point of his philosophical development. The argument from incongruent counterparts convinced Kant that the metaphysical foundations of his pre-critical cosmology were based on an untenable concept of space. It attacked his own pre-critical relational theory of space, by arguing in favour of *some* concept of absolute space, in a Newtonian or in another sense, but certainly not in favour of Kant's 1755 account. Thus he came to believe that the analytic method had led him astray in deriving the foundations of cosmology. The method defended in the *Prize Essay*, when applied to our idea of space, did not give rise to adequate real definitions, but rather to the definitive destruction of the metaphysical foundations of his pre-critical system.

3.4.1 Criticizing Leibniz's Analysis Situs

It is not accidental that Kant's 1768 paper begins by criticizing Leibniz's mathematical program of an *analysis situs*, the project of deriving Euclidean geometry from spatial relations as primitive concepts. Leibniz's project aims at establishing Euclidean geometry by means of two primitive relations only: equality, i.e., length identity, and similarity, i.e., angle identity. A third central concept of geometry, congruence, is not primitive but derived. It results from their combination.[11] Kant counters that a right hand and a left hand are incongruent, although identical in lengths and angles. His analysis of the idea of space aims at demonstrating that Leibniz's project of an *analysis situs* leaves the conceptual basis of geometry underdetermined, with fatal consequences for a cosmological concept of space.

---

found in van Cleve and Frederick 1991, but not this one. The claim "that Kant had no good argument" (George and Rusnock 1994, 466) is related only to mathematics, and misses the cosmological aspects of the argument.

[11] De Risi (2007, 284–286) gives a very detailed account of Leibniz's (two) definition(s) of incongruence and their relation to Kant's 1768 argument.

From a mathematical point of view, two remarks have to be made. First, Kant's observation strengthens Euler's mathematical criticism of a Leibnizean relational view of geometry (see Sect. 3.4.1.2). Euclidean 3-space has an affine structure not captured by Leibniz's (merely topological) foundations of geometry (Euler 1748); in addition, however, it is orientable (van Cleve and Frederick 1991). Second, the lack of this conception in Leibniz's *analysis situs* may be cured by adding a third primitive relation, namely mirroring symmetry. But this is not what Kant wants to show.

3.4.1.1 The Ultimate Ground of Geometrical Concepts

Kant's 1768 argument is metaphysical rather than mathematical. Due to the missing elaboration of the program of an *analysis situs* by Leibniz, he sees himself entitled to take it up for a metaphysical exposition of the concept of space:

> At any rate, it looks as if a certain mathematical discipline, which *Leibniz* called *analysis situs* [...] – it looks as if this discipline was never more than a thought in *Leibniz's* mind. [...] But to judge by the meaning of the term, what I am seeking to determine philosophically here is the ultimate ground of the possibility of that of which *Leibniz* was intending to determine the magnitudes mathematically. (2:377)

According to this passage, his goal is to determine the "ultimate ground", or sufficient reason, of the geometrical concepts that are subject to Leibniz's *analysis situs*. In contradistinction to Leibniz, whose project aimed at the mathematical foundations of geometry, he wants to determine this ground "philosophically", not "mathematically" (2:277), given that for him the concept of space is at the interface of mathematics and metaphysics (see Sect. 3.3.2).

Therefore, the "possibility" of the "analysis situs" cannot *only* be the *logical possibility* of mathematical concepts. According to the *Prize Essay*, mathematical concepts are constructed according to the synthetic method. Their construction results in arbitrary combinations of concepts and in nominal definitions. In contrast, the possibility of philosophical or metaphysical concepts is *real possibility*. Hence, Kant's task of 1768 is to give an exposition of the real possibility, i.e., the real ground of possibility, of physical space. His method for finding such an exposition is the analytic method, which the *Prize Essay* had declared to be the only adequate philosophical method.

Against this background it is clear that Kant's 1768 paper reinterprets Leibniz's "analysis situs" in terms of physical geometry. In his view, the concepts of relation and position are not sufficient for determining the "ultimate ground", or the metaphysical foundations, of geometry:

> For the positions of the parts of space in reference to each other presuppose the direction [Gegend] in which they are ordered in such a relation. In the most abstract sense of the term, direction does not consist in the reference of one thing in space to another – that is really the concept of position – but in the relation of the system of these positions to the absolute space of the universe. (2:377).

How does "absolute space" come into play here? According to Leibniz, physical space is the order of all co-existing phenomena, and geometrical space is the sum total of all geometric locations obtained by abstracting from the physical relata of spatial relations. In the *Metaphysical Foundations of Mathematics*, Leibniz (1715, 669) calls this abstract space "absolute space" and defines it as the "fullest locus, or the locus of all loci", i.e., as a topological manifold. Of course, this is not the physical space of cosmology. Kant, on the other hand, does not have an abstract topological manifold in mind, but rather the three-dimensional Euclidean space of Newtonian physics, as applied in the real use of reason. He asks for the ultimate ground of the determination of geometric positions and associates this with the "direction", that is, the relation of the system of all positions to "absolute space". According to the cosmology of 1755/1756, "absolute space" is identical to the system of all substantial relations of coexistence of the physical monads, and the ultimate ground or *ratio essendi* of this system is the mediation between the monads by the "scheme of the divine understanding" (1:412). Since the *Dreams*, this theory had no longer seemed tenable to Kant. In the 1768 article, he asks

> [...] whether there is not to be found in the intuitive judgements about extension, such as are to be found in geometry, clear proof that: *Absolute space, independently of the existence of of all matter and as itself the ultimate foundation [Grund] of the possibility of the compound character of matter, has a reality of its own.* (2:378).

According to the *Prize Essay*, to ask for the relations between the sum total of all geometrical positions and cosmological space is a matter of the interface between geometry and metaphysics, i.e., the metaphysical foundations of geometry, which have to be clarified by means of the analytical method. The respective conceptual analysis applies to the empirical idea of space, the clear and distinct marks of which are given by geometrical phenomena.

3.4.1.2 Euler's Influence

In order to distinguish this procedure from attempts to define the concept of space *a priori* in metaphysics, Kant now refers to Euler:

> Everybody knows how unsuccessful the philosophers have been in their efforts to place this point once and for all beyond dispute, by employing the most abstract judgements for metaphysics. Nor am I familiar with any attempt to attain this end so as to speak *a posteriori* [...] apart, that is, from the treatise of the illustrious *Euler* the Elder, which is found in the *Proceedings of the Berlin Royal Academy of Sciences* for the year 1748. (2:378)

Commentators disagree on Kant's reception of Euler's work. Laywine (1993, 17–34) and Friedman (1992, 4 and 16–17) emphasize an earlier influence of Euler on Kant's 1755/1756 writings, given that the *Physical Monadology* referred to the monadology debate at the Berlin Academy. Of these writings, however, only *On Fire* mentions Euler's theory of light, in a defence of the Cartesian view that light is not an entity on its own but the propagation of pressure within the ether (1:378). This theory of light propagation also appears in the *New Elucidation* as Kant's example for the distinction between epistemic and ontic grounds (1:393; see Sect. 1.4.2.1).

Buroker (1981, 50–51) suggests that Euler's 1748 treatise drew Kant's attention to the phenomenon of incongruent counterparts, but she does not explain in which way. Walford (1999), who gives a detailed discussion of Euler's influence on Kant's 1768 argument, claims against Buroker that

> [...] her suggestion [...] only succeeds in obscuring the fact that Kant had long recognized the importance of directionality, for it plays a very significant role in his astronomy (in the 1755 *Naturgeschichte*), on meteorology (in the 1756 *Theorie der Winde*), in the analysis of matter (in the 1756 *Monadologia*), and in the explanation of real opposition (in the 1763 *Versuch*). (Walford 1999, 314)

If this claim is correct, why did Kant not take incongruent counterparts into account before 1768? In my view, Buroker's suggestion is essentially correct, but she is missing a key puzzle piece (as the other Anglo-American commentators do). Waschkies (1987, n. 101, 514–515) proposes that Kant only took note of the German translation, given that he was not well versed in French. Indeed, he first explicitly mentions Euler's 1748 treatise in the *Negative Magnitudes* of 1763, the publication year of the German translation of Euler's treatise (Euler 1763). As George and Rusnock (1994, 466) observe, referring to the *Metaphysik Herder*, Kant mentions spatial incongruence for the first time in 1763:

> Aequalia et similia congruunt non nisi in plano. (AA 28:15)

The passage refers critically to Baumgarten's *Metaphysica*, §70 (AA 17.042). Around the same time, in the *Prize Essay*, Kant criticizes the use of analytical definitions in mathematics, referring to Wolff's concept of similarity:

> Mathematicians [...] sometimes have offered analytic definitions. But it must also be said that for them to do so is always a mistake. It was in this way that *Wolff* considered similarity in geometry: he looked at it with a philosophical eye, with a view to subsuming the geometrical concept of similarity under the general concept. But he could have spared himself the trouble. [...] The general definition of similarity is of no concern whatever to the geometer. It is fortunate for mathematics that [...] in the end nothing is actually inferred from such definitions. [...] Otherwise this science would be liable to exactly the same wretched discord as philosophy itself. (2:277)

This remark, however, is no "first attempt at a resolution" (George and Rusnock 1994, 466) of the later incongruence problem of 1768. In the *Prize Essay*, Kant is concerned with the systematic separation of the mathematical (i.e., synthetic) and philosophical (i.e., analytic) method, and he criticizes the use of the analytic method in mathematics. Kant only faces the incongruence problem of 1768 when he seeks for the metaphysical foundations of physical geometry, and applies the analytic method to the cosmological concept of space.

There are more hints that the German translation of Euler's treatise might have been crucial for Kant's 1768 argument. An anonymous Wolffian edited the book titled *Vernünftige Gedanken von dem Raume, dem Orth, der Dauer und der Zeit* (Reasonable Thoughts on Space, Position, Duration and Time). In addition to the translation, it included correspondence between Georg Venzky (1704–1757), rector of the secondary school of Prenzlau, and Euler, on which the editor critically commented, and a brief account of the Leibniz–Clarke debate. In the second letter to Euler, Venzky explicitly mentions the geometrical difference in altitude, width, and depth:

> I have such a concept of space that it is an indefinite extent, in which many things can coexist. If I am to demonstrate the correctness of this concept against my opponents, I would like to point out that we live and move in a part of space; that we know from experience that this space and extent extends in length, width, depth and height. (Euler 1763, 77)[12]

He then explains the difference between space and time via characterizing space by extension in *several* directions or dimensions, and time by mere length (Euler 1763, 79–80). Euler, who does not respond to this point at all in his answer, does not realize its significance. The anonymous editor does neither. Concerning the different ways in which space and time extend, he just notes:

> [...] it will soon be understood that the usual explanations [...] of the immeasurable extent [...] are only fraudulent concepts, and contradictory in themselves. [...] The first extends in length, the other in width or extension. We do not yet understand anything of this. Whether others will understand the author better, we have good reason to doubt. But don't we know that there is also a length in the extent, or in the space? And how do we distinguish length from width or extension?[13]

Kant, who presumably understands Venzky's point better, takes the latter question up by asking for the "difference of the directions in space". In doing so, he has a more specific point in mind than the three dimensions of space, namely the difference between right and left. This question is *not* found in Euler's writing of 1748.[14] There, Euler only argues based on the laws of inertia and force in favour of absolute space. He points out that a relational concept of space is not sufficient to express the claim that a body keeps or changes its direction, according to the laws of inertia and force, given that it is impossible to define the concept of the *direction* in a relational theory of space:

---

[12]"Ich habe einen solchen Begriff von Raum, daß er ein unbestimmter Umfang sey, darin vieles nebeneinander seyn kann. Soll ich die Richtigkeit dieses Begriffs noch wider die Gegner darthun: so gebe ich zu bedenken, daß wir ja in einem Theil des Raums leben und uns bewegen; daß wir aus der Erfahrung wissen, dieser Raum und Umfang erstrecke sich in die Länge, Breite, Tiefe, Höhe." My translation.

[13]"[...] so wird man bald begreifen, dass die gewöhnlichen Erklärungen [...] des unermeslichen Umfangs [...] nur betrügerische Begriffe, und an sich widersprechend seien. [...] Die erste erstreckt sich in die Länge, die andere in die Breite oder Ausbreitung. Wir verstehen hieraus noch nichts. Ob andere den Verfasser besser verstehen werden, daran zweifeln wir mit gutem Grund. Aber weis man denn nicht, dass in dem Umfang, oder in dem Raum, auch eine Länge sey? Und wie unterscheidet man die Länge von der Breite oder Ausbreitung?" My translation.

[14]Euler considers the three dimensions of space only in the *Letters to a German Princess*. The French version of the first and second part appeared in 1768, the German translation in 1769. The *Letters* only influenced Kant after 1768, and he quotes them in his *Dissertation* of 1770 (2:414, 2:419).

> From this it is evident that the identity of direction, which is a truly essential circumstance in the general principles of motion, could not be explained absolutely by the relation or the order of the coexisting bodies. Therefore, there must still be some other real thing apart from a body to which the idea of 'same direction' relates, and there is no doubt that it is to space, which we have just established to be real. (Euler 1748, §XVII)[15]

From a modern point of view, Euler complains that Leibniz's relational spatial concept only defines a topological but not an affine structure. Kant is looking for a stronger argument. In the *Directions in Space*, he attributes to Euler the merit of arguing "so to speak *a posteriori*", or by means of applying metaphysical principles "in concreto", in favour of the reality of absolute space, based on an analysis of the use of geometric concepts in mechanics. For him, Euler's argument is not satisfactory, because it is only aimed at the physical concept of space, without taking into account the metaphysical presuppositions of the properties of Euclidean space used in mechanics. From Kant's point of view, the inadequacy of a relational concept of space for cosmology is already evident in Euclidean geometry, prior to Newton's mechanics. The relational concept of space is too weak to provide a real ground for the distinction between the directions of three-dimensional Euclidean space. The thoughts of Venzky, forgotten today, expressed in his letters to Euler, are more likely to have brought Kant to this point than Euler's own.

3.4.1.3 A Closer Look at Kant's 1768 Argument

Kant develops his argument in several steps, intended to counter different aspects of the reduction of the cosmological concept of space to a relational concept. First he points out that we need the distinction between right and left, as it is defined relative to our body, in order to orient ourselves according to the geographic directions:

> [...] the most precise map of the heavens [...] would not enable me [...] to infer from a known direction, for example, the north, on which side of the horizon I ought to expect the sun to rise. The same thing holds of geographical and, indeed, of our most ordinary knowledge of the position of places. Such knowledge would be of no use to us unless we could also orientate the things thus ordered [...] by referring them to the sides of our body. (2:379–380)

This passage may be read as an objection to one of Leibniz's arguments against Clarke's defence of absolute space, which probably did not convince Kant. In his third letter to Clarke, Leibniz claims that there is no sufficient reason to distinguish the actual state of the world from its state obtained by "changing east into west":

> [...] that it is impossible there should be a reason why God, preserving the same situations of bodies among themselves, should have placed them in space after one certain particular

[15]"D'où il est evident, que l'identité de direction, qui est un circonstance fort essentielle dans les principes généraux du mouvement, ne sauroit absolument être expliquée par la rélation, ou l'ordre des corps coëxistants. Donc il faut qu'il y ait envore quelque autre chose de réell, outre les corps, à laquelle se rapporte l'idée d'une même direction; & il n'y a aucun doute, que ce ne soit l'espace, dont nous venon d'etablir la réalité."

> manner and not otherwise—why everything was not placed the quite contrary way, for instance, by changing east into west.[16] (Leibniz and Clarke 1715/1716, 15)

It is possible to interpret the exchange of east and west in two ways: either as a rotation of 180°, or as a spatial reflection. De Risi (2007, 289–290) argues that Leibniz meant the exchange in the latter sense, accepting the radical consequence that the world and its incongruent counterpart are indiscernible. For Kant, to claim invariance of the world under mirroring certainly goes too far.[17] De Risi, however, claims that Kant's attack on Leibniz's metaphysics misses the point, emphasizing that

> Kant's true argument against Leibniz should target the principle of the identity of indiscernibles [...]. (De Risi 2007, 288)

According to my interpretation of the *New Elucidation* and the *Directions in Space*, he indeed does so. His argument from incongruent counterparts is in line with his 1755 criticism of the principle of indiscernibles. In 1768, Kant presumably considers the assumption that the world and its incongruent counterpart are indiscernible as an illegitimate, unjustified real use of the principle of sufficient reason, which is on a par with applying the principle of indiscernibles to identical objects (see Sects. 1.4.2.2 and 2.2.2.2). In the *New Elucidation*, his next step was to show that this illegitimate use of the principle is at odds with empirical evidence:

> Nor it is proper here to suppose that there is some hidden difference which escapes the senses [...], for that would be to search for knots in a bullrush. (1:409–410)

In *Directions in Space*, too, he cites empirical instances of objects in nature that agree in all relational properties but nevertheless are different:

> Indeed, there is a well-known characteristic distinctive of certain natural phenomena, which [...] consists in the particular direction in which the order of the parts is turned. In virtue of this distinctive characteristic, two creatures may be distinguished from each other, even though they may be exactly the same in respect of size, proportion and even the relative position of their parts. (2:380)

Here he refers to the existence of left-handed and right-handed objects, which are incongruent with their counterpart obtained by spatial reflection. He considers this phenomenon to be in conflict with Leibniz's principle of indiscernibles, which he had implicitly used ten years earlier in order to support his relational account of space. According to him, chirality, the respective spatial property of hands and screws, is *not* a relational property in the sense of Leibniz's *analysis situs*. In his view, handedness can only be defined in terms of a relation to the "directions" in space, that is, to the distinctions of right and left, above and below, in front of and behind. These "directions" are defined in relation to "absolute space" as an abstract

[16]"[...] qu'il est impossible qu'il y ait une raison, pourquoy Dieu, gardant les mêmes situations des corps entre eux, ait placés le corps dans l'espace ainsi et non pas autrement, et pourquoy tout n'a pas été mis àss rebours (par exemple) par un échange d l'orient et de l'occident."

[17]Weyl (1952, 21) understands him in this way, too.

totality of positions (2:377), or, as we would say today, relative to a given coordinate system with a given left- or right-handed orientation. His mathematical point is that the chirality of an object is only defined relative to an already oriented coordinate system. Hence, Kant's argument of 1768 already implies a purely mathematical criticism of Leibniz's program of the *analysis situs* (for details, see Buroker 1981, Ch. 3–5, and Buroker 1991, 315–326). In his view, the chirality of an object cannot be determined independently of the geometrical space to which the object belongs.

However, Kant's argument does not reduce to this mathematical issue. Mathematically, his quest for the "absolute" chirality of a spatial structure does not make sense. The clear and distinct metaphysical concept of space, which Kant has in mind in the context of the analytic method of his *Prize Essay*, has two features. Space is three-dimensional: this feature corresponds to the number of linearly independent basis vectors of a vector space. And it is orientable: this corresponds to a left-handed or right-handed arrangement of the basis vectors. From a mathematical point of view, this arrangement is not an absolute property of space, but a mere convention, which enables us to distinguish a left-handed coordinate system *relative* to a right-handed coordinate system, and vice versa. In contrast, Kant claims that a definite handedness, or orientation, is an "absolute" property of an object, which can only be explained in terms of a relation to absolute space, as some kind of background space of physical geometry. In his view, the absoluteness of this property is due to our idea of space, i.e., to the orientation of our internal spatial coordinate system. His crucial claim is that the chirality of a right-handed or a left-handed object is an *absolute property* that derives from its relation to absolute space. He describes the goal of the *Directions* as follows:

> What we are trying to demonstrate, then, is the following claim. The ground of the complete determination of a corporeal form does not depend simply on the relation and position of its parts to each other; it also depends on the reference of that physical form to universal absolute space, as it is conceived by the geometers. (2:381)

Only absolute space, and not relational space, provides the "ultimate ground", or sufficient reason, for determining the shape of spatial bodies. Kant employs the empirical example of the difference between the left and the right hand:

> It is apparent from the ordinary example of the two hands that the shape of the one body may be perfectly similar to the shape of the other, and the magnitudes of their extensions may be exactly equal, and yet there may remain an inner difference between the two, this difference consisting in the fact, namely, that the surface which encloses the one cannot possibly enclose the other. (2:382)

Kant then claims that the "inner difference" of the left and the right hand "must [...] rest upon an inner ground":

> Since the surface which limits the physical space of the one body cannot serve as a boundary to limit the other, no matter how that surface be twisted and turned, it follows that the difference must be one which rests upon an inner ground. (2:382)

Hoefer (2000, 239 and 254, n. 5) notes that Kant's reference to an "inner ground" is cryptic here, given that the ground of the difference is the relation to absolute space. The quoted passage does not make clear whether handedness is an intrinsic or a relational property of objects, i.e., the relation to a surrounding background space.

But the other text passages clearly indicate that Kant understands absolute space as a given reference frame with an oriented coordinate system, relative to which the chirality of an object is definite. Probably Kant uses the expression "inner ground" just as an elliptical phrase for the "ultimate ground of the inner difference". He continues:

> This inner ground cannot, however, depend on the difference of the manner in which the parts of the body are combined with each other. For, as we have seen from our example, everything may in this respect be exactly the same. (2:382)

He then suggests a thought experiment that gives a speculative causal argument for the reality of this reference space:

> [...] imagine that the first created thing was a human hand. That human hand would have to be either a right hand or a left hand. The action of the creative cause in producing the one would have of necessity to be different from the action of the creative cause producing the counterpart. (2:382–383)

Here, he employs Leibniz's causal reading of the principle of sufficient reason in order to emphasize that the difference between right- and left-handed objects must have a real ground. He then contrasts this difference with the relational conception of space of the Leibniz–Wolff school and his own 1755/1756 cosmology, pointing out that according to the concept of space "of many modern philosophers, especially German philosophers" (2:383), the space emerging from this act of creation would only exist in the inner relations of the parts of the first created hand. Finally, he attempts to reduce the relational concept of space to absurdity:

> However, there is no difference in the relations of the parts of the hand to each other, and that is so whether it be a right hand or a left hand; it would therefore follow that the hand would be completely indeterminate in respect of such a property. In other words, the hand would fit equally well on either side of the human body; but that is impossible. (2:383)

The very last conclusion is an obvious *non sequitur*, if one reads "impossible" in a *logical* sense of impossibility, i.e., if one assumes that Kant intends a *logical reductio ad absurdum*. Understood in this way, the argument seems to presuppose what needs to be proven here, i.e., the reality of an external reference space that determines the difference between right and left for our halves of the body absolutely, and not merely relative to the hand created at first. Only the absolute indeterminacy of the orientation of the single hand, in a relational account of space, follows from his premises; but not the claim that this indeterminacy is logically impossible. A defender of relationalism may continue Kant's thought experiment as follows. In the next step, God might create the rest of the world in accordance with the orientation of the single hand he had first created. If he did so, he would end up either with our universe, or with its incongruent counterpart. From Leibniz's relationalist point of view, there is no internal difference between both possible worlds. Hence, according to his principle of indiscernibles, both are not only physically equal but *identical*. This conclusion may be *counterintuitive*, but it is *not logically impossible*.

To accuse Kant of committing such a crude logical error, however, misses the point. One has also to take the historical context of the argument into account. In 1768, Kant still is indebted to a Leibniz–Wolffian logic and semantics. In addition, he is grappling with a cosmological problem which lies at the interface of physics and metaphysics. For him, "impossible" in the passage quoted above can only mean that the conclusion of his thought experiment has *no real possibility*. Therefore I propose to reconstruct the logical structure of Kant's 1768 argument as follows:

- *Premiss 1*: According to our idea of space, a single hand imagined as a first piece of creation is necessarily a left or right hand. That is, in relation to our idea of space it has definite chirality.
- *Premiss 2*: According to relationalism, a hand on its own is undetermined with respect to the property of being a left or right hand. That is, it has no definite chirality.
- *Conclusion*: From a relationalist point of view, a single hand created on its own can be arbitrarily embedded in our idea of space. That is, in a creation resulting in creating our bodies it should fit on both sides of the body, which (according to our idea of space) is absurd. Therefore, relationalism has *no real possibility*.

The main point of this argument is: The spatial properties of incongruent counterparts show that relationalism is incompatible with our idea of space. Our idea of space is *not* relational, it is *absolute* as far as it determines the orientation of hands or screws. This point is fatal for Kant's pre-critical cosmology. Taken together, a relational concept of space and our idea of space result in incompatible criteria for the individuation of objects. According to relationalism, a left hand on its own is identical with a right hand on its own. According to our idea of space, they are not. Hence, a relational concept of space does not suffice to individuate the left or respectively right hand, which we imagine in our idea of space.

Note, however, that Kant's argument only applies to a hand on its own which makes up the whole universe. Solely under this assumption, i.e., if objects *within* the world and objects which *make up* a world are underpinned by the same individuation criteria, it turns out that the pre-critical cosmology is based on an incoherent theory of space. In the critical period, i.e., in the *CPR*, Kant obviously no longer shares this assumption. But in 1768, Kant faces a puzzle in view of the argument from incongruent counterparts and of Leibniz's well-established arguments against absolute space. He is still convinced that there are good reasons for a relational theory of space, but the thought experiment about the orientation of a single hand created by God discredits relationalism.

As a result of the 1768 argument, Kant considers *both* contemporary candidates for an "objective" account to be refuted. His analysis of our idea of space gives rise to a serious objection against his pre-critical Leibnizean relational theory of space. But he cannot simply reject it in favour of Newton's absolute concept of space, given that he had good (Leibnizean) reasons for rejecting the latter, too. The real predicates which, for good reasons, he had attributed to space in his pre-critical system, had turned out to be logically incompatible. That is, he faced the problem that the concept of space employed in his pre-critical cosmology had no real possibility.

3.4.2 Two Problems with Kant's 1768 Argument

We should note, however, that Kant's 1768 argument only looks convincing against the background of his pre-critical cosmology and the analytic method of the *Prize Essay*. At this point, the suspicion may arise that the puzzle of 1768 is a self-generated problem of Kant's pre-critical cosmology. Indeed, a closer look at it reveals two problems that make Kant's conclusion debatable even in its historical context. First, even though from an epistemological point of view his use of Leibniz's principle of indiscernibles in the writings of 1755, 1758, and 1768 looks coherent, an incoherence in his attitude towards the symmetries of nature remains. Second, one has to ask to what extent Kant's conclusion of 1768 is epistemologically tenable, given that it results from his pre-critical account of real possibility and is still indebted to the conceptual background of the Leibniz–Wolff school.

3.4.2.1 An Incoherent Account of Symmetries

In his 1758 defence of relationalism, Kant relies on Leibniz's famous invariance arguments. According to them, it makes no sense to ask for the *position* of the universe in space and time, or to assume an empty space or time in which the universe is located, given that it is impossible to distinguish them (2:23). In Section 3.3 I argued that Kant here adopts Leibniz's principle of indiscernibles in the sense of a verificationist criterion of meaning, in a way which is compatible with his 1755 defence of atomism. In the *New Elucidation*, he emphasizes that objects with identical properties such as atoms or crystals are possible, but they are individuated in terms of position alone. Hence, their permutation does not give rise to any observable effect, and no real ground or sufficient reason is required to explain the constellation of atoms or crystals in the world (1:409; see Sect. 1.4.2.2). In this way, Kant objects to Leibniz's permutation argument in the Fourth Letter to Clarke and supports it in addition by empirical evidence (1:409–410).

From an epistemological point of view, Kant's 1755 defence of atomism and his 1758 defence of relationalism look coherent, even though the former rejects Leibniz's principle of indiscernibles, whereas the latter employs it. Both lines of reasoning have in common that in Kant's view no real ground, or sufficient reason, is required in order to explain the respective symmetry of the phenomena, i.e., the invariance of the universe under translations and rotations, and the invariance of a system of atoms or parts of a crystal under permutations.

In a certain sense, his 1768 argument, too, fits in with this line of reasoning. Considered as two possible worlds, a handed universe and its incongruent counterpart are indiscernibles, and Leibniz's argument against the existence of atoms is easily transferred to them. According to Leibniz, God would not have created one rather than the other, given that he had no sufficient reason to distinguish them. Kant, however, insists that we should ask for the *absolute orientation* of a single hand which makes up a world. He rejects the Leibnizean conclusion that a handed universe and its incongruent counterpart would be identical. For him, the universe and its mirror image are different, with their respective relation to our idea of space as their distinguishing mark. The main difference from his defences of relationalism and atomism is that left-handed and right-handed objects can*not* be transformed into each other by any symmetry operation within three-dimensional Euclidean space. They are mirroring-asymmetric, incongruent counterparts. In accordance with a verificationist criterion, the distinction between them has empirical meaning and so raises the quest for a real ground, or cause, of the difference.

The problem is that Kant generalizes the difference between left-handed and right-objects *within* nature to the universe *as a whole*. It has been noted that he does so in a certain parallel to the *Scholium* on space and time in the *Principia*, in which Newton generalizes the observed inertial effects acting on rotating bodies from his famous bucket experiment to a rotating system of two connected bodies in empty space (Newton 1726, 412–413):

> Just as Newton considered it absurd that for a rotating system in an otherwise empty universe inertial forces no longer occur, Kant considers it absurd that a single hand in the universe has no specific handedness.[18] (Lyre 2005, 11)

Newton's thought experiment in the *Scholium* and Kant's thought experiment of 1768 have in common that they are counterfactual. From a verificationist point of view, both generalizations are illegitimate. Newton made a counterfactual claim about the inertial effects of absolute space on rotating bodies. Kant, who in 1758 rejects Newton's absolute space based on a Leibniz-oriented verificationist criterion of meaning, in 1768 makes a counterfactual claim about a single hand that makes up the universe. His generalization from incongruent counterparts within nature to the handed universe is obviously *not* in accordance with his verificationism of 1758.

Comparing the arguments of 1755, 1758, and 1768, we note that Kant adopted Leibniz's principle of indiscernibles for continuous spatio-temporal symmetries, but he rejected it for discrete symmetry transformations such as permutations or mirroring. From the point of view of twentieth-century physics, his pre-critical uses of Leibniz's principle of indiscernibles turn out to be incoherent. They appeal to an unjustified difference between continuous and discrete symmetries. That the universe is invariant under continuous spatio-temporal symmetry transformations seemed possible to Kant in the sense of real possibility. That the universe should also be invariant under mirroring, did not.

---

[18]"So wie nun Newton die Möglichkeit für abwegig hielt, dass für ein rotierendes System in einem ansonsten leeren Kosmos keine Trägheitskräfte mehr auftreten, hält Kant es für abwegig, dass ein einziges händiges Objekt im Kosmos keine spezifische Händigkeit besitzt." My translation.

3.4.2.2 How Compelling is Kant's Conclusion of 1768?

From an epistemological point of view, the recent discussion on incongruent counterparts, in so far as it focuses on the question of whether handedness commits us to substantivalism or not, falls back into positions that Kant no longer defended after his critical turn. He concluded from his 1768 argument that *neither* relationalism *nor* substantivalism have real possibility. But is this the only option?

At this point, we have to ask how compelling his conclusion is. As shown above, it is based on his 1763 distinction between the logical and real grounds of reason. Before 1770, within the framework of a Leibniz–Wolffian doctrine of ideas, Kant had a modified Leibnizean concept of real possibility, according to which a concept can only have real possibility if its real definition is non-contradictory. Leibniz's own account of real possibility was stronger. For Leibniz, real possibility means the logical compossibility of substances, that is, the condition that the complete concepts of all things in the world must be compatible with each other. For Kant, who criticized Leibniz's theory of monads in the *New Elucidation*, only a substantially weaker condition remained. According to the *Prize Essay*, the real definition of a metaphysical concept must not give rise to contradiction, and it has to agree with the phenomena found in our mind (see Sect. 2.3.2.3). Hence, in his pre-critical philosophy, the real possibility of a concept only implies its logical possibility plus its adequacy.

In 1762, Kant still thought that the analytic method could give rise to non-contradictory real definitions of metaphysical concepts. In 1768, he was convinced that the analytic method, or respectively the analysis of our idea of space *plus* Leibniz's arguments against Newton's absolute space, gives rise to a contradictory concept of space if applied to a handed universe: according to Leibniz's invariance arguments, space is not absolute but relational; according to the argument from incongruent counterparts, in contrast, it cannot be relational but must be absolute. In his terms of his notion of real possibility, Kant's immediate conclusion was: for a handed universe, the non-coincidence of incongruent counterparts shows that relationalism has no real possibility. Against the background of the analytical method of his *Prize Essay* and his pre-critical concept of real possibility, this conclusion seemed inevitable to him. But Newton's account of absolute space, too, still seemed untenable. The text of the *Directions* indicates that the conclusion of 1768 is *not* intended to prove the reality of *Newton's* absolute space. Kant does not mention Newton at all, and he characterizes absolute space as an epistemic rather than a metaphysical concept:

> Finally, our considerations make the following point clear: absolute space is not an object of outer sensation; it is rather a fundamental concept which first of all makes possible all such outer sensation. (2:383)

Anticipating here to a certain extent his critical turn towards the transcendental conditions of the possibility of experience, he summarizes as follows:

> A reflective reader will not, therefore, dismiss the concept of space, as it is construed by geometers and as it has also been incorporated into the system of natural science

> by penetrating philosophers, as a mere figment of thought [Gedankending],[19] though the concept is not without its difficulties. Such difficulties reveal themselves when the attempt is made, employing the ideas of reason, to understand the reality of space, which is intuitive enough for inner sense. (2:383)

Here, Kant emphasizes the difficulties of conceiving of absolute space, which is more than a mere subject of thought (*bloßes Gedankending*) but less than a real entity, even if its reality is "intuitive (*anschauend*) enough for inner sense". These remarks demonstrate that the 1768 argument paves the way to the critical conception of space and time as forms of intuition, which Kant advocates from the *Dissertation* of 1770 on. Kant's conclusions of 1768 anticipate to a certain extent the 1770 theory of space and time as pure forms of intuition—yet exactly to what extent they do is debatable.[20]

This, then, is my interpretation of Kant's position. Now comes the problem: If space is only an epistemic concept, the proof resulting from the 1768 argument turns out to be substantially weaker than claimed in *Directions in Space*. The conclusion that a single hand first created by God has no definite orientation is simply incompatible with our intuitive idea that it must either be a right or a left hand. The indefiniteness of its orientation violates the requirement that a hand must be unambiguously embeddable into our idea of space. Accordingly, one should replace the term 'impossible' in the conclusion of 1768 by 'implausible',

---

[19]The Cambridge translation "figment of the imagination" is not correct here. The German expression "bloßes Gedankending" indicates that Kant here does *not yet* anticipate his 1770 theory of intuition. The translation "figment of the imagination" suggests that Kant speaks of an *ens imaginarium*, a mere fiction of perception. But the expression "bloßes Gedankending" is much closer to Kant's critical account of an *ens rationis*, i.e., a figment of *thought*. The distinction between an *ens rationis* or "empty concept without object" and an *ens imaginarium* or "empty intuition without concept" (*CPR*, B 348) presupposes Kant's theory of space and time as forms of pure intuition, which is first mentioned in Kant's 1769 notes on metaphysics. Even if Kant here already anticipates his later transcendental conditions of the possibility of experience, the idea of space for him is still a *concept*, not an intuition. His use of the terms "idea of reason" and "intuitive" ("anschauend") is not yet specific; see below and next note. His distinction between concepts and intuitions would finally become his way out of the 1768 puzzle (see Sect. 3.4.3).

[20]See Friedman (1992, 29) as well as the objections against identifying the 1768 result with Newton's concept of absolute space in von Wolff-Metternich (1995, 88–90) and Kauark-Leite (2017). Kaulbach (1960, 98) suggests that Kant in 1768 argues that our idea of space is bound to our body. De Risi (2007, 291) points out that Kant's source for doing so, as for his examples of incongruent counterparts, is Buffon's *Histoire Naturelle*, mentioned at the beginning of *Directions in Space* (3:277). Kaulbach (1960) and Kauark-Leite (2017) identify the 1768 result with the 1770 conception of space as a form of intuition. But in 1768, Kant did not yet have his later account of space as a *repraesentatio singularis*. See the wavering remarks on space in the notes on metaphysics of 1768–1769 (see Sect. 3.4.3). De Risi (2007) stresses the difference between Kant's 1768 version of the argument and his later accounts based on the critical theory of space and time. He, however, interprets Kant's 1768 argument as an argument in favour of Newton's absolute space and concludes that it "fails to achieve its goal" (De Risi 2007, 292), whereas in the versions of 1770 and later, "Kant's argument on the incongruents would prove nothing at all—as it would assume from the start the ideality it would want to deduce" (De Risi 2007, 286). I hope to show here that the state of affairs is much more complicated than any of these authors realized.

'unimaginable', or 'counter-intuitive'. If we understand the latter term, counter-intuitive, in the sense of violating the conditions of intuition, then Kant's argument here would amount to his critical theory of space as a form of intuition, which he first presented in 1770. In 1768, however, Kant did not yet have this theory. My conclusion concerning this problem is that in 1768 he could not strictly distinguish between counter-intuitiveness and real impossibility.

In the end, two ways of interpreting Kant's conclusion remain. According to a stronger reading, the argument proves the real (though not logical) impossibility of a relational concept of space. According to a weaker reading, it only proves that our idea of space is a given epistemic conception that precedes the cosmological use of geometric concepts. The stronger reading corresponds to the proof result obtained against the conceptual background of the Leibniz–Wolff school, while the weaker reading corresponds to the proof result obtained with Kant's later theory of intuition. Only the latter is still convincing from a modern logical point of view. The former is not. The weaker reading is also coherent with Kant's later versions of the argument of 1770, 1783, and 1786, which cite the phenomenon of incongruent counterparts as additional evidence for his critical theory of space as a pure form of intuition (see Sect. 3.4.4).

But what, then, does the "real possibility" of the lone hand and its relation to absolute space in the 1768 argument mean? The *critical* answer of 1781 to this question is: Only concepts which are representable in pure intuition have real possibility (respectively objective reality). In 1768, Kant does not yet differentiate between concepts and intuitions as two different kinds of representations. Without this distinction, he apparently confuses two semantic requirements: (i) the *intuitive* requirement that a concept should be representable in space and time; and (ii) the *logico-conceptual* requirement that a concept should have real possibility in the sense of admitting of a non-contradictory (and adequate) real definition. That is, in 1768 he replaces the metaphysical concept of space by an epistemic concept, without realizing the consequences of doing so. The conclusion that relationalism has no real possibility now diminishes to the substantially weaker claim that it is implausible, counter-intuitive, or unimaginable.

According to Kant's views after 1768, the puzzle of incongruent counterparts commits us to dispense with metaphysical realism. This conclusion may be too strong. In the context of his transcendental philosophy, there is indeed no longer any puzzle; but this only proves that transcendental idealism is *sufficient* for resolving the puzzle. It does not prove that Kant's critical turn is a *necessary* condition for it.

3.4.3 Solution of the Puzzle

In sum, then, the 1768 argument made Kant think that neither Leibniz's relationalism nor Newton's absolute space have real possibility. From the argument from incongruent counterparts he concluded that relationalism is untenable, but he was not willing to put Leibniz's arguments against Newton aside. He would resolve the puzzle by suggesting that absolute space is an epistemic concept. But in 1768 he was still unclear about the significance of this epistemic turn. In 1769, he had the crucial additional insight that space is a pure form of intuition, in contradistinction to other metaphysical concepts, such as ground, substance, etc., which are pure forms of thought. In his *Inaugural Dissertation* of 1770, he finally presented his new theory of space and time as pure forms of intuition. This epistemic theory of space and time sheds new light on the 1768 argument from incongruent counterparts in terms of the weaker conclusion indicated above (in Sect. 3.4.2.2). The details of these developments are investigated in the next chapter (see Sects. 4.2.1 and 4.2.2): for now, I sketch how they affected Kant's pre-critical concept of real possibility and his 1768 argument from incongruent counterparts.

If our idea of space is a pure intuition rather than a concept, the argument from incongruent counterparts seems no longer to be fatal for cosmology. Intuition enables us to distinguish left-handed and right-handed objects. Intuition also enables us to imagine a possible world which consists of a single hand only, and it gives rise to the conclusion that it is "impossible" to imagine a single hand that "would fit equally well on either side of the human body" (2:383). Obviously here we are neither dealing with logical possibility, nor with real possibility in the sense of asking for a real definition of physical space, but only with our subjective impossibility of imagining a lone hand without definite chirality.

In the course of the critical turn, the 1770 distinction between concepts and intuitions indeed led to *two* crucial semantic shifts in Kant's account of real possibility. The first comes into effect in the 1770 dissertation, when Kant still believed that unrestricted metaphysical knowledge, i.e., pure intellectual knowledge of the world, is possible (see Sect. 4.2.3). The second is made explicit in the *CPR*. In 1781, Kant denies any cosmological cognition that exceeds the limitations of possible experience (see Sect. 6.2.2 and Chap. 5), that is, he rejects the objective reality of any metaphysical concepts which cannot be represented in space and time as forms of pure intuition.

In 1770 Kant is already distinguishing between concepts and intuitions as two different kinds of representations (Sect. 4.2.2), but he is not yet at the point of identifying real possibility with representability in space and time. In the *Dissertation* of 1770, he critically mentions that "*unrepresentable* and *impossible* are commonly treated as having the same meaning" (*Dissertation*, 2:388). At that time he still thinks that symbolic intellectual knowledge of an actually infinite world is possible in cosmology, even though such a concept is not representable in our idea of space (see Sects. 4.2.3 and 4.4). The *Dissertation* emphasizes that there is no intuitive knowledge of intellectual things, but only symbolic cognition (2:396). That is, in 1770 he is still convinced that the "real use of reason" in metaphysics gives rise to a symbolic concept of the world as a sum total of all substances. Accordingly, the *Dissertation* of 1770 makes a sharp distinction between the real possibility of symbolic metaphysical concepts on the one hand, and the representability of spatio-temporal concepts in intuition on the other. Eleven years later, Kant no longer distinguishes between real possibility and spatio-temporal representability. In the *CPR*, real possibility (or objective reality) is bound to intuition. To be more precise:

in contrast to the 1768 argument on incongruent counterparts, it is no longer *tacitly* but rather *explicitly* bound, on the condition of being representable in space and time.

3.4.4 Incongruent Counterparts After the Critical Turn

On the basis of Kant's theory of intuition, whether that of 1770 or from 1781 onwards, his pre-critical thought experiment about the creation of a single hand, or the orientation of a possible world, is no longer a puzzle. From a critical point of view, it reveals the non-conceptual character of space as pure intuition. The internal difference of incongruent counterparts cannot be depicted with concepts. This indicates that space is an intuition. In accordance with this result, it has been concluded that the existence of incongruent counterparts commits us to admitting irreducible non-conceptual aspects of empirical properties (Hanna 2008; Landy 2013) or non-extensional features of physical quantities (Falkenburg 2006).

In the 1770 dissertation, incongruent counterparts reappear as distinct singular representations that indicate that space is a pure intuition. Incongruent counterparts are "perfectly similar and equal" but they do not coincide (2:403). That is, they fall under the same relational concepts (of Leibniz's project of an *analysis situs*, see Sect. 3.4.1), even though they exhibit a difference. The difference between a left hand and a right hand cannot be expressed by means of concepts, it can only be intuitively grasped:

> Which things in a given space lie in one direction and which things incline in the opposite direction cannot be described discursively nor reduced to characteristic marks of the understanding by any astuteness of the mind. Thus, between solid bodies which are perfectly similar and equal but incongruent such as the left and right hands [...] there is a difference, in virtue of which it is impossible that the limits of their extensions should coincide [...] It is, therefore, clear that in these cases the difference, namely the incongruity, can only be apprehended by a certain pure intuition. (2:403)

The *Dissertation* cites the existence of incongruent counterparts as additional evidence that space is a pure form of intuition rather than an entity. Here, the difference of incongruent counterparts ranks among the features of our idea of space that do not derive from a universal concept of space, but can only be seen *in concreto* in space:

> Thus, between solid bodies which are perfectly similar and equal but incongruent such as the left and right hands (in so far as they are conceived only according to their extension), or spherical triangles from two opposite hemispheres, there is a difference, in virtue of which it is impossible that the limits of their extension should coincide – and that, in spite of the fact that, in respect of everything which may be expressed by means of characteristic marks intelligible to the mind through speech, they could be substituted for one another. It is, therefore, clear that in these cases the difference, namely, the incongruity, can only be apprehended by a certain pure intuition. (§15 C, 2:403).

The *Prolegomena* of 1783 and the *MFSN* of 1786 also cite the phenomenon of incongruent counterparts as an "affirmative argument" for the subjectivity of the concept of space. In the *Prolegomena*, Kant emphasizes that incongruent counterparts

> are surely not representations of things as they are in themselves, and as the pure understanding would cognize them, rather, they are sensory intuitions, i.e., appearances, whose possibility rests on the relation of certain things, unknown in themselves, to something else, namely our sensibility. Now, space is the form of outer intuition of this sensibility, and the inner determination of any space is possible only through the determination of the outer relation to the whole space of which the space is a part (the relation to outer sense) [...]. We can therefore make the difference between similar and equal but nonetheless incongruent things [...] intelligible through no concept alone, but only through the relation to right-hand and left-hand, which refers immediately to intuition. (*Prolegomena*, 4:285–286)

The *MFNS* mentions the "internal difference" of incongruent counterparts as follows, in the context of dealing with the direction of motions:

> But what is here the side towards which the motion is directed? This has a kinship with the following question: On what rests the inner difference of snails, which are otherwise similar and even equal, but among which one species is wound rightward, the other leftward; or the winding of the kidney bean and the hop, where the first runs around its pole like a corkscrew, or, as sailors would express it, against the sun, whereas the second runs with the sun? This is a concept which can certainly be constructed, but, as a concept, can in no way be made clear in itself by means of universal characteristics and in the discursive mode of cognition, and can yield no thinkable difference in the inner consequences in the things themselves [...], but is nevertheless a genuine mathematical, and indeed inner difference, which is connected with, although not identical to, the difference between two circular motions that are otherwise equal in all parts, but differ in direction. (4:483–484)

Then Kant refers to the passage from the *Prolegomena* quoted above:

> I have shown elsewhere that, since this difference can certainly be given in intuition, but can in no way be captured in clear concepts, and thus cannot be rationally explicated (*dari, non intelligi*), it supplies a good confirming ground of proof for the proposition that space in general does not belong to the properties or relations of things in themselves, which would necessarily have to be reducible to objective concepts, but rather belongs merely to the subjective form of our sensible intuition of things or relations [...]. (4:484)

In the *CPR*, Kant does not mention incongruent counterparts. This is due to the fact that the *CPR* develops the principles of transcendental philosophy according to the synthetic method, in contradistinction to the analytic presentation in the *Prolegomena*, as Kant explains there:

> Here then is such a plan subsequent to the completed work, which now can be laid out according to the analytic method, whereas the work itself absolutely had to be composed according to the synthetic method, so that the science might present all of its articulations, as the structural organization of a quite peculiar faculty of cognition, in their natural connection. (4:263)

The "completed work" which is "composed according to the synthetic method" is the *CPR*; and its plan subsequently outlined "according to the analytic method" is given in the *Prolegomena*. Kant further explains:

> In the Critique of Pure Reason I worked on this question synthetically, namely by inquiring within pure reason itself, and seeking to determine within this source both the elements and the laws of its pure use, according to principles. [...] Prolegomena should by contrast be preparatory exercises; they ought more to indicate what needs to be done in order to bring a science into existence if possible, than to present the science itself. They must therefore rely on something already known to be dependable, from which we can go forward with confidence and ascend to the sources, which are not yet known [...]. The methodological procedure of prolegomena [...] will therefore be analytic. (4:274–275)

Here, Kant leaves no doubt that his critical view of the analytic and the synthetic method still agrees with the traditional meaning of the regressive and progressive method tracing back to Pappus (see Appendices A.2 and A.2.1). After the collapse of his pre-critical system, however, the analytic method for him is no longer a method of deriving metaphysical concepts (as suggested in the *Prize Essay*), but rather a method of presenting his critical doctrine of transcendental idealism in a way that makes it plausible by proceeding from "something already known" to "the sources, which are not yet known" (see also Hatfield 1997, xix–xx). Given that the argument from incongruent counterparts employs the analytic rather than the synthetic method, it reappears in the *Prolegomena*, but not in the Transcendental Aesthetic of the *CPR*. Most Kant scholars miss this point (for an exception, see Severo 2005, 2007).

Kant mentions incongruent counterparts, or left and right hands, for the last time in his work *To Orient Oneself* of 1786 on the orientation in thinking. There, he again takes up the problem of geographical orientation, in remarks similar to those of the 1768 essay, and reinterprets them in subjective terms of *feeling* the directions of space only via their relation to our body:

> [...] to orient oneself means to use a given direction (when we divide the horizon into four of them) in order to find the others—literally, to find the sunrise. Now if I see the sun in the sky and know it is now midday, then I know how to find south, west, north, and east. For this, however, I also need the feeling of a difference in my own subject, namely, the difference between my right and left hands. [...] Thus even with all the objective data of the sky, I orient myself geographically only through a subjective ground of differentiation; and if all the constellations, though keeping the same shape and position relative to one another, were one day by a miracle to be reversed in their direction, so that what was east now became west, no human eye would notice the slightest alteration on the next bright starlit night, and even the astronomer—if he pays attention only to what he sees and not at the same time to what he feels—would inevitably become disoriented. (8:136)

The context of this passage is that Kant wants to show how this subjective concept of spatial orientation can be extended by analogy, to

> [...] thinking in general [...] extending itself beyond all the bounds of experience [...] solely to bring its judgments under a determinate maxim according to a subjective ground of differentiation in the determination of its own faculty of judgment.*
>
> * Thus to orient oneself in thinking in general means: when objective principles of reason are insufficient for holding something true, to determine the matter according to a subjective principle. (8:134–135)

These remarks not only demonstrate once more how important analogical reasoning was for Kant, but also indicate how decisive the argument from incongruent counterparts was for his critical project.

3.5 Conclusions: From Metaphysics to Epistemology

Kant's pre-critical metaphysics only lasted from establishing its foundations in the 1755/1756 trilogy to defending the underlying analytic method in the *Prize Essay*. When the latter was published, the pre-critical system had already begun to collapse. The distinction between logical and real grounds introduced in the *Only Possible Argument* and the *Negative Magnitudes* first resulted in delimiting logic from metaphysics, and later in making the principles of the *New Elucidation* seem suspicious. Kant's increasing distrust in the foundations of his pre-critical system culminated in the *Dreams* and the *Directions in Space*.

The collapse of the pre-critical cosmology resulted in the epistemic turn of his metaphysics, towards which the *Prize Essay* had already made a first and decisive step. There, Kant defended the conceptual analysis of metaphysics along the lines of Newton's analysis of the phenomena, making explicit the analogy between inner and outer experience on which his 1755/1756 uses of analytic methods had implicitly relied. The analogy gave rise to a crucial problem neglected in the *Prize Essay*: What are the criteria for adequate metaphysical concepts? How can we know whether our inner, subjective experience of metaphysical ideas agrees with the objects of outer experience, or the objective world? The *Dreams* testifies to Kant's discomfort about this question, in view of the similarities between Swedenborg's spiritism and his own 1755 theory of the interactions of substances. Hence the quest for clarifying the epistemic conditions of the real use of reason, which became manifest in the claim that "metaphysics is the science of the boundaries of human reason" (3:368).

The *Dreams* shed additional light on the *Directions in Space*. Kant's criticism of Leibniz's project of an *analysis situs* and his argument from incongruent counterparts only became completely comprehensible against the background of the *Dreams*. The *Dreams* defeated the pre-critical theory of the interactions of substances mediated by God. With this defeat, the grounds for combining the existence of indiscernible atoms with a relational account of space vanished, and the latter became suspicious. Criticizing Leibniz's relational concept of space, Kant asks for the ultimate ground, or sufficient reason, of the difference between incongruent counterparts. Then, he identifies this ultimate ground with absolute space, not in Newton's sense of a real entity, but rather in the sense of an epistemic idea. In terms of the *New Elucidation*, the ultimate ground or sufficient reason of the absolute orientation of a lone hand or any other left- or right-handed object is an *epistemic* rather than an ontic ground (see Sect. 1.4.2.1). The epistemic concept of absolute space suggested in the *Directions in Space* perfectly conforms to the epistemological turn suggested in the *Dreams*.

According to Kant's solution to his 1768 puzzle, the argument from incongruent counterparts commits us to dispensing with metaphysical realism. This conclusion may be too strong, however. In the context of his transcendental philosophy, obviously no puzzle any longer remains. But this way out only proves that transcendental idealism is *sufficient* for resolving the puzzle. It does not prove that Kant's critical turn is a *necessary* condition for doing so. In the end, it is little wonder that the debate on Kant's argument from incongruent counterparts has continued up to the present day.

References

Allison, Henry E. 2012. *Essays on Kant*. Oxford: Oxford University Press.

Beiser, Frederick C. 1992. Kant's intellectual development. In *The Cambridge Companion to Kant*, ed. P. Guyer, 26–61. Cambridge: Cambridge University Press.

Buroker, Jill V. 1981. *Space and Incongruence. The origin of Kant's idealism*. Dordrecht: Reidel.

Buroker, Jill V. 1991. The Role of Incongruent Counterparts in Kant's Transcendental Philosophy. In van Cleve et al. 1991, 315–319.

Byrd, Jeremy. 2008. A remark on Kant's argument from incongruent counterparts. *British Journal for the History of Philosophy* 16: 789–800.

De Risi, Vincenzo. 2007. *Geometry and Monadology. Leibniz's Anaslysis Situs and Philosophy of Space*. Birkhäuser: Basel.

Earman, J. 1989. *World Enough and Spacetime. Absolute versus Relational Theories of Space and Time*. Cambridge: MIT Press.

Euler, Leonhard. 1748. Reflexions sur l'espace et le temps. In: *Memoires de l'academie des sciences de Berlin* 4, 1750, 324–333. Engl. transl. by M. Saclolo and P. Wake: Reflections on Space and Time. http://eulerarchive.maa.org/docs/translations/E149tr.pdf.

Euler, Leonhard. 1763. *Vernünftige Gedanken von dem Raume, dem Orth, der Dauer und der Zeit*. Quedlinburg: Gottfried Heinrich Schwans Wittwe.

Falkenburg, Brigitte. 2006. Intuition and cosmology: The puzzle of incongruent counterparts. In: *Intuition and the Axiomatic Method*, ed. E. Carson and R. Huber, 157–180. Dordrecht: Springer.

Friedman, Michael. 1992. *Kant and the Exact Sciences*. Cambridge: Harvard University Press.

George, Rolf, and Paul Rusnock. 1994. Snails rolled up contrary to all sense. *Philosophy and Phemenological Research* 54: 459–466.

Grier, M. 2001. *Kant's Doctrine of Transcendental Illusion*. Cambridge: Cambridge University Press.

Guyer, Paul. 1987. *Kant and the Claims of Knowledge*. Cambridge: Cambridge University Press.

Hanna, Robert. 2008. Kantian non-conceptualism. *Philosophical Studies* 137: 41–64.

Hatfield, Gary. 1997. *Introduction to the translation of Kant's "Prolegomena"*. Cambridge: Cambridge University Press 1997, revised ed. 2004.

Hoefer, Carl. 2000. Kant's hands and Earman's pions: Chirality arguments for substantival space. *International Studies in the Philosophy of Science* 14: 237–256.

Kanzian, Christian. 1993. Kant und Crusius 1763. *Kant-Studien* 83: 399–407.

Kauark-Leite, Patricia. 1993. On the epistemic status of absolute space: Kant's directions in space read from the standpoint of his critical period. Kant-Studien 108: 175–194.

Kaulbach, Friedrich. 1960. *Die Metaphysik des Raumes bei Leibniz und Kant*. Köln: Univ.-Verlag.

Kaulbach, F. 1982. *Immanuel Kant. Zweite, durchgesehene Auflage*. Berlin: de Gruyter.

Landy, David. 2013. What incongruent counterparts show. *European Journal of Philosophy* 21: 527–527.

Laywine, A. 1993. *Kant's early metaphysics and the origins of the critical philosophy*. Atascadero: Ridgeview Publishing.

Leibniz, Gottfried Wilhelm. 1715. Initiarerum mathematicarum metaphysica. In: *Mathematische Schriften*, ed. C.I. Gerhardt, Berlin, 1849–1860, repr.: Hildesheim: Olms 1971, vol. 7, 17–29. Engl. (in extracts): The metaphysical foundations of mathematics. In *Gottfried Wilhelm Leibniz*, L.E. Loemker. Philosophical Papers and Letters, 666–674. Dordrecht: Reidel 1969.

Leibniz, Gottfried Wilhelm, and Samuel Clarke. 1715/1716. Correspondence. In: *Die philosophischen Schriften*, ed. C.I. Gerhardt, Berlin, 1875–1890, repr.: Hildesheim: Olms 1965, vol. 7, 345–440. Quoted after the Engl. transl. and repr.: Edited, with Introduction, by Roger Ariew. Indianapolis & Cambridge: Hackett Publishing Co. 2000.

Longuenesse, Béatrice. 2001. In *Kant und die Berliner Aufklärung: Akten des IX. Internationalen Kant-Kongresse*. Band 1: Hauptvorträge, ed. V. Gerhardt, R.P. Horstmann, R. Schumacher. Berlin: de Gruyter, 66–85.

Lyre, Holger. 2005. Metaphysik im "Handumdrehen": Kant und Earman, Parität und moderne Raumauffassung. *Philosophia Naturalis* 42: 49–76. Quoted after the digital version: http://www.lyre.de/kanthand.pdf. Access: 23 Mar 2020.

Mühlhölzer, Felix. 1992. Das Phänomen der inkongruenten Gegenstände aus Kantischer und aus heutiger Sicht. *Kant-Studien* 83: 436–453.

Nerlich, Graham. 2009. Incongruent counterparts and the reality of space. *Philosophy Compass* 4/3: 598–613.

Newton, Isaac. 1726. *Principia mathematica philosophiae naturae*, 3rd edn. Quoted after: *The Principia: A New Translation* by I.B. Cohen and A. Whitman, 1999. Berkeley: University of California Press.

Reich, Klaus. 1958. *Einleitung zu: Kant, Immanuel: Über die Form und die Prinzipien der Sinnen- und Geisteswelt*. Lat.-dt. Hamburg: Meiner, VII–XVI.

Rusnock, Paul, and Rolf George. 1995. A last shot at Kant and incongruent counterparts. *Kant-Studien* 86: 257–277.

Schönfeld, M. 2000. *The Young Kant*. Oxford: Oxford University Press.

Severo, Rogério Passos. 2005. Three remarks on the interpretation of Kant on incongruent counterparts. *Kantian Review* 9: 30–57.

Severo, Rogério Passos. 2007. A puzzle about incongruent counterparts and the Critique of Pure Reason. *Pacific Philosophical Quarterly* 88: 507–521.

van Cleve, James, and Robert E. Frederick, ed. 1991. *The Philosophy of Right and Left. Incongruent Counterparts and the Nature of Space*. Dordrecht: Kluwer.

von Wolff-Metternich, Brigitta-Sophie. 1995. *Die Abkehr vom Mathematischen Methodenideal*. Berlin: De Gruyter.

Walford, David. 1992. Introduction to: *Immanuel Kant: Theoretical Philosophy* 1755–1770. Translated and edited by D. Waldford in collab. with R. Meerbote. New York: Cambridge University Press, xxxv–lxxxi.

Walford, David. 1999. The aims and method of Kant's 1768 *Gegenden im Raum* essay in the light of Euler's 1748 *Réflexions sur L'Éspace*. *British Journal for the Philosophy of Science* 7: 305–332.

Walford, David, and Ralf Meerbote. 1992. *Immanuel Kant: Theoretical Philosophy 1755–1770*. Translated and edited by DẆaldford in collab. with R. Meerbote. New York: Cambridge University Press.

Waschkies, Hans-Joachim. 1987. *Physik und Physikotheologie des jungen Kant. Die Vorgeschichte seiner Allgemeinen Naturgeschichte und Theorie des Himmels*. Amsterdam: B. Grüner Publishing Company.

Weyl, Hermann. 1952. *Symmetry*. Princeton: Princeton University Press.

Chapter 4
A Closer Look at the Critical Turn

Initially I saw this doctrine as if in twilight. I tried [...] to prove propositions and their opposite, not in order to establish a skeptical doctrine, but rather because I suspected I could discover in what an illusion of the understanding was hiding. The year '69 gave me a great light. (18:69)

In a broad sense, Kant's critical turn is the transition from his pre-critical system of 1755/1756 to his transcendental philosophy of 1781. Kant's pre-critical metaphysics aimed at reconciling a metaphysical system in Wolff's style with the principles of Newtonian physics. We have seen in the preceding chapters how complex this project was (Chap. 1); how he attempted to realize it by means of analytic methods (Chap. 2); how his *Prize Essay* established an analogy between the analytic method of Newtonian science and the use of conceptual analysis in metaphysics (Sect. 2.3); and how he became increasingly aware of the problems of his 1755/1756 system, by distinguishing the logical and the real use of reason (Chap. 3). The critical turn in the above broad sense began with the *Dreams* of 1766 and the *Directions in Space* of 1768. The *Dreams* indicate an epistemic turn for metaphysics, according to which "metaphysics is a science of the *limits of human reason*" (3:367–368) (Sect. 3.2). The *Directions* suggests an epistemic concept of absolute space, in view of the puzzle of incongruent counterparts (Sect. 3.4). This epistemic turn also becomes evident in his notes on metaphysics dated to 1769 (phase κ), which contain several remarks on the subjective nature of metaphysical concepts. Kant's famous note 5037 on metaphysics of course relates to these insights and what he made of them after 1768.

In a narrower sense, however, it is only at this point that Kant's critical turn begins. It spans the silent decade of the 1770s and ends with the publication of the *Critique of Pure Reason* of 1781. Its beginnings are usually dated 1769, the year of the "great light" mentioned in note 5037 (18:69), or 1770, the year of the *Dissertation*. In the latter, Kant first presents his epistemic theory of space and time as pure intuitions, laying thus the grounds for his transcendental philosophy, but

B. Falkenburg, *Kant's Cosmology: From the Pre-Critical System to the Antinomy of Pure Reason*, European Studies in Philosophy of Science 12,
https://doi.org/10.1007/978-3-030-52290-2_4

appending to it the outline of a metaphysics which compared to the *New Elucidation* only shows a few epistemological aspects. In the *CPR*, the pre-critical metaphysics is finally replaced by the Transcendental Dialectic and its cosmology part by the doctrine of the antinomy of pure reason.

To shed more light on the questions of why and when Kant's pre-critical cosmology was replaced by the critical doctrine of the antinomy, this chapter investigates in detail the development of Kant's thought after 1768. According to the traditional view of the critical turn, Kant already had the antinomy of pure reason before 1770, the year when he allegedly resolved it in the *Dissertation*. I will argue that from 1766 to 1770 Kant did *not* yet assume that reason must become entangled in a cosmological antinomy. At that time, indeed he was still far from a statement of the later antinomy of pure reason. Before 1769 he saw his critical doctrine of transcendental idealism just "as if in twilight" (18:69), according to note 5037. In addition, he states there that he attempted

> to prove propositions and their opposite, not in order to establish a skeptical doctrine, but rather because I suspected I could discover in what an illusion of the understanding was hiding. (N. 5037, 18:69)

Many Kant scholars have understood this enigmatic dictum in the sense of Hume's skepticism and the antinomy of pure reason. Kant, however, clearly distinguishes the skeptical method proposed in N. 5037 from Hume's skepticism. After the *Dreams* and the *Directions*, he no longer considered the analytical method as a reasonable approach to establishing a system of metaphysics. He was therefore looking for a new philosophical method, but did not yet have the critical method of the *CPR*. The cosmological antinomy, in the strict sense of an internal contradiction in which reason entangles itself, must have emerged later. In 1769 he had the skeptical method which he employed in the apagogic proofs of the antinomy of 1781, but not yet the antinomy itself. In fact, the 1770 *Dissertation* claims the peaceful coexistence of the sensible and the intelligible worlds, rather than an antinomy. The cosmological antinomy arose from his later inversion of the arguments of 1770, as a closer look at Kant's notes on metaphysics and his correspondence of the early 1770s shows.

In order to understand the origin of the cosmological antinomy, Kant's critical resolution of the antinomy should also be taken into account. In the *CPR*, Kant gives a twofold resolution. For the objects of the "dynamic" antinomy concerning freedom and the sufficient reason of the existence of the world, the sensible and the intelligible worlds may coexist. For them, the 1770 account of metaphysics in 1781 still seemed tenable to Kant. In contrast, for the "mathematical" antinomy concerning the spatio-temporal extent of the universe and the existence of atoms, it did not. From Kant's critical point of view, the more substantial problem is the mathematical antinomy, that is, the cosmological question of whether the universe and the number of its parts are finite or infinite. Therefore, Kant's notion of the infinite is also decisive for understanding how he arrived at the cosmological antinomy.

4.1 What is Wrong with the Traditional View?

Kant's critical turn has been the focus of research for more than a hundred years. Kant scholars agree that the critical turn in a narrow sense was initiated with the *Dissertation* of 1770, in which he for the first time presented his doctrine of space and time as subjective forms of intuition, laying the grounds for the transcendental aesthetics of the *CPR*. This doctrine does indeed mark the decisive point at which he turned against the 1755/1756 system, according to which space and time are metaphysical concepts with objective meaning.

There is controversy in Kant research, however, concerning the motives that prompted him to develop his space-time theory of 1770 and about the significance of the *Dissertation*. Since the end of the nineteenth century, and up until today, influential Kant scholars have claimed that the doctrine of the antinomy is the oldest part of the *CPR*. Their view is based on the claim that Kant had already known of the cosmological antinomy since the late 1760s and resolved it with the 1770 *Dissertation*. In contrast, other studies identify the development of the transcendental deduction of the categories in the early 1770s as the actual motive behind Kant's transition to transcendental idealism. In the following I argue that both views miss the point. The former retrojects Kant's later cosmological antinomy into the puzzle of 1768, which referred to the concept of space at the interface of mathematics and metaphysics, but not yet to the cosmological concept of the world as a totality, and which was overcome with the 1770 theory of space and time. The latter neglects the significance of epistemological problems related to cosmological cognition for Kant's development in the silent decade.

Since the emergence of neo-Kantianism, the following pattern of misconstruing Kant's philosophical development has been preserved, in ever new variations:[1]

1. In the 1760s, Humes' skepticism woke Kant out of his "dogmatic slumber".[2]
2. The use of the skeptical method led him to discover the cosmological antinomy in which transcendental idealism is rooted.
3. The "great light" of 1769 consisted in the insight that the antinomy can be resolved by the 1770 theory of space and time as subjective forms of intuition.

---

[1]It was first presented by Fischer (1860, 1909). Erdmann (1884), Adickes (1897), and Riehl (1924) were particularly influential. See also the overview by Reich (1958, VII–XII). More recently, Kreimendahl (1990) has continued this long-established misconstruction of Kant's development; see below. In contrast, Beiser (1992, 36–46, 55) points out that Crusius was more important for Kant's development than Hume's skepticism, and he emphasizes Rousseau's influence on Kant's critical attitude towards metaphysics expressed in the *Dreams*.

[2]After Paulsen (1875), Kuno Fischer (1909, 212) emphasized Hume's influence on Kant. The relevant passages stem from the third edition (1882) of Fischer's book; see the explanations in the appendix to Fischer (1909, 659–660). The dispute about the influence of Hume is as old as the attempts to reconstruct Kant's development (Fischer 1909, 683; Kreimendahl 1990, 15).

Kant himself laid the foundation for this interpretation, through his later remarks in the *Prolegomena* and the famous letter to Garve of September 21, 1798 (12:257–258; see below). In addition, the often-quoted note 5037 on metaphysics was taken as evidence for a direct relation between the epistemic turn of 1769 and the antinomy of pure reason. It is one of the notes stemming from the late seventies, in which Kant looked back at the development of his critique of reason shortly before the final wording of the *CPR*. The full text reads:

> If I only achieve as much as being convincing that one must suspend the treatment of this science until this point has been settled, then this text will achieve its purpose.
>
> Initially I saw this doctrine as if in twilight. I tried quite earnestly to prove propositions and their opposite, not in order to establish a skeptical doctrine, but rather because I suspected I could discover in what an illusion of the understanding was hiding. The year '69 gave me a great light. (18:69)

The year 1769 was the period between publishing the *Directions in Space* and the *Dissertation*. Many Kant scholars have identified Kant's attempts "to prove the propositions and their opposite" with the skeptical method, which makes the conflict between theses and counter-theses a touchstone of reason, and associated it with Hume's skepticism. Whereas the former claim is undisputed and correct, the latter is debatable, given that Kant repeatedly accuses Hume of a too-pronounced tendency to doubt (see Sect. 4.3.2). The "doctrine" initially seen "as if in twilight" has correctly been equated with transcendental idealism, which, from Kant's later point of view, invalidates the doctrines of his former metaphysics. Merging the skeptical method, Hume's influence, and the shadowy new doctrine then resulted in scholars linking the "light" of 1769 to the antinomy of 1781.

According to this interpretation of the critical turn, Kant would have formulated the antinomy of pure reason before the *Dissertation*, and this antinomy was the impulse for his subjective, epistemic concepts of space and time, which Kant then in turn employed for resolving that very antinomy. Should he really be suspected of such a circular approach? Nevertheless, Kant's own later remarks fed such an interpretation. According to the *Prolegomena*, "the remembrance of David Hume [...] first interrupted my dogmatic slumber" (4:260). And the letter to Garve of September 21, 1798 contains the famous passage which relates this remark to the discovery of the cosmological antinomy (12:257–258). According to the *Prolegomena* and this letter, either the traditional view is correct despite all the objections just noted, or else Kant awoke stepwise from his "dogmatic slumber" and merged the different stages of awakening much later. However, the passage in the *Prolegomena* refers to the principle of causality, and even over a century ago, Adickes (1897, 113, 122–123) had emphasized that the remarks in the letter to Garve are questionable. Every historian of science would agree that the later autobiographical remarks of the protagonists of scientific revolutions are not very reliable.[3]

[3]See the discussion of the way in which Kant refers to Hume's skepticism in Thöle (1991, 25–29). Klemme (1996, 38–44), too, convincingly opposes the traditional view.

Klaus Reich (1958) is the author who has most decisively opposed the traditional pattern of interpretation. He argues that Kant's writings of 1766, 1768, and 1770 address one and the same problem from different points of view, namely the difference between sensibility and understanding considered as two distinct faculties of human cognition. Referring on the one hand to Kant's own cautious formulations, and on the other to the problems that Kant dealt with in the years before and after the *Dissertation*, he dates the discovery of the cosmological antinomy to the early 1770s, that is, to a period after the *Dissertation* of 1770. Norbert Hinske (1970, 106) tried to mediate between the two points of view by distinguishing different stages of the "antinomy problem". In conformity with Reich, he dates the last stage, i.e., the discovery of the cosmological antinomy proper as a conflict of pure reason with itself, to the early 1770s.[4] Reich's sketchy proposal for an interpretation of Kant's development towards the antinomy has unfortunately never been followed up in detail by other leading Kant scholars. Instead, more recently Kreimendahl (1990) again took up the long-established view of Riehl, Erdmann, and Adickes, qualifying it by recourse to Hume's influence:

1. Kant found the antinomy problem in Book I of Hume's *Treatise*, and this was the decisive impulse for developing his transcendental philosophy.
2. He discovered the cosmological antinomy in the 1760s and this was the reason for establishing his new doctrine of space and time.
3. The insight into the subjectivity of space and time revealed that the antinomy is only an apparent conflict of reason with itself, and this was the "light" of 1769.

Kreimendahl's interpretation is based on the topics of the third and the fourth antinomy, that is, the questions of whether the causality of nature is compatible with a causality of freedom, and whether the substances in the world are necessary or contingent. Kreimendahl considers these to be decisive for the genesis of the doctrine of the antinomy:

> [...] the idea of the antinomy, as a contradiction of reason with itself, initially develops particularly in the field of those cosmological questions which Kant later deals with under the third and fourth antinomy. This is hardly accidental. For it is the area to which Kant's interests were directed from an early age, and in it the principle of causality, on which Hume had concentrated his skeptical remarks, plays a decisive role.[5] (Kreimendahl 1990, 191)

The notes on metaphysics dated to the years 1764–1768 (phases η, θ, and ι) indeed contain several remarks on topics of the third and fourth antinomy of the *CPR*, yet however do not yet mention any contradiction of reason with itself. But these topics are only half of the story. The above approach neglects the significance of the 1768 puzzle for Kant's views about space and time (see Sect. 3.4) as well as his later account of the cosmological antinomy as an unavoidable internal conflict

[4]Engelhard (2005, 280) follows Hinske in suggesting that Kant's remarks have been overstated, but were not completely mistaken.

[5]My translation.

of pure reason (see Sect. 5.2). Whoever backdates the antinomy before 1769, and its resolution to 1770, assumes that in 1769 Kant already had his 1781 concept of reason.

After the *Dreams*, Kant indisputably pursued the project of a critique of reason, which concerned the limits of the real (and not just logical) use of metaphysical principles. He was in particular concerned with the legitimate use of the principle of sufficient reason, continuing the critique of Leibniz in the *New Elucidation* (see Sect. 1.4.2). But in his pre-critical writings, as well as in the notes on metaphysics until 1768, he deals only with the logical principle of contradiction and the distinction between logical and real opposition.

The first notes on metaphysics in which the logical concept of contradiction applies to cosmological problems stem from 1768–1769. They are related to the epistemic turn of metaphysics initiated by the *Dreams* and to the skeptical method (see below Sects. 4.2.1 and 4.3.2). N. 3912 (last note of phase ι) deals with the subject of the first antinomy, anticipating the "apagogic" proof of the thesis that the world has a beginning. N. 3942 (phase κ) expresses the insight that human freedom can only be compatible with the laws of nature if the principles of metaphysics are subjective; otherwise they give rise to contradiction. Many Kant scholars associated these and related notes with the "antinomy problem". However, a contradiction produced by the objective use of metaphysical concepts, which can be avoided by considering them as subjective, certainly does not yet have the meaning of the later cosmological antinomy, as an unavoidable conflict of reason with itself.

Hence, there are three strong reasons why the later cosmological antinomy should not be rashly identified with the "contradictions" resulting from an objectively intended use of metaphysical concepts, such as stated in N. 3942 (and similarly, in N. 3936):

1. Kant never mentions an antinomy or antithetic of pure reason in the *Dissertation*, or before. Until 1770 he refers at most to "difficulties" to be resolved and "contradictions" to be avoided, but never to an "antithetic" or "antinomy". He uses the expression 'antithetic' for the first time in the year after writing the *Dissertation*, in the sense of a philosophical method, namely the skeptical method. Around 1773–1775 at earliest he explicitly uses the expression 'antinomy' (in N. 4275 and N. 4742; see Sects. 4.3.2 and 4.3.3).

2. One may object that Kant had recognized the problem of the cosmological antinomy long before he explicitly used the expressions 'antithetic' and 'antinomy'. This objection sees the core of the later antinomy in the difficulties and contradictions of the real use of metaphysical concepts, which he detected after the *Dreams* and the *Directions in Space* and to which his notes on metaphysics of 1768–1769 refer.[6] However, this reading is not compatible with

[6]This is the strategy adopted by Ertl (2002, 620). He explains the notion of antinomy in the literal sense, i.e., as a contradiction of laws, and claims that "we must treat Kant's usage of the term 'antinomy' and the corresponding notion as two closely related, but different issues. Plainly, a notion can be present before the appropriate term has been coined for it." The literal sense of the

the precise meaning of the cosmological antinomy, as explained in the *CPR*. There, Kant characterizes the cosmological antinomy as an inevitable, albeit apparent, conflict of reason with itself (see Sect. 5.2). The "contradictions" at issue in the notes of 1768–1769 seemed anything but inevitable to Kant. Indeed, then, he already had the solution when conceiving of the difficulty. In a certain sense, this observation holds even for the puzzle of 1768, given that in the *Directions in Space* he argues in favour of an epistemic concept of space, and *not* in favour of Newton's absolute space (see Sects. 3.4.2.2 and 3.4.3). Moreover, the antinomy is a conflict that emerges from the cosmological concept of the world. It results neither from the mere concepts of space and time nor from the mere real use of metaphysical concepts such as 'cause' or 'substance'.

3. According to the resolutions to the four antinomies in the *CPR*, the conflict of reason with itself arises in two variants, one of which seemed definitely more fatal to Kant than the other. The dynamic antinomy of causality of nature or freedom (third antinomy), as well as of the absolute contingency or necessity of all that exists in the world (fourth antinomy), does less damage to metaphysics, in Kant's view. The *CPR* resolves it in such a way that thesis and antithesis are both true and the conflict is simply avoided by distinguishing the intelligible from the sensible world. This resolution was in fact known to Kant before 1769, and goes back to Leibniz's distinction between *phaenomena* and *noumena*, which makes it possible to consider the events and things in the world in different respects as subject to free decisions or determined by the laws of nature, and as absolutely necessary or contingent. For Kant, the true scandal was the mathematical antinomy, which refers to the beginning of the world in space and time (first antinomy) and to the ultimate constituents of the substances in the world (second antinomy). The *CPR* resolves it by claiming that thesis and antithesis are both wrong, so that neither the statement of a finite nor that of an infinite size of the world, or number of its parts, can be maintained. As late as 1772 Kant still thought that the mathematical conflict between the opposite principles concerning the beginning of the world in time could be resolved in the same way as the dynamic conflict, that is, by reconciling opposite sensitive and intellectual grounds (NN. 4616–18, dated phase ζ; see Sect. 4.3.3). Later, however, he came to see the antithetic claims about the spatio-temporal structure of the world as a totality as an unexpectedly vicious problem, which results from incompatible propositions about the extension of the spatio-temporal world and the constitution of the substances in it. The cosmological antinomy in the strict sense of an unavoidable conflict of reason with itself therefore does not arise from the subjects of the dynamic antinomy, but only from those of the mathematical antinomy.

---

notion matches Kant's expression 'antithesis' (N. 4275), i.e., his skeptical method (see Sect. 4.3.2), but it does not match Kant's later, critical concept of the antinomy, in the sense of an inevitable cosmological conflict of reason with itself. The latter can be found neither in the notes 3922, 3942, 3999, 4007, 4225, on which Ertl relies, nor in the Berlin papers (N. 3716 and 3717) which he (most convincingly) dates to 1768–1769.

As will be shown below (in Sects. 4.3.3 and 4.4), Kant was only able to discover an inescapable antinomy on the basis of his 1770 doctrine of space and time and his pre-critical cosmological concept; and he did so not earlier than 1772. This confirms Reich's interpretation of Kant's development, as well as Hinske's plea for a more differentiated use of the term 'antinomy'. However, this only becomes visible by examining in detail the cosmological and methodological problems which Kant faced before and after 1770.

4.2 Metaphysical Cognition at the Crossroads

Kant did not explicitly relate the results of the *Dreams* to those of the *Directions in Space*. Nevertheless, his sharp criticism of dogmatic metaphysics in the former and his criticism of relationalism in the latter follow the same line of argument. Both writings suggest an epistemic turn for metaphysics (see Sects. 3.2 and 3.4.3), which gave rise to a radical departure from the analytic method of justifying metaphysical principles, as defended in the *Prize Essay*. His notes on metaphysics after 1766 bear witness to his wavering, tentative views about metaphysical cognition. To make the best of it, however, they also give first hints to a new and more reliable philosophical method, the skeptical method.

The *Dissertation* of 1770 is ambiguous with respect to the epistemic turn for metaphysics. It adds the new epistemic theory of space and time to the outline of a rationalist cosmology, in a dualistic approach to the sensible world of phenomena in space and time and the intelligible world of traditional metaphysical objects. The question of how Kant manages to combine both cosmological concepts is decisive for understanding his philosophical development at the beginning of the silent decade.[7]

4.2.1 The Epistemic Turn of 1769

Kant's notes on metaphysics of 1768–1769 reveal two important new features of his approach to metaphysics as compared to the *Prize Essay*. On the one hand, he now begins to qualify all metaphysical concepts as epistemic and subjective. On the other, he is now in search of a new touchstone for the tenability of metaphysical principles, and finds it in attempts to refute propositions by *reductio ad absurdum*.

The note N. 3892 on metaphysics, dated to 1766–1768, expresses Kant's doubts about Newton's account of space as follows:

[7]Beiser (1992, 55) suggests that Kant in hindsight came to identify these dogmatic claims of 1770 as his "dogmatic slumber" famously interrupted by Hume. See, however, Kant's letter to Lambert of September 2, 1770 (10:98), quoted below in Sect. 4.3.

> Whether there is a *spatium absolutum* or *tempus absolutum*, would mean to say as much as whether between two things in space everything [. . .] in between can be destroyed and yet the specific empty gap would remain, and whether, if [. . .] for a whole year all motions and changes would stop, the following could not begin, so that **in between an empty year** would have happened. We do not resolve these difficulties, but respond to our opponents by retorts [retorsion], because their view has these very difficulties. (N. 3892, phase θ; 17:330)[8]

As shown above (see Sect. 3.3.1), Kant here follows Leibniz's arguments against Clarke. In addition, he emphasizes that he employs this objection against Newton's absolute space and time "by retorts", that is, by using what after 1770 he will call the skeptical method, in a first approach to a new reliable philosophical method after the defeat of the analytic method (see Sect. 4.3.2). The note indeed anticipates the "apagogic" proof of the antithesis of the first antinomy given in the *CPR*, which is based on a variant of this argument (see Sect. 5.4.1.2). In view of this note, however, even the 1768 argument from incongruent counterparts may be considered as an argument "by retorts", or an early demonstration of the skeptical method.

The very first note in which Kant explicitly links the term 'contradiction' to a cosmological problem, N. 3912, is closely related. Its subject is the beginning of the world in time. Kant proves the thesis that the world has a beginning by *reductio ad absurdum* of the opposite assumption:

> The world has a beginning, i.e., a state which is not a consequence of another state: *terminum a priori; non est aeternus a parte ante.* For if each state of the world were a consequence of another state of the world, all states of the world would have a different state of the world before them; so one state of the world would be different from all states. which *contradictorium.*[9] (N. 3912, ι; 17:340)

The note anticipates to a certain extent the "apagogic" proof of the thesis of the first antinomy given in the *CPR*. The apagogic proof figure is nothing else than the response "by retorts" mentioned before in N. 3892. This means that N. 3912, too, is related to the skeptical method which Kant employs from 1770 onward. In 1768–1769, Kant does not yet associate the method of retorts with the later antinomy of pure reason and its proofs (nor does he so associate the skeptical method, until at least 1772). At this date, it is still linked to considering metaphysical concepts as subjective, as the notes N. 3936 and N. 3942 dated to 1769 clearly indicate:

---

[8]"Ob es ein *spatium absolutum* oder *tempus absolutum* gebe, würde soviel sagen wollen, ob man zwischen zwey Dingen im Raume alles [. . .] dazwischen liegende vernichten könne und doch die bestimte leere Lücke bleiben würde, und ob, wenn [. . .] ein gantzes Jahr Bewegungen und veränderungen überhaupt aufhöreten, nicht das folgende Anheben könne, so dass ein **leeres zwischen Jahr** verlaufen wäre. Wir lösen diese Schwierigkeiten nicht auf, sondern antworten unseren Gegnern durch die retorsion, weil ihre Auffassung eben diese Schwierigkeiten hat." My translation (note not contained in the Cambridge edition).

[9]"Die Welt hat einen Anfang, d.i. einen Zustand, der keine folge aus einem andern Zustande ist: *terminum a priori; non est aeternus a parte ante.* Den wenn ein ieder Zustand der Welt eine Folge aus einem andern Zustand der Welt wäre, so würden alle Zustände der Welt einen andern Zustand der Welt vor sich haben; also würde ein Zustand der welt von allen Zuständen unterschieden seyn. welches *contradictorium.*" My translation (note not contained in the Cambridge edition). It is the last note of phase ι, dated phase ι^2–κ^3, hence probably not earlier than 1768.

The *principia subiectiva*, if they are considered objectively, contradict each other. E.g., Everything has a ground, contradicts this sentence: nothing is necessary by itself etc etc. (N. 3936, κ; 17:355)[10]

A cognition is true which is in agreement with the constitution of the object. [...] But since the metaphysical concepts of ground, substance, etc., are not properly speaking representations of the objects, [...] these concepts are not objective; therefore in the axioms of them everything is subjective. [...]

The surest proof that they are not objective is that they stand in evident contradiction [Wiederspruch].[11] (N. 3942, κ; 17:357)

Here, he refers to the skeptical method in order to support the epistemic turn of metaphysics initiated by the *Dreams*. This use of the skeptical method in favour of his new, epistemic approach to metaphysics fits in well with N. 5037 of the late 1770s, which emphasizes that in 1769 the doctrine of transcendental idealism began to dawn on him when he tried

to prove propositions and their opposite, not in order to establish a skeptical doctrine, but rather because I suspected I could discover in what an illusion of the understanding was hiding. (N. 5037, 18:69)

The notes on metaphysics of 1768–1769 show that he did so for the concepts of space, time, ground, substance, and other metaphysical concepts, considering all of them at that time as subjective. But he differentiated space and time on the one hand, and the other metaphysical concepts, on the other, in terms of the logical distinction between "coordination" and "subordination":

Metaphysical concepts pertain 1. merely to the relation of coordination: *absolutum et relativum*, whole, part, *continuum, discretum*, one, all (the first, the last, a single one); 2. or to that of subordination in the logical sense: universal or particular; 3. to subordination in the real sense: ground, consequence, cause, effect. Whence arises the concept of the first cause, the final consequence, the cause of everything, of anything. 4. to existence: necessary, contingent, possible; 5. Substance (subject, predicate), simple, composite, *actio, passio (vis, receptivitas) spontanea, iners*. A whole of substances. World. (N. 3941, κ; 17:356)

In 1769, too, he begins to associate space and time with *a priori* conditions of sensibility, in contrast to the other metaphysical concepts mentioned here. The distinction on which transcendental idealism relies, between intuitions and concepts as two different kinds of ideas or representations, first appears in the notes NN. 3957 and 3958 on metaphysics dating from 1769, in terms of "intuitive" and "rational forms", or "intuitive concepts" and "concepts of reason". It is worth quoting them at some length:

[10]"Die *principia subiectiva*, wenn sie obiectiv erwogen werden, wiedersprechen sich. Z.E. Alles hat einen Grund, wiederspricht diesem satz: nichts ist nothwendig durch sich selbst etc etc." My translation (note not contained in the Cambridge edition).

[11]The Cambridge edition translates "Wiederspruch" (contradiction) as "self-contradiction". However, compare this with N. 3936 quoted above, which claims that certain opposite metaphysical principles "contradict each other"; see above.

> All human cognitions can be divided into two main genera: 1. Those which arise from the senses and are called empirical; 2. those which are not obtained by means of the senses at all, but rather have their ground in the constant nature of the [crossed out: cognitive power] thinking power of the soul, and can be called pure representations. [...] Now we have a twofold form for cognitions: the intuitive and the rational form. The former occurs only in the immediate cognition of individual things, the latter in general representations; the former I will call intuitive concepts, the latter concepts of reason. Now in all empirical cognition we can look first merely to the matter, and this consists of sensation; second, to the form of intuition; third, to the form of reason in concepts. The form of appearances rests solely on space and time, and these concepts do not arise through the senses or sensation, but rather rest on the nature of the mind, in accordance with which the various sensations can be represented under such relations. (N. 3957 (κ), 17:364–365)

> All cognitions from experience (empirical) belong either to sensation and contain the matter of empirical cognition, or to appearance and contain at the same time the form, or to the concept and contain what is general in different sensations or appearances. Sensation represents individual objects insofar as they stimulate the senses, e.g., red, black, sweet, hard, warm, etc., consequently only the matter of empirical cognition. The form of objects is thought in accordance with space and time. The form of empirical cognition is that of coordination; the form of rational cognition is that of subordination. (N. 3958 (κ), 17:366)

From now on Kant distinguishes between the pure forms of intuition, space and time, and the pure forms of the understanding, such as real ground (or cause), substance, etc.[12] Space and time are treated in parallel. According to N. 4077, the skeptical method shows that they are not the absolute entities which Newton assumed, but only the subjective conditions of sensible objects or phenomena:

> Space and time precede things; that is entirely natural. Both, namely, are subjective conditions, under which alone objects can be given to the senses. Taken objectively, this would be absurd. Hence the difficulty about the location of the world and time before the world. Yet in absolute time no location is determined without actual things, hence absolute time cannot yield any ground for the explanation of the *phaenomenorum*. (N. 4077 (κ), 17:405–406)

4.2.2 Space and Time as Pure Intuitions

The *Dissertation* of 1770 lays the grounds for transcendental idealism, establishing the principles of the Transcendental Aesthetic of the *CPR*. These present Kant's definite epistemic solution to the puzzle of incongruent counterparts. The way out of the puzzle of 1768 is: *No* objective metaphysical concept of space has real possibility (see Sect. 3.4.3). Using the skeptical method, Kant transfers this result to the concept of time and conceives of space and time as subjective forms of pure

[12]The Berlin papers (NN. 3716 and 3717), which Ertl (2002) convincingly dates to 1768–1769, also seem to belong in this context: "The principium of the form of all experiences is space and time. The principium of the form of all judgments of pure reason: identity and contradiction. The principium of the form of all *a posteriori* judgments of reason: ground and force" (N. 3717, 17:255).

intuition (see Sect. 4.2.1). Accordingly, the *Dissertation* distinguishes intuitions and concepts as two different kinds of ideas or mental representations, which correspond to the cognitive capacities of sensibility and understanding. The intuitions of space and time are sensible grounds of cognition. In contrast, the logical and real uses of the concepts and principles of metaphysics are based on intellectual grounds of cognition.

Concepts are general representations, whereas intuition is a singular representation. Concepts contain things *under* themselves, whereas space and time embrace all things *within* themselves (2:399). Space and time are singular representations *a priori*. Kant argues that space and time are *a priori*, even though "without any doubt" they are acquired rather than innate; they have "*been acquired*, not, indeed, by abstraction [...], but from the very action of the mind" (2:406).[13] In this way Kant replaces his pre-critical objective, metaphysical account of space and time as relations of substances, by the subjective, epistemic theory of space and time as pure intuitions.

In the *Dissertation*, Kant attributes the following characteristics to the idea of space (2:402–405):

A. it is not obtained by abstraction from outer sensations;
B. it is an unrestricted singular representation which embraces all limited singular representations;
C. it is a pure intuition, in which the axioms, postulates, and problems of geometry as well as the three dimensions of space and the difference of incongruent counterparts can be represented *in concreto*;
D. it is not objective and real, but subjective and ideal, that is, a scheme of coordination or correlation of sensory perceptions;
E. it is the formal condition of the possibility of our sensory perceptions; and thus the absolutely ultimate formal ground of empirical truth.

The characteristics of time are analogous (2:398–400). The explanations of the nature of space and time in the *Dissertation* are parallel to the corresponding passages of the Transcendental Aesthetic of the *CPR*, and to the classification of the kinds of representation in the logic lectures from 1772 on. Comparison of the *Dissertation* with the parallel passages in the *CPR* demonstrates that (A.) and (B.) are sufficient to determine space as a pure intuition (C.), and that (D.) and (E.) are not logically independent of (A.) and (B.). In the two editions of the *CPR*, the characteristics corresponding to (D.) and (E.) appear as "conclusions" from the conceptions of space and time.[14] Hence, the dissertation already attributes to

[13]Concerning the aspect that they are not obtained by abstraction, Kant follows Euler (1748, § XIV) and Euler's second letter to Venzky (Euler 1763, 103–104).

[14]*CPR*, A 26/B 42; A 32–34/B49–51. Edition B of the *CPR* distinguishes the *metaphysical exposition* and the *transcendental exposition* of the conceptions of space and time and adds as an independent new mark that space and time are given infinite quantities. See my more detailed discussion in Falkenburg (2000, 131–134). As Carson (1998) aptly notes, this is an independent

the pure forms of sensibility the following primary marks. Space (2:402) and time (2:398–399) are:

(1) not abstracted from sensory impressions or perceptions, but presupposed in all sensory perceptions; and
(2) not general, but singular representations.

From these two characteristics Kant immediately concludes that space and time are intuitions rather than concepts. According to (2), they are singular; according to (1), they are *a priori*; and singular representations *a priori* are pure intuitions:

> 2. *The idea of time is singular* and not general. [...] 3. Therefore, *the idea of time is an intuition*. And since, in so far as it is the condition of the relations to be found in sensible things, it is conceived prior to all sensation; it is not a sensory but a pure intuition. (2:399)

> *The concept of space is thus a pure inituition*, for it is a singular concept, not one which has been compound from sensations, although it is the fundamental form of all outer sensation.[15] (2:402–403)

According to Kant's logic, too, intuitions are singular representations:

> All cognitions, that is, all representations related with consciousness to an object, are either intuitions or concepts. An intuition is a singular representation (*repraesentatio singularis*), a concept a universal (*repraesentatio per notas communes*) or *reflected* representation (*repraesentatio discursiva*). (§ 1, 9:91)

In a certain sense, Kant's theory of space and time is a step away from his pre-critical Leibnizean account of space as a real sum total of relations back to the authentic Leibniz, who considered space and time as ideal, too. In the *Metaphysical Foundations of Natural Science* (*MFNS*), Kant correspondingly claims:

> A great man, who has contributed perhaps more than anyone else to preserving the reputation of mathematics in Germany, has frequently rejected presumptuous metaphysical claims to overturn the theorems of geometry concerning the infinite divisibility of space by the well-founded reminder that space belongs only to the appearance of outer things. But he has not been understood. This proposition was taken to be asserting that space appears to us, though it is otherwise a thing, or relation of things, in itself, but that the mathematician considers it only as it appears. Instead, it should have been understood as saying that space is in no way a property that attaches in itself to any thing at all outside our senses. (4:507)[16]

Nevertheless, Kant's theory of space and time as ideal, subjective forms of intuition differs in a crucial point from Leibniz's theory. For Kant, the intuition of space

---

argument that is by no means the starting point of Kant's reflections on the character of space and time. This argument is not needed to show that space and time are singular representations *a priori*. However, it shows in addition that Kant's criticism of Leibniz's theory of the complete concept and the corresponding concept of the actually infinite enforces the assumption that space is a pure intuition.

[15]Here Kant uses the expression 'concept' more generally in the sense of 'idea'.

[16]According to n. 26 to the text in the Cambridge Edition, "Leibniz seems to be the most probable" candidate for the "great man" (Friedman 2004, 44). Gerlach (1998), however, suggests that the "great man" may have been the Leibnizian Ploucquet, who in 1748 submitted a prize essay to the monadology debate of the Berlin.

makes it possible to individuate objects with identical properties. In this way, he integrates his 1755 criticism of Leibniz's principle of indiscernibles into his 1770 solution to the puzzle of incongruent counterparts.

4.2.3 The Formal Cosmology of 1770

Despite the epistemic turn in metaphysics, the *Dissertation* of 1770 still gives an outline of dogmatic metaphysics that is substantially unchanged. As the title of the writing indicates, Kant wants to achieve this goal in 1770 through a dualistic approach, by distinguishing two cosmological concepts, the concepts of the sensible and the intelligible world. The objects of both worlds are given by two different cognitive faculties, sensibility and understanding, which give rise to intuitions and concepts as two different kinds of representation. The approach of 1770 differs in particular from Wolff's epistemology, which only logically distinguishes between sensitive and intellectual ideas (in terms of more confused vs. distinct ideas), as Kant emphasizes (2:395).

It has been debated whether Kant's dualism of 1770 is ontological or epistemological (Grier 2001, 52–54, vs. Guyer 1987, 13–24). And his approach is indeed ambiguous, given that in 1770 he has no theory of how intellectual concepts relate to their objects. According to the *Dissertation*, the sensible world of phenomena and the intelligible world of noumena are epistemically and ontologically disjoint, and Kant is not clear about the question of how the understanding can have any access at all to intelligible substances. He claims on the one hand that the intelligible substances are related to one another by real coordination (2:390), and on the other that human beings have only symbolic cognition of them (2:396). In this way, he suggests a formal ontology of logical individuals with "*real* and objective" relations (2:390), again merging the logical and the real use of reason, as he did in the *New Elucidation* (see Sect. 2.2.2.3).

The dualism of 1770 serves to avoid the puzzle of 1768 with regard to the concept of space, but also preserves the pre-critical project of cosmology as part of a system of metaphysics. Kant still thought that this strategy for establishing the principles of cosmology was feasible without contradiction.[17] Around 1770 he neither saw an inescapable cosmological contradiction, nor had an elaborate theory of cognition that differentiated between understanding and reason and made it possible to diagnose an inner conflict of reason.

---

[17] Baum (1994) also suggests this interpretation. According to him, in 1770 Kant is still "firmly on the side of Platonism in matters of rational cosmology (A 471/B 499), i.e., on the side of the thesis in the antinomy of the *CPR*, or on the side of 'dogmatic rationalism' of Dohn's metaphysics lecture" (Baum 1994, 188). In contrast to the present work, however, Baum (1994, 193) sees the origin of the later antinomy of pure reason in "the theme of the third antinomy taken for itself" (my translations). Grier (2001, 109–111) suggests that the doctrine of the antinomy as an inevitable conflict of reason with itself emerged from the very distinction between understanding and reason, during the silent decade.

The intellectual cosmological concept of the *Dissertation* establishes a non-empirical theory of the world as a whole. In 1770, Kant still attributed some cognitive content to it, albeit with some epistemological reservations (2:409–410). It has to be noted that Kant's 1770 theory is free of antinomy not only according to his fragmentary epistemology of 1770, but also from the point of view of his elaborated critical epistemology of 1781. After the critical turn, Kant only denies that a purely intellectual cosmological concept has objective reality and refers to the spatio-temporal world. The antinomy of 1781 is epistemic since it concerns the cognition of the spatio-temporal world as a totality (see Chap. 5).

The notion of the intelligible world of 1770 is defined as an intellectual concept, or concept of the pure understanding, which is not subject to the conditions of sensibility. Its definition presupposes the strict separation of the sensitive and intellectual grounds of knowledge. What is specific about the intelligible in contrast to the sensible world is the intellectual origin of its concept, which is created by the understanding and not given by the senses. The composition of intellectual ideas from particular ideas is discursive and not intuitive, i.e., obtained by logical composition and subordination and not by co-ordination (see N. 3941 quoted above in Sect. 4.2.1). In addition, the *Dissertation* distinguishes the real use of intellectual concepts from their logical use, which only concerns their logical subordination:

> [...] the use of the understanding [...] is twofold. By the first of these uses, the concepts themselves, whether of things or relations, are *given*, and this is the REAL USE. By the second use, the concepts, no matter whence they are given, are merely subordinated to each other, the lower, namely, to the higher (common characteristic marks), and compared with one another in accordance with the principle of contradiction, and this use is called the LOGICAL USE. (2:393)

The real use concerns metaphysical objects, the concepts of which are not given by sensibility. One obtains them by disregarding all empirical ideas. They operate on an abstract domain which is, however, not given by abstraction from sensible objects, but generated by the pure understanding. Kant calls them 'pure ideas':

> As for that which belongs strictly to the understanding, and in the case of which the *use of the understanding is real*: such concepts, whether of objects or of relations, are given by the very nature of the understanding: they contain no form of sensitive cognition and they have been abstracted from no use of the senses. [...] Hence, a concept of the understanding *abstracts* from everything sensitive, but it is *not abstracted* from what is sensitive. Perhaps, a concept of the understanding would more rightly be called *abstracting* rather than *abstracted*. For this reason, it is more advisable to call concepts of the understanding '*pure ideas*', and concepts which are only given empirically '*abstract* concepts' (2:394)

This "abstracting" domain of the intellectual cosmological concept and the concepts and relations operating on it are not accessible to intuition, but only to symbolic cognition (2:396–397). Hence, one may understand Kant's cosmology of 1770 as a formal theory in a Leibnizean sense of symbolic cognition. Its concepts operate on the "abstracting" domain of the "world". This domain is not identical to the domain of mathematical objects. According to the *Dissertation*, the objects of geometry are given in space, and the objects of arithmetic, i.e., numbers, by performing operations in time and space (2:397). Kant thus related mathematics to space and time as pure forms of intuition as early as 1770. In contradistinction to mathematical objects, the "abstracting" domain of the intelligible world of 1770 is not given by intuition.

Today we would say that Kant's intellectual cosmological concept of 1770 establishes a formal ontology which is based on uninterpreted logical relations and axioms, and which may be expressed in terms of a symbolic calculus of individuals. The scope or extension of the intellectual cosmological concept is a domain of abstract individuals ("substances") for which, from today's perspective, at least an axiom of extensionality should hold. Instead of such an axiom, Kant uses an abstract concept of coexistence, which however has no spatial interpretation. His formal cosmology of 1770 has the following structure:

1. The concept of the world is defined in terms of the whole–part relation, as a whole that is not a part of a more comprehensive whole. The world is thus a mereological maximum. Kant gives this definition implicitly, in terms of analysis and synthesis:

   > In the case of a substantial compound, just as analysis does not come to an end until a part is reached which is not a whole, that is to say a SIMPLE, so likewise synthesis does not come to an end until we reach a whole which is not a part, that is to say a WORLD. (2:387)

2. The "matter" of the world, i.e., its parts, are formal, symbolic substances. The intellectual "form" of their part–whole relations is not the subordination of cause and effect, or principles and their consequences, but the co-ordination of intelligible objects, i.e., coexistence (2:390–391).

   Here, coexistence is the formal relation of the arrangement of substances. It is subject to a relational logic that includes uninterpreted relational predicates and operates on a domain of logical individuals. Coexistence is a discursive and not an intuitive relation, given that the empirical relation of spatial coexistence is neglected.

3. Coexistence should not be thought of as an ideal, but rather as a real relation; as a formal correlation of real intelligible objects (2:390). This claim is compatible with the 1770 account of space as a pure form of intuition, given that according to the *Dissertation* spatial relations are ideal. In which sense these intelligible objects (i.e., logical individuals) are real Kant does not explain.

   In addition, however, he associates the coexistence of the substances with a "principle of the *possible influences* of the substances which constitute the world" (2:390), which is the real ground of their connections. This assumption is obviously a residue of his theory of physical influence defended in the *New Elucidation* of 1755.

4. The world as "*absolute* totality of its component parts" (2:391) is actually infinite with regard to the coexisting substances and their successive states, even though Kant already expounds the problems of this assumption here.

According to this cosmological theory, the world is a mereological composite of infinitely many substances. Each substance has an infinite number of inner states;

the states of one and the same substance have a formal total order; this total order is preserved in the composition of substances. Hence, the substances constitute a totally ordered domain of individuals in the sense of modern logic, on which two formal relations hold:

(i) a part–whole relation (coexistence) with a mereological maximum, the world, and infinitely many mereolological atoms, the simple substances or monads;
(ii) a total order (succession) of infinitely many components, which constitute each monad as well as their sum total, the world.

Kant's *Dissertation* of 1770 thus takes a step back from the philosophy of Wolff's school, returning to the authentic Leibniz not only with regard to the ideality of space and time, but also with regard to the intellectual cosmological concept.[18] This return to Leibniz is also indicated by Kant's remarks on the real use of metaphysical concepts in terms of symbolic cognition (2:395–396). If the cosmological concept is conceived in the sense of Leibniz's logic, the claim that symbolic cognition of the intelligible world is possible obviously does not give rise to contradiction. Its structure can be expressed by means of symbolic logic and reconstructed as a formal ontology.

4.2.4 Phenomena and Noumena

How, in 1770, does Kant conceive of the connection of this intelligible world to the world of sensory phenomena in space and time? I suggest that he does so largely in the sense of a modern theory of abstraction. The concepts of the intelligible world, its noumenal substances, and their formal relations are obtained by abstracting from the sensible world, the phenomena that constitute it, and their spatio-temporal relations. But the abstract cosmology of 1770 is more than just a formal ontology. It still aims at a dogmatic cosmology as part of a system of metaphysics. Kant combines his abstraction theory with metaphysical dualism. His intellectual cosmological concept refers to a domain of *noumena* as a correlate of the sensible world of phenomena.

The connection of the abstract cosmology of the *Dissertation* with the sensible world is conceived after the model of Leibniz's monadology. The phenomena of the sensible world are founded upon intelligible substances. These substances are parts of the intelligible world and have logical relations with a real correlate. The appearance of these relations in the sensible world is based on the forms of intuition, space and time, and is therefore ideal. The forms of intuition give a spatio-temporal interpretation to the logical or intellectual relations between the substances in the world. This interpretation is determined by the form of sensibility, hence it is ideal and subjective.

[18] According to Gerlach (1998), Kant here followed the Leibnizian Ploucquet.

In contrast to Leibniz, however, Kant still attributes real, objective relations to substances or monads. Here he does not mean spatio-temporal relations, given that these are ideal according to the *Dissertation* of 1770. But he still attempts to preserve the theory of physical influence of 1755/1756, albeit in an abstract version, for which he assumes that it possesses a real, intelligible, non-sensible correlate which is at least accessible to symbolic cognition. Thus he thinks of the intelligible world as some kind of logical correlate of the sensible world. His approach of 1770 is to follow Leibniz's metaphysics, at least to a certain extent. The sensible world is founded upon an intelligible world, the logical structure of which is reflected in the phenomena. The logical relation between the sensible world and the intelligible world does not give rise to contradiction, given that the phenomena are strictly separated from the domain of intelligible substances.

The most important difference between the *Dissertation* and Leibniz's epistemology and metaphysics is that Kant now identifies space and time with intuitions instead of concepts. He attributes them to sensibility, considered as a cognitive faculty that has strictly to be distinguished from understanding or reason. This is a rupture with the entire rationalist theory of cognition from Descartes to Leibniz and Wolff, up to Kant's own pre-critical system. It signifies a genuine scientific revolution of 18th philosophy. Kant himself, when he put it into effect as a first and decisive step towards his Copernican turn of 1781 (*CPR*, B xiv), saw its consequences only "as if in twilight", according to N. 5037 (18:69). The revolution took place in 1769, as the notes on metaphysics quoted above show (see Sect. 4.2.1), and is closely related to the "great light" of 1769. To quote Beiser (1992, 51): "That great light was most probably the distinction between reason and sensibility".

4.3 In Search of a New Philosophical Method

In the 1770 *Dissertation*, the theories of sensitive and intellectual cognition stand side by side, partly unconnected, partly merged. Kant was well aware of these eclectic features. He had written the *Dissertation* in view of the necessity to prepare an academic essay, under lack of time and in a poor state of health. It was incoherent and methodologically immature, in striking contrast to his lifelong intention to establish a systematic metaphysics through an appropriate cognitive method.

In a letter to Lambert of September 2, 1770, Kant expresses his own discontent with the *Dissertation*. He regards the explanations of the cosmological concept and the form of the intelligible world as insignificant. Only the sections on the distinction between sensibility and understanding, on space and time as well as on the method of metaphysics, seem promising to him, albeit in strong need of further elaboration:

> The first and fourth sections can be scanned without careful consideration; but in the second, third, and fifth, though my indisposition prevented me from working them out to my satisfaction, there seems to me to be material deserving more careful and extensive exposition. (10:98)

The fact that he recommends readers to disregard his intellectual cosmological concept does not mean that he wrote these parts of the *Dissertation* against his better knowledge, as defenders of the traditional view of the critical turn tend to assume.[19] It was not Kant's custom to defend any thought contrary to his own views, as he emphasized in his letter to Mendelssohn from April 8, 1766, concerning the *Dreams*:

> Although I am absolutely convinced of many things that I shall never have the courage to say, I shall never say anything I do not believe. (10:69)

4.3.1 The Critique of Subreptic Axioms

In 1770, Kant is not much interested in the dogmatic features of his intellectual cosmological concept, but rather in the epistemological principles according to which it can be separated from the principles of sensibility. At that time, the principle of separating the sensitive and intellectual grounds of cognition plays the role of a methodological principle for establishing metaphysics as a symbolic cognition of the intelligible world. In the *Dissertation*, he emphasizes that we human beings do not have an intuition of the intelligible world, and that our understanding is only capable of symbolic cognition by the use of abstract universal concepts, not by concrete singular concepts (2:396–397).

The further passages of the letter to Lambert of September 2, 1770, show that Kant now plans a kind of propaedeutics of metaphysics, which deals with separating the contents of metaphysics from all empirical or sensitive cognitive conditions, following the approach of the *Dissertation*. In a letter to Marcus Herz of June 7, 1771, he also reports on a planned work "The Limits of Sensibility and Reason", which was to be based on the idea of separating the cognitive faculties of the understanding and sensibility (10:122–123). It would partly fulfill and partly carry on the program of the *Dissertation*. As he emphasizes in his famous next letter to Herz of February 21, 1772, the theoretical part of the planned work would be divided into a "general phenomenology" and explanations on "metaphysics, but this only with regard to its nature and method" (10:129). Due to the question of how both parts should be connected, he finally saw the approach of 1770 fail. In particular, he was not clear about the question of how the concepts of metaphysics relate to their objects, or, vice versa, how the concepts of metaphysical objects come to our mind:

---

[19]According to Kreimendahl (1990, 213), in the *Dissertation* Kant presented the outline of a consistent dogmatic cosmology and metaphysics in which he no longer believed, because he already knew the antinomy of pure reason including its resolution. This hypothesis seems to be conceived ad hoc, given that it serves to save the traditional interpretation of Kant's critical turn, without taking seriously Kant's struggle with cosmological problems and his search for a coherent theory of cognition to overcome these problems. See also Baum 1994.

> As I thought through the theoretical part, considering its whole scope and the reciprocal relations of all its parts, I noticed that I still lacked something essential, something that in my long metaphysical studies I, as well as others, had failed to consider [...]. I asked myself this question: What is the ground of the relation of that in us which we call "representation" to the object? (10:129–130)

According to the fifth section of the *Dissertation*, metaphysics needs a specific method which guarantees the real use of metaphysical principles, and which differs from the method of all other sciences. Kant describes this method as "wholly unknown":

> Hence, since the method of this science may not be well known at the present time, apart, that is, from the kind which logic teaches generally to all the sciences, and since the method which is suited to the particular character of metaphysics may be wholly unknown, it is no wonder that those who devoted themselves to this enquiry seem, hitherto, to have accomplished scarcely anything at all [...]. (2:411)

The method "which logic teaches generally to all the sciences" is the traditional method of analysis and synthesis, which since the Port-Royal Logic (Arnauld and Nicole 1685) had been part of academic teaching, as Kant's logic lectures also demonstrate (see Sect. 2.1 and Appendix A.2). Kant here distinguishes it from the required specific method of metaphysics, in sharp contrast with the *Prize Essay* of 1764 (see Sect. 2.3). Indeed, this method had even gone astray for physical geometry. As the 1768 argument from incongruent counterparts showed (see Sect. 3.4), Kant not only saw his pre-critical system failing, but also the analytic method for the real use of reason in cosmology. Now he was looking for a new specific philosophical method which would be able to reestablish metaphysical cognition.

The methodological part of the *Dissertation* does not yet have much to offer for this new method of metaphysics. It is restricted to the critique of "subreptic axioms" that result from applying the conditions of sensibility to intelligible objects, and which declare everything that is merely incomprehensible to intuition as impossible (2.412–417; for many instructive details, see Grier 2001, 57–64). At this point, one may wonder why this critique should not also apply to Kant's 1768 conclusion that a relational account of space is impossible, or unimaginable, in view of incongruent counterparts? (See Sect. 3.4.1.3). However, according to the fragmentary epistemology of 1770 the "subreptic" axioms erroneously suggest that unimaginable concepts of the pure understanding cannot grasp the essence of metaphysical objects. They misleadingly claim that objects which escape our intuition are impossible if they are unimaginable—that is, if their concept does not meet the conditions of sensibility:

> For this lack of accord between the *sensitive* faculty and the faculty of the *understanding* [...] points only to the fact that *the abstract ideas which the mind entertains when they have been received from the understanding very often cannot be followed up in the concrete and converted into intuitions*. But this *subjective* resistance often creates the false impression of an *objective* inconsistency. And the incautious are easily misled by this false impression into taking the limits, by which the human mind is circumscribed, for the limits within which the very essence of things is contained. (2:389)

In 1770 Kant still claimed that the "very essence of things" could be determined without contradiction and grasped by pure ideas. According to his fragmentary epistemology of 1770, objective symbolic cognition of the intelligible world is possible, whereas the subjective conditions of intuition tend to lead us to conflate intellectual and sensitive epistemic grounds, making us believe that the cosmological concept is logically impossible. Due to this conflation, the "*subjective* resistence" to imagining an intellectual cosmological concept that goes beyond intuition gives rise to an the impression of an "*objective* inconsistency". This results, for example, in the "subreptic" principle:

> *Whatever is, is somewhere and somewhen.* (2:413)

From the perspective of 1770, such "subreptic axioms" generate metaphysical pseudo-problems. In particular, they generate "idle questions about the places in the corporeal universe of immaterial substances" (2:413), or "absurd questions [...], for example, why did not God establish the world many centuries earlier?" (2:424). Here, Kant reassesses Leibniz's arguments against absolute space and time (see Sect. 3.3.1) as follows. Any claim about the position of the world in space and time, or about the position of immaterial substances within the world, is based on the "subreptic" application of sensitive grounds of cognition to intelligible objects, which are not subject to experience and incomprehensible to intuition. However, how could any cosmological cognition of intelligible objects be possible; and what could justify the certainty of the intellectual cosmological concept, as compared to the certainties of logic, mathematics, or natural science? In 1770, Kant "silently passed over" this problem, as he admitted two years later, in his letter to Marcus Herz:

> In my dissertation I was content to explain the nature of intellectual representations in a merely negative way, namely, to state that they were not modifications of the soul brought about by the object. However, I silently passed over the further question of how a representation that refers to an object without being in any way affected by it can be possible.[20] (10:130–131)

Against the background of the *Dreams*, according to which metaphysics is a "science of the *limits of human reason*" (3:367–368), Kant would now be expected to search for the method of metaphysics in the means of determining the limits of human cognition. The approach of the *Dissertation*, however, does not go beyond the bare demarcation between sensibility and understanding as our respective cognitive faculties of intuition or concepts. Paradoxically, this demarcation leads to him establishing a dogmatic cosmology through criticizing the "subreptic axioms" of sensibility.

Kant was aware that his methodological approach of 1770 was ambiguous. The sought-after method of metaphysics ought to lead to symbolic cognition of the

[20]The problem named here does not refer to the topic of Kant's later deduction of the categories (as assumed in Carl 1989), but primarily to the intellectual idea of the world as the key cosmological concept of the *Dissertation*.

intelligible world, in a way which was uncoupled from all cognitive conditions of sensibility. Yet, at the same time, it ought to be achieved by abstraction (see Sect. 4.2.3). From a methodological point of view, Kant's formal cosmology of the *Dissertation* is partly obtained by the synthetic method of mathematics and a symbolic logic in Leibniz's style, and partly by an abstraction method based on the results of conceptual analysis in metaphysics. In this way, the approach of 1770 once again remains eclectic. In 1770 Kant could not resolve this methodological ambiguity: he could only present the method "which logic teaches generally to all the sciences" (i.e., the traditional analytic-synthetic method) as necessary but not sufficient for metaphysics, emphasizing that in addition a "wholly unknown" method is required for metaphysics (2:411).

4.3.2 The Skeptical Method

The unifying principle underlying the system of 1755 was the quest for the best reasons *in favor of* two opposing doctrines (see Sects. 1.3.1 and 1.3.2). This requirement is contrary to the *skeptical method* which Kant now develops, that is, the requirement to find the strongest reasons *against* a philosophical position. The skeptical method counters a position with arguments to either disprove it or confirm it by dispelling all conceivable objections. It is a touchstone of the tenability of metaphysical principles. Kant realized in 1766 that he was in need of such a touchstone. The wavering rhetoric of the *Dreams* (see Sect. 3.2) has been considered as a first probe of the skeptical method (Reich 1958, XII; Kaulbach 1982, 100), while the *Directions in Space* gave a further demonstration of it. The 1768 argument made skeptical use of the analytic method, inferring from the phenomenon of incongruent counterparts and its inductive generalization to the whole universe that a relational concept of space is untenable.

Kant's further search for a reliable method of metaphysics gave rise to differentiating between the "dogmatic", "skeptical", and "critical" methods, the latter finally being the sought-after method (See *Logic*, 9:83–84; Vienna logic, 24:884–885). The dogmatic method and the critical method are positive methods of finding metaphysical principles. The skeptical method, in contrast, is only negative. The dogmatic method of traditional metaphysics gives rise to metaphysical judgments. The skeptical method suspends these judgments by confronting them with all conceivable counter-arguments. Finally, the critical method consists in examining the conditions of the possibility of objective cognition in order to arrive at judgments against which there is no longer any objection and which can be considered certain.

After abandoning the analytic method of metaphysics and before developing the critical method of transcendental philosophy, however, only the skeptical method was available to Kant. He started to use it not later than 1768, applying it to the conceptions of space and time in the notes on metaphysics (see Sect. 4.2.1). At that time, he claimed to "respond to our opponents by retorts" (N. 3892, θ; 17:330), suggesting apagogic proofs via *reductio ad absurdum* (N. 3912, last note of ι;

17:340). In the light of notes N. 3892 and N. 3912 (dated to 1766–1768), the argument on incongruent counterparts also looks like an argument by "by retorts", which presents a *reductio ad absurdum* of the relational concept of space. In the notes on metaphysics from 1770 onwards, Kant characterized the skeptical method as an antithetic procedure or "antithesis":

> Antithesis: a method of reason for discovering the *oppositionem* of **subjective laws**, which, if it is taken **for objective** *per vitium subreptionis, is skepticismus (in sensu obiectivo)*; if, however, it is only a [*crossed out*: critique of the subject] propaedeutic, then it is a *methodus skepticus* for the determination of the subjective laws of reason. *Antithesis subiectiva.* (N. 4275, μ (1770–71); 17:492)

Here, Kant relates this antithetic procedure to skepticism, if understood in the sense of the "subreptic axioms" of the *Dissertation*, which conflate the subjective and objective conditions of cognition. In contrast, if used for demonstrating that conflicting principles are merely subjective, he calls it the skeptical method (see also N. 3936 and N. 3942, quoted in Sect. 4.2.1). A first elaboration is found in the Blomberg logic, which probably stems from 1771. According to it, Kant attributes the origin of skepticism to Socrates (24:207) and identifies "skepticism" with "the method of skeptical doubt", emphasizing that philosophy

> has gained far more from the *sceptici* than from the proud dogmatists [...]. *Skepticism*, however, or the method of skeptical doubt, where one establishes a distrust in oneself, considers the grounds for and against the cognition one has, and in this way strives to come to complete certainty concerning it, this is the *kathartikon*, reason's best means of purgation. (24:208)

Contrary to popular belief, Kant did not at all associate the skeptical method with Hume's skepticism. As the famous reflection 5037 indicates, around 1770 he did not use the skeptical method "to establish a skeptical doctrine", but rather because he suspected he could "discover in what an illusion of the understanding was hiding" (N. 5037, 18:69). This passage is best understood as distinguishing Kant's own skeptical method from Hume's "skeptical doctrine". The Blomberg logic supports this interpretation, harshly putting Hume's skepticism down:[21]

> Skeptical method is a true investigating of truth by means of postponemen)t [...] We are really not at all inclined by our nature to postponement. In our times people are called skeptics [...] such as are not real *academici* but instead merely display a skeptical method in itself and, as it were, affect it, e.g., a Hume. (24:210)
>
> *David Hume* is especially known as a *skepticus* who had an overwhelming, indeed, a somewhat extravagant inclination to doubt. (24:217).

On the same line is N. 4469 (about 1772), which emphasizes that the skeptical method does not result in universal doubt:

---

[21] This harsh criticism of Hume's skepticism is in striking contrast to Kreimendahl's interpretation, according to which this very passage "clearly shows why Kant was interested in Hume. It is the skeptical procedure that Hume handled more virtuously than anyone before and which Kant has advanced into the 'skeptical method' [...]" (Kreimendahl 1990, 12, my translation).

The skeptical method is the best and only one for beating back objections by means of retorts. Does there then arise from it a universal doubt? No, but the presumptions of pure reason with regard to the conditions of the possibility of all objects are thereby beaten back. All judgments of healthy reason with regard to the world and the practical receive thereby their great reputation. Healthy or practical reason can never be persuaded that there is no God, if only subtle reason does not seek to gain status from it. (N. 4469, phase ξ; 17:563)

4.3.3 Reversing the Arguments of 1770

After 1770, Kant realizes that he has not yet sufficiently aligned his dogmatic cosmological concept with his new doctrine of space and time as subjective forms of intuition. But at that time he does not yet reject cosmology as a metaphysical doctrine, and he continues to search for a positive method to establish cosmology and metaphysics. As he summarizes in N. 5116:

I have not always judged this science thus. [. . .] In some parts I believed myself able to add something to the common store, in others I found something to improve, but always with the aim of thereby acquiring dogmatic insights. For the doubt that was so boldly stated seemed to me so much to be ignorance without reason that I gave it no hearing. [. . .] It took a long time before in such a way I found the whole dogmatic theory to be dialectical. Yet I sought something certain, if not with regard to the object, then still with regard to the nature and boundaries of this sort of cognition. I gradually found that many of the propositions that we regard as objective are in fact subjective, i.e., they contain the *conditiones* under which alone we can have insight into or comprehend the object. But while I thereby certainly became careful, I was still not instructed. For [. . .] I still believed I could find a method for extended dogmatic cognition through pure reason. For this I now required insight into how a cognition *a priori* is possible in general. (18:95–96; υ)

The notes on metaphysics and letters from 1772 show how Kant only gradually departs from dogmatic cosmology. He now begins to seriously ask himself how cognition that is not bound to the conditions of sensibility could ever be achieved; and he becomes increasingly convinced that metaphysics, at least in the field of theoretical cognition, is not possible as a dogmatic doctrine, but only in the sense of a critique of reason. Accordingly, he writes in N. 4445:

The employment of metaphysics with regard to the theoretical is merely negative; it does not open up the cognition of things and is not **dogmatic**; for where should it get the cognition of things without the senses[?] (17:552; ξ)

The departure from dogmatic cosmology is closely linked to the considerations on the relation of our ideas to cognitive objects, the significance of which Kant first emphasized in his letter to Marcus Herz of February 21, 1772 (10:129–135). This problem leads him to proceed from the project of considering the limits of human reason of 1766 to a more specific critical program, according to which human cognition is bound to the conditions of our sensibility to guarantee comprehensible adequacy conditions (see Carl 1989). Kant at this time does not yet differentiate between understanding and reason. The explanations in the above letter concern the real use of pure concepts of the understanding such as ‘substance’, ‘reason’, or 'cause', which finally give rise to the problem of a transcendental deduction of the pure concepts of the understanding. But they also concern the concept of the intelligible world of the *Dissertation*, the reference to which is no less problematic. From 1772 Kant also had to think about the adequacy conditions of the cosmological concept and its two versions, 'sensible world' and 'intelligible world', which gave the *Dissertation* its title. His efforts to clarify the adequacy conditions of this two-fold cosmological concept finally gave rise to the cosmological antinomy of 1781.

In 1772, Kant no longer attributes cognitive content to a cosmological concept which is not subject to the conditions of sensibility, given that there are no criteria of how purely intellectual ideas should be related to their objects. In contrast to his approach of 1770, he no longer associates the intellectual cosmological concept with the real use of reason, but merely with its logical use.

In the terminology of the *CPR*, he now assumes that it has no objective reality, given that it does not satisfy the conditions of possible experience. According to the distinctions of the amphiboly chapter of the *CPR*, this only means that the cosmological concept of the *Dissertation* is the empty concept of an *ens rationis*. In contrast, the cosmological antinomy of 1781 refers to the contradictory concept of a *nihil negativum*, which is formed from incompatible predicates and has no instance, on pain of contradiction (*CPR*, B 349/A 292).

Based on the considerations of 1772, Kant finally began to move from the non-antinomic approach of the *Dissertation* to the doctrine of the antinomy. N. 4525 is the first note on metaphysics which states that the conception of the spatio-temporal world as a totality is contradictory:

> The world is the **absolute** whole of possible experience. We may well conceive of an absolute world-whole, but not in space and time. The absolute-whole in appearance is a contradiction. (18:95–96; $\xi - -\omega$?; not before 1772)

The "contradiction" mentioned here makes the origin of the cosmological antinomy known for the first time. Kant is now coming to realize that the 1770 concept of the intelligible world cannot result in objective cognition. In contradistinction to the assumptions of the *Dissertation*, the cosmological concept cannot be a pure intellectual concept. If it is to have any cognitive content and belong to the real use of reason, it must also be contaminated, so to speak, with the sensible conditions of space and time. According to the *Prolegomena* of 1783, too, the cosmological antinomy results from a contradiction inherent to a cosmological concept which satisfies the cognitive conditions of objects in space and time (see 4:341–342 and Sect. 5.2.3). The cosmological antinomy of 1781 is thus not prompted by the intellectual world concept of a "whole which is not a part" defined in the *Dissertation* (2:387). It "does not concern the understanding's pure concept of a whole of things in general", as he puts it in the *CPR* (B 443/A 416). The dogmatic doctrine of the world as a whole, which the intelligible cosmological concept of 1770 establishes, is non-antinomic not only according to the rudimentary epistemology of 1770, which Kant started to revise in 1772, but also from the critical point of view of 1781. However, according to the latter it is not a cosmology proper; that is, Kant no longer considers it as a discipline of the real use of reason. The antinomy of 1781 refers to the cosmological concept of the spatio-temporal world. It emerges from the concept of the sensible world as the sum total of all empirical phenomena. According to the theory of objective cognition developed by Kant after 1772, we cannot objectively conceive of the intelligible world without conflating it with the whole of all spatio-temporal phenomena, that is, the concept of the sensible world, and vice versa.

The Transcendental Dialectic of the *CPR* emphasizes this point. The cosmological antinomy is an inevitable conflict of reason with itself, which results from the conception of an absolute totality of phenomena in space and time. The cosmological idea derives from the intellectual concept of nature as the sum total of all sensory phenomena under laws (see Sect. 5.2.3). It results from the concept of the "absolute totality in the series of conditions for a given appearance" (B 398/A 340), i.e., from the concept of an extensional and nomological completion of nature as the sum total of phenomena under laws. Kant accordingly emphasizes that the cosmological idea

> concerns nothing other than the exposition of appearances, hence it does not concern the understanding's pure concept of a whole of things in general. (B 443/A 416)

This diagnosis indicates that the 1781 doctrine of the antinomy presupposes not only the space-time theory of 1770, but also some additional key elements of the epistemology developed after 1772. The cosmology of 1770 relates to the cosmological antinomy of the 1781 as follows. The proof of an inevitable inherent conflict of reason in the Transcendental Dialectic of the *CPR* is exactly opposite to the 1770 strategy of criticizing subreptic axioms in order to provide sound foundations for cosmology. From 1772 onwards Kant regarded this strategy of avoiding contradictions as meaningless, insofar as it deprived cosmology of the spatio-temporal world as its object of cognition. The antinomy of 1781 then arises from reversing the strategy of 1770. The reversal is based on the insight that an objective real use of reason in general, and objective cosmological cognition in particular, is inevitably bound to the structure of our cognitive faculties, including the formal conditions of sensibility.

As a result of this reversal, in the doctrine of the antinomies of 1781 Kant argues as follows. The very structure of our cognitive faculties necessarily pretends the objective reality of the cosmological ideas, without taking the conditions of possible cognition into account. The conflict between both is inevitable and can only be resolved by realizing that the horizon of our cognition is limited.

How did Kant arrive at the claim of an inevitable contradiction in the cosmological concept, reversing the arguments of 1770? The *Dissertation* avoids any conflict between the intellectual and the sensitive cosmological concepts by strictly separating the domains of both. In 1772 Kant begins to realize that the separation of the sensitive and intellectual grounds of cognition does not give rise to a convincing epistemology, or that symbolic cognition alone cannot give rise to a real use of reason. Now, he begins to see that cosmological claims similar to the "subreptic axioms" criticized in the *Dissertation* are inevitable, because the very nature of our cognitive faculties gives rise to a conflation of the sensitive and the intellectual grounds of cognition.

With regard to the antinomy of the *CPR*, the structural difference between the "mathematical" and the "dynamical" antinomies has to be noted. Both variants of the cosmological antinomy are based on different relations between the substances in the world. The "dynamical" antinomy of the world as a totality of law-like connections is based on logical relations of ground and consequence, which he gives a causal or modal interpretation. For dealing with relations of ground and consequence, in the 1781 account Kant still employs his 1770 strategy of seeking to avoid any conflict between the sensitive and the intellectual grounds of cognition. In contrast, the "mathematical" antinomy of the world as a spatio-temporal totality is based on the relations of coexistence and succession. For them, the solution of 1770 fails, given that in order to apply to the real world (in contrast to a merely logical universe), the symbolic relations of the intellectual cosmological concept have to be interpreted in spatio-temporal terms.

For the critical Kant, the true scandal of metaphysics is therefore the mathematical antinomy, which refers to the beginning of the world in space and time and to the ultimate components of the substances in the world. In order to understand how the doctrine of the antinomies emerged, two questions need to be clarified.

1. How does Kant come to the conclusion, by reversing the argument of 1770, that the cosmological concept is not just empty, but contradictory?
2. What is the source of the mathematical antinomy, and why does Kant no longer accept the solution of 1770 for its subject?

The key steps towards the cosmological antinomy have to be reconstructed from the notes on metaphysics. Shortly after writing the *Dissertation*, Kant begins to use the term "antithesis" in the sense of the skeptical method (see N. 4275 quoted in Sect. 4.3.2). He first does so based on his 1770 approach of criticizing "subreptic axioms", which claim an objective conflict between principles which in fact are only subjective. In N. 4616 and N. 4618, Kant still proposes to resolve the antithetic conflict regarding the beginning of the world on the grounds of the *Dissertation*:

> [N. 4616:] The world considered *intellectualiter* has no beginning; not because it has endured for an infinite time, for then it would be considered *sensitive*, but rather because in this respect it is not considered in time at all.
>
> The *sensibilis mundus* has no beginning because a first beginning is impossible. He who considers the world *sensitive* cognizes no boundaries in it. (17:610; ξ–ω)

> [N. 4618:] If we consider the world as the sensible world, we will find no beginning in the *regressu* of appearances. But these appearances are not actual things; thus one does not actually say: the world is in itself without a beginning, but rather: the appearance (in relation to us) has no beginning. In appearances there is no absolute boundary of increase or diminution. The world has a beginning, not, however, as appearance, but rather as being in itself. (17:610; ξ or τ)

N. 4743 once more takes up the topics of the Leibniz–Clarke debate, in a similar line of argument:

> **Cause** and **beginning** are this sensitive and that intellectual. The beginning is only in the world, but not of the world. Why did God not create the world earlier? God is in no [abs] relationship against absolute empty time.[22] (17:694; ρ–σ)

These arguments remain on the lines of the reasoning in the *Dissertation*. Their reversal to the later antithetic of pure reason, according to which there is an objective conflict of reason within itself, must have taken place later, after the letter to Herz of February 21, 1772. One of the first notes which anticipates the apagogic proof structure of the first antinomy of the *CPR* is N. 4617 (dated to 1772–1776):

> The world cannot be conceived as having a beginning in accordance with grounds of reason, and in accordance with grounds of sensibility it cannot be conceived as not having a beginning. (18:210; ξ–τ)

In the early 1770s, the apagogic proof structure, which Kant then calls the method "by retorts" (N. 3892; 17:330) or "antithesis" (N. 4275; 17:492), belongs to the use of the skeptical method (see Sects. 4.2.1 and 4.3.2). The term "antinomy" is not explicitly found before 1773–1775 (if the dating of the notes is reliable). One of the first occurrences is in N. 4742:

> Nothing absolutely first is to be encountered among appearances, but it may well be in the *synthesis* of the understanding. Thus there is to be sure no first beginning, but there may be a first cause, part, action, etc. Something first as a *phaenomenon* would appear as the boundary of nothing. The antinomy of reason is therefore nothing other than the difference between principles of reason insofar as the *data* are sensible, i.e., dependent on objects, or intellectual, i.e., given by the mind itself, which is certainly consistent with regard to particular possible experiences but not with regard to the whole of them. Hence all the actions of human beings considered *a posteriori* are empirically determined, but *a priori* are undetermined and free. (18:694; ρ–σ)

This note, too, makes parallel references to the subjects of the first and the third antinomy, the "first beginning" of the world and the "actions of human beings".

It has to be noted that here Kant identifies the "antinomy of reason" only vaguely with "the difference between principles of reason" that are applied to sensible or intellectual ideas. He identifies "sensible" with "dependent on objects" and "intellectual" with "given by the mind itself". In contrast to sensible ideas such as "beginning", he assigns the "first" principles of cognition to the intellectual concepts of "cause", "part", or "action". Here, he lays the grounds for associating the later thesis or antithesis claims in the doctrine of the cosmological antinomy to the positions of dogmatism or empiricism, respectively. The note is found in the preliminaries to drafts on a "dialectic of sensibility" and a "dialectic of the understanding", located in the notes on metaphysics of phase σ dated to 1775–1777 (NN. 4756–4760; 18:699–713). There, Kant conceives of the antinomy in the immediate context of a "transcendental theory of appearance" and a "transcendental theory of experience", without yet making a clear distinction between the contents

[22]"**Ursache** und **Anfang** sind ienes intellectuell, dieses sensitiv. Der Anfang ist nur in der Welt, aber nicht von der Welt. Warum Gott die Welt nicht eher erschaffen? Gott ist in keinem [abs] Verhaltnis gegen die absolute leere Zeit." My translation; not contained in Cambridge edition.

of the transcendental aesthetics, analytics, and dialectics of 1781. What is more, he still speaks of the "antithetic or apparent Antinomy of pure reason" (N. 4757, 17:704; see also Guyer 1987, 391–392). There is still no mention of an antinomy in the sense of an inevitable, provable contradiction.

Hence, when Kant began to conceive of the antinomy as a central component of a "transcendental dialectic", around 1775, he did not yet explicitly characterize its subject as an inevitable conflict of reason with itself. And he did not yet have a clear account of the elements of his later transcendental philosophy. Only when he reversed his argument of 1770 with regard to the infinity of the world, did he discover the inevitability of this conflict and claim a provable contradiction concerning the cosmological cognition of the world as a whole. This reversal is closely related to his conceptions of the mathematical and the metaphysical infinite of 1770, which became subject to an epistemological transformation in the mid 1770s.

4.4 The Mathematical and the Metaphysical Infinite

Let us have a closer look at Kant's conception of the infinite. In the pre-critical period before 1770, he uses the expressions 'infinite' and 'unlimited' in the sense of 'immeasurable' and 'immeasurable in space and time', respectively. In the *Theory of the Heavens*, he considers the structure of the universe as something that "contains in itself everything that can exist, that adopts no measure, in short, that is infinite" (1:310); and he distinguishes the actual infinity of God's thought, the "concept of infinity that is in his mind all at once", from "the concept of another infinity in a *spatially combined connection*", according to which the "extent of the world" is "without limits" (note on 1:310).

For the formal cosmological concept of the *Dissertation* of 1770, he uses a formal concept of the actual infinite, which lacks any spatial or temporal interpretation. In note N. 4197 (dated to 1770 or later), he distinguishes this metaphysical concept from the mathematical conception of the infinite as follows:

> The mathematical infinite is without ending according to the laws of sensible appraisal, the metaphysical infinite is without any ending at all, i.e., without limits.[23] (17:452; λ or later)

The note refers to § 248 of Baumgarten's *Metaphysica* (17:81). It attributes the metaphysical infinite to the intelligible world, and the mathematical infinite to the sensible world. The former is actually infinite, the latter only potentially infinite, as N. 4195 indicates (which also refers to Baumgarten's § 248):

> In the infinite, the difficulty is to reconcile the totality with the impossibility of a *synthesis completae*. Consequently, the difficulty is subjective. On the other hand, the *potentialiter*

[23]"Das mathematisch unendliche ist ohne Ende nach Gesetzen der sinnlichen schätzung, das metaphysisch unendliche ist ohne Ende überhaupt, d.i. ohne Schranken." Note not contained in Cambridge edition; my translation.

> *infinitum (infinitum coordinationis potentialis)* is very well comprehensible, but without totality.[24] (17:452; λ or later)

The metaphysical infinite differs from the mathematical infinite in that the latter underlies the subjective conditions of sensibility, whereas the former does not. According to the *Dissertation* of 1770, mathematical objects must be constructed by restricting space. Hence the mathematical infinite is unlimited, though not given as an actually infinite multitude or class of mathematical objects. Nevertheless, Kant emphasizes that the concept of an actual mathematical infinity should not be rejected. It is worth looking at the respective passage of the *Dissertation* in detail, because Kant's 1770 treatment of infinity raises the very problem to which he finally reacts with the mathematical antinomy of 1781.

In his remark to § 1 of the *Dissertation*, Kant distinguishes the concepts of the world and its ultimate substantial compounds as "abstract concepts of the understanding", on the one hand, and their representation "in the concrete by a distinct intuition", on the other (2:387). Then, he explains that for a continuous or an infinite magnitude the latter cannot be completely given:

> [...] in the case of a *continuous magnitude*, the *regression* from the whole to the parts, which are able to be given, and in the case of an *infinite* magnitude, the *progression* from the parts to the given whole, have in each case *no limit*. Hence it follows that, in the one case, complete analysis, and, in the other case, complete synthesis, will be impossible. Thus, in the first case, the whole cannot, according to the laws of intuition, be thought completely as regards *composition* and, in the second case, the compound cannot be thought completely as regards *totality*. (2:388).

Here he clearly uses the concepts of analysis and synthesis in the sense of the traditional analytic-synthetic or resolutive-compositive method (see Sect. 2 and Appendix A.2). Then, he emphasizes that the impossibility of representing continuous or infinite magnitudes in intuition does not preclude that they may exist in the world:

> From this it is clear how, since *unrepresentable* and *impossible* are commonly treated as having the same meaning, the concepts both of the *continuous* and the *infinite* are frequently rejected. For, indeed, *according to the laws of intuitive cognition*, any representation of these concepts is absolutely impossible. Now, although I am not pleading here a case for these concepts∗ [...] nonetheless it will be of the greatest importance to have given a warning that the people who use such a perverse method of arguing are guilty of the gravest errors. (2:388–389).
>
> ∗ Those who reject the actual mathematical infinite do not exactly make the task difficult for themselves. [...] For them, the *infinite* is *that magnitude than which a greater magnitude is impossible*; and the mathematical infinite is for them that multiplicity (of a unit which can be given) than which a larger multiplicity is impossible. They then substitute *largest* for *infinite*, and, since a largest number is impossible, they readily conclude against an infinite, which they themselves have constructed. Alternatively, they call an infinite multiplicity an

[24]"Im Unendlichen ist die Schwierigkeit, die totalitaet mit der unmöglichkeit einer *synthesis completae* zu vereinbaren. folglich ist die Schwierigkeit subjectiv. Dagegen ist das *potentialiter infinitum (infinitum coordinationis potentialis)* sehr wohl begreiflich, aber ohne totalitaet." Note not contained in Cambridge edition; my translation.

> *infinite number*, and declare that an infinite number is absurd, which it obviously is. But they are fighting with figments of their own imagination. (n. on 2:388)

In 1781, in his remark to the thesis proof of the first antinomy, he still rejects this conception of the infinite as a maximum magnitude, seeing as an erroneous argument:[25]

> I could also have given a plausible proof of the thesis by presupposing a defective concept of the infinity of a given magnitude, according to the custom of the dogmatists. A magnitude is **infinite** if none greater than it (i.e., greater than the multiple of a given unit contained in it) is possible. [...] only this concept does not agree with what is usually understood by an infinite whole.

Kant himself proposes to

> conceive of the mathematical infinite as a magnitude which, when related to a measure treated as a unit, constitutes *a multiplicity larger than any number* [...]. (n. 2:388)

Here, he does not distinguish between 'magnitude [quantum]', 'multitude', and 'number'. By 'magnitude' he means a measured number of concrete units. According to his definition, the mathematical infinite is not a number, but a magnitude that exceeds any (finite) number. From a modern perspective, this corresponds to an ordinal concept of infinity. In contrast to modern mathematics, however, Kant does not consider the mathematical infinite as a mathematical object. According to his definition, the infinite is not a number, but an immeasurable magnitude, i.e., a limiting concept of mathematics or arithmetic. In terms of Leibniz's theory of cognition, the object of this limiting concept is only accessible by symbolic cognition. It is not intuitively given, but must be represented by a symbol such as ∞.

In the *Dissertation* of 1770, the concept of the infinite is ambiguous—that is, either metaphysical or mathematical, depending on its intellectual or sensible origin. As a mathematical limiting concept, the infinite is given by intuition; in 1770 Kant considers space and time to be represented as infinite given magnitudes, just as in 1781 (*CPR*, A 25/B 39):

> [...] following the paradigm of sensitive intuition, it is rather the case that *the infinite contains the ground* of each *part* which can be thought [...] For it is only when both infinite space and infinite time are given that any definite space and time can be specified by *limiting*. [...] Therefore, all the fundamental properties of these concepts lie beyond the limits of reason [...] (2:405)

In particular, any restriction of space that is not a point represents the continuum as a given (i.e., actually) infinite quantity. Nevertheless, the concept of the mathematical infinite does not refer to an actually infinite magnitude. It does not refer to an infinite number or multitude of given elements, because the mathematical infinite is not completely representable in intuition as a given magnitude. The understanding can only successively construct this magnitude as a potential infinite, the scope of

---

[25] Accordingly, in the first antinomy of 1781 the proof of the thesis employs Kant's concept of the infinite of 1770 (see below Sect. 5.4.1.1).

which is generated as an incompletable sequence of mathematical objects. In note N. 4079 (dated to 1769), Kant attributes this cognitive restriction to the nature of the human understanding, which can conceive of an infinite totality only as successively generated, but not as given:

> The difficulty of imagining a *quantum simultanei* as infinite is due to the nature of the human understanding, which can only think of a *totum* synthetically according to its possibility, i.e. *successive addendo unum uni*. The synthesis, however, which is to go into infinity, is never complete.[26] (17:406, phase κ)

Kant's 1770 account of the mathematical infinite gives rise to an obvious problem. The potential infinite corresponds to the ordinal definition 'greater than any finite number', which constitutes his concept of the mathematical infinite. However, since Kant does not preclude the possibility of an actual mathematical infinite, there is a gap between his definition of the infinite as an ordinal limiting concept, which the understanding can only successively generate, and the infinite as a cardinal magnitude given by the nature of intuition. Hence, Kant's account of the mathematical infinite is ambiguous too. Concerning the pure form of intuition, it is actually infinite; but regarding the nature of the human understanding, it is only potentially infinite.

Nevertheless, in 1770 Kant still considers it possible to conceive of the metaphysical infinite in the symbolic terms of an intellectual cosmological concept. Defending the possibility of the actual mathematical infinite, he cannot clearly separate this concept and that of the metaphysical infinite, which he conceives as an actually given infinite multitude of substances in the world. He seems to consider the metaphysical infinite as a cardinality larger than any number of objects given in intuition. However, this conception comes very close to his account of the mathematical infinite. Indeed, he concludes his support of the actual mathematical infinite with an appeal to an intuitive non-human understanding, which can grasp a given multitude of infinite cardinality:

> For there could be an understanding, though certainly not a human understanding, which might distinctly apprehend a multiplicity at a single glance, without the successive application of a measure. (1:388 n.)

From 1772 on Kant began to question the cognitive conditions of the intellectual cosmological concept (see Sect. 4.3.3). This led him to exclude the actual infinite from cosmology as epistemically inaccessible. In note N. 4707 (dated about 1773, referring to § 246 of Baumgarten's *Metaphysica*), he accordingly asks himself:

> The question is whether a given quantity is infinite. The condition is that it is given.[27] (17:682, phase ρ)

---

[26]"Die Schwierigkeit, sich ein *Quantum simultaneum* als unendlich vorzustellen, beruhet auf der Natur des Menschlichen Verstandes, der ein *totum* seiner Möglichkeit nach nur synthetisch denken kann, d.i. *successive addendo unum uni*. Die synthesis aber, die ins unendliche gehen soll, ist niemals complet." Note not contained in Cambridge edition; my translation.

[27]"Die Frage ist, ob eine gegebene Größe unendlich sey. Die Bedingung ist, dass sie gegeben sey." Note not contained in Cambridge edition; my translation.

In N. 4780 (dating back to 1775 at the earliest, and referring to § 255 of Baumgarten's *Metaphysica*), Kant explicitly links the definition of 'infinite' to the impossibility of completing the synthesis of an object of knowledge:

> Infinitude is the absolute impossibility of a complete synthesis (not of the completeness of the object) of the composition or decomposition of a given object. (17:725, σ–υ)

This is an epistemic definition of infinity, given that Kant had used the term 'synthesis' since 1772 in the sense of a cognitive operation. The concept of infinity is no longer defined here as a limiting concept of mathematical quantities, but as a limiting concept of the synthetic activities of the human understanding. What is infinite is any object that in principle cannot be given completely by successive composition or decomposition of the corresponding representation. To apply this epistemic definition of infinity to mathematical objects results in the 1770 concept of the mathematical infinite, according to which it is absolutely impossible to generate the mathematical infinite as a given magnitude successively by the composition or decomposition of given finite magnitudes.

Hence, the above epistemic definition of 1775 (or later) does not affect Kant's mathematical concept of infinity. Space as a given infinite magnitude is actually infinite; but the class of mathematical objects obtained from restricting space is potentially infinite. However, this epistemic definition affects the metaphysical use of the concept. It restricts the metaphysical concept of infinity to the potentially infinite. To attribute the property of being infinite to the world as a whole implies two claims:

(i) The world is a totality, i.e., it is completely given.
(ii) The world is infinite, i.e., it is impossible to complete the mathematical synthesis of its magnitude.

For the intellectual cosmological concept of 1770, Kant considered both claims to be compatible by assuming an actually infinite totality of intelligible substances, which in principle are beyond our cognition. According to the cognitive conditions which Kant has been seeking for since 1772, however, one does not in this way arrive at cognition of the world. To apply the mathematical concept of infinity to the world in space and time, which is subject to the cognitive conditions of our understanding, means to claim that:

(i) The world is completely given as a spatio-temporal object.
(ii) It is impossible to complete the spatio-temporal synthesis of the world.

To retain both claims is impossible on pain of contradiction, because it would be to claim of the world that it is actually infinite and at the same time that it is not. This new insight obviously points towards the antinomy of pure reason. Kant expresses it immediately after giving the above epistemic definition of infinity, in note N. 4780:

> The [crossed out: synthesis of] appearance is infinite, and its division proceeds to the infinite. [...] By contrast, in the intellectual the synthesis is complete, but the condition for cognizing this completeness *in concreto* is sensible [...]. Reason therefore demands

independence from the sensible, but the determination of its concept can only be sensible (antinomy). (17:725)

To apply the epistemic concept of infinity of 1775 to the concept of the world as a whole in its spatio-temporal extension ("mathematical *synthesis*") and its causal connections ("dynamical [. . .] *synthesis*") thus results in an antinomy with regard to the actual infinite. The mathematical account of space and time as infinite given magnitudes, which are completely given by pure intuition, does not give rise to such an antinomy. Only the metaphysical use of the mathematical infinite, i.e., the cosmological concept of the spatio-temporal world, does so. Accordingly, in 1781 Kant does not use the mathematical definition of the infinite of 1770 for the thesis proof of the first antinomy, but rather the epistemic definition of 1775. In the thesis proof of the first antinomy of the *CPR*, he defines the infinite as follows:

But now the infinity of a series consists precisely in the fact that it can never be completed through a successive synthesis. (*CPR*, A 427/B 455)

In the 1781 doctrine of antinomies, this epistemic definition of the infinite is again applied to the cosmological concept of the world in space and time. As in the 1770 cosmology, the cosmological concept concerns the metaphysical use of the mathematical infinite. In contrast to 1770, however, on pain of contradiction, the metaphysical infinite can no longer be conceived as the actual scope of the concept of all intelligible substances, but only as a potentially infinite composition or decomposition of the spatio-temporal world, which can in principle not be completed by the synthetic operations of the understanding. Any attempt to do so in a cosmological concept of the spatio-temporal world conflates the mathematical concept of the infinite and its metaphysical use. This leads to a self-contradictory cosmological concept, from which the cosmological antinomy of 1781 results (see Sect. 5.2.3).

4.5 Conclusions: Reconsidering the Critical Turn

From the *Dissertation*, the notes on metaphysics written in the years before and after 1770, the Blomberg logic, and Kant's correspondence in the early 1770s, the following picture of Kant's philosophical development emerges. The long-established traditional interpretation of the critical turn is erroneous above all in that it hastily identifies the "contradictions" that are mentioned in the notes on metaphysics around 1769, with precisely those subjects later dealt with in the doctrine of the antinomies of the *CPR*. Even if Kant did indeed construct the cosmological antinomy out of this material, he only obtained it by reversing the arguments of the *Dissertation*.

Around 1769–1770, he certainly did not yet believe in the existence of unavoidable contradictions in the field of cosmology. From 1768 to 1770 he does not refer to any cosmological antinomy nor its resolution. The puzzle of 1768 is not an antinomy in the later sense of an unavoidable conflict of reason with itself, but a consistency problem concerning the concept of space at the interface of physics and metaphysics (Sect. 3.4) discovered "by retorts" (N. 3892), i.e., by the skeptical method (Sect. 4.3.2).

After the *Dreams* of 1766 and the *Directions in Space* of 1768, Kant was convinced that the analytic method defended in the *Prize Essay* of 1764 had gone astray as a reliable method for establishing the foundations of metaphysics. Both writings, the *Dreams* as well as the *Directions in Space*, suggested an epistemic turn for metaphysics. In addition, Kant started to make use of the skeptical method as a new, critical tool for probing metaphysical principles. In parallel, he proposed that the traditional metaphysical concepts are subjective rather than objective, and that considering them as objective gives rise to contradictions, which however could easily be avoided by taking their subjective nature into account. The *Dissertation* of 1770, and in particular the criticism of "subreptic" axioms in the fifth part, is based on a strategy of avoiding contradictions. It differentiates between intuition and concepts, and the sensible and intelligible world, without perceiving any conflict between the intellectual and the sensitive cosmological concepts. It still aims at a dogmatic cosmology, namely at a (contradiction-free) theory of the intelligible world, which would comprise symbolic cognition in a Leibnizian sense (Sects. 4.2.3 and 4.3.1). From 1770 Kant was in search of a new specific method of metaphysics, but at that time his only means were the principle of strictly distinguishing the sensitive and intellectual grounds of cognition, and the skeptical method of invoking all conceivable arguments and counterarguments.

It was not before 1772 that the later doctrine of the antinomy began to take shape, emerging during the 'silent decade' in the course of developing the critical theory of objective cognition. Now Kant began to abandon the project of a dogmatic cosmology. He started to link the possibility of objective cognition to the conditions of sensibility, and to deny the cognitive content of the symbolic concept of the intelligible world of the *Dissertation*. This led him to reverse the contradiction-avoidance strategy of 1770 into the later proofs of the cosmological antinomy (Sect. 4.3.3). Dating the 'critical turn' back to the years 1769–1770 misses the point that the most important systematic step towards the antinomy, namely the reversal of the arguments of 1770, did not take place before 1772–1775.

According to the *CPR*, the inevitable conflict of reason with itself arises primarily for the topic of the mathematical antinomy. The notes on metaphysics of the 1770s show that this problem does not stem from a revision of the mathematical concept of infinity, but from criticizing its metaphysical use (Sect. 4.4). Between 1772 and 1775 Kant arrived at the conclusion that the cosmological concept of the spatio-temporal world as a totality gives rise to an inevitable contradiction, due to the need to apply the mathematical concept of infinity to cosmology.

Kant's development in the 1770s clearly demonstrates that the crucial breakthrough of 1769 was not to discover or even resolve the later antinomy of pure reason, but rather to dispense with the Leibniz–Wolff theory of ideas and with the traditional analytic-synthetic methods of establishing the foundations of metaphysics. The "great light" of 1769 was the insight that concepts and intuitions are two different kinds of mental representations, which belong to the distinct cognitive faculties of understanding and sensibility, as emphasized by Beiser (1992, 51). After all, for him to abandon the rationalist theory of ideas was a genuine revolution in philosophy. In 1769, Kant started to perceive the principles of transcendental idealism "as if in twilight" (N. 5037; 18:69). The skeptical method (which, however, must *not* be misunderstood in terms of Hume's skepticism!) bore fruit a few years later, leading to the discovery of the cosmological antinomy in a strict sense. Hence, Kant's insights in the year before writing his *Dissertation* of 1770 are often overestimated. The "great light" of 1769 still left much in the dark, as Reich (1958, XVI) put it.

This reconstruction fits in well with Reich's interpretation of the critical turn. Two objections against it remain, however, and so call for refinement. The famous N. 5037 from the late seventies refers to the "great light" of 1769 together with attempts "to prove propositions and their opposite", albeit not "in order to establish a skeptical doctrine" (18:69). In some sense Kant at that time must have already been concerned with antithetical propositions and their proofs. The notes on metaphysics of that time show that he was thus testing the skeptical method. The traditional interpretation of Kant's critical turn is based on this vague evidence. In addition, N. 5116 (which dates from the same period as N. 5037) demonstrates that the discovery of the subjectivity of the real use of metaphysical concepts in 1769 was only a first step towards the critique of cognition. Retrospectively, writing in the late 1770s, Kant considered the insights which allowed him to definitely abandon the project of a dogmatic metaphysics to have been more decisive:

> It took a long time before in such a way I found the whole dogmatic theory to be dialectical. Yet I sought something certain, if not with regard to the object, then still with regard to the nature and boundaries of this sort of cognition. I gradually found that many of the propositions that we regard as objective are in fact subjective, i.e., they contain the *conditiones* under which alone we can have insight into or comprehend the object. But while I thereby certainly became careful, I was still not instructed. [...] for I still believed I could find a method for extended dogmatic cognition through pure reason. (18:95–96)

Here, Kant emphasizes three points:

(1) He did not discover the internal conflict of reason at one stroke; it took a long time until he recognized the formation of metaphysical theories as being dialectical.
(2) The insight into the subjectivity of cognitive conditions made him increasingly critical of the claims of dogmatic metaphysics, but it did not yet amount to the decisive breakthrough to his critical philosophy.
(3) When he realized the subjectivity of the conditions of our cognition, around 1769, he still continued to search for a method of establishing dogmatic metaphysics. At that time he was still far from demonstrating that such attempts inevitably give rise to a self-contradictory cosmological concept.

My reconstruction of the "great light" of 1769, attempted above, must therefore be modified at least in two respects. On the one hand, it neglects the importance of methodological questions for the critical turn. Kant was dissatisfied with the skeptical method, and yet was committed to it around 1769. Therefore he used it ruthlessly to examine the concepts and principles of metaphysics, hoping to find an error in the dogmatic approach which would give him hints to a new methodology. The insight that there could no longer be a new dogmatic method was still to come, and would result only from discovering the cosmological antinomy as an unavoidable contradiction, using the skeptical method and the critical method developed from 1772 onwards, and then reconsidering the use of the mathematical infinite in metaphysics.

On the other hand, my above reconstruction perhaps overestimates the insights of 1769. If there was a decisive breakthrough in 1769, it only concerned the break with Leibniz's and Wolff's doctrine of ideas, or with the Cartesian tradition of clear and distinct ideas. Compared to the entire rationalist and empiricist tradition from Descartes to Locke, Leibniz, Wolff, and Berkeley to Hume, it is not the insight into the subjectivity of our cognitive conditions that is fundamentally new about Kant's theory of 1770. What is new is his distinction between intuitions and concepts as two kinds of ideas or representations, which as distinct cognitive capacities are substantially different as regards their structure and origins, and not just in respect of their degrees of clarity. This distinction gave rise to a revolution in metaphysics, understood as a science of the principles of human cognition, the consequences of which Kant himself was by no means able to realize in 1770. Like a scientific revolution, Kant's Copernican turn did not happen in one leap, but rather by many small steps.

References

Adickes, Erich. 1897. *Kants Systematik als systembildender Faktor*. Berlin: Mayer und Müller.

Arnauld, Antoine, and Pierre Nicole. 1685. *La logique, ou lŠart de penser*. Quoted after the Engl. transl.: *Logic or the Art of Thinking*. Transl. and ed. by J. V. Buroker. Cambridge University Press 1996.

Baum, Manfred. 1994. Kants kritischer Rationalismus. Zur Entwicklung des Vernunftbegriffs nach 1770. In *Vernunftbegriffe in der Moderne*, ed. H.F. Fulda and R.-P. Horstmann, 184–198. Stuttgart-Bad Cannstatt: Clett-Cotta.

Beiser, Frederick C. 1992. Kant's intellectual development. In *The Cambridge Companion to Kant*, ed. P. Guyer, 26–61. Cambridge University Press.

Carl, Wolfgang. 1989. *Der schweigende Kant. Die Entwürfe zu einer Deduktion der Kategorien vor 1781*. Göttingen: Vandenhoeck & Ruprecht.

Carson, Emily. 1998. Kant on intuition in geometry. *Canadian Journal of Philosophy* 27: 489–512.

Engelhard, Kristina. 2005. *Das Einfache und die Materie. Untersuchungen zu Kants Antinomie der Teilung*. Kantstudien-Ergänzungshefte 148. Berlin: de Gruyter.

Erdmann, Benno. 1884. Die Entwicklungsperioden von Kants theoretischer Philosophie. In *Reflexionen Kants zur Kritik der reinen Vernunft. Aus Kants handschriftlichen Aufzeichnungen*, ed. B. Erdman, Vol. 2. Reprint Stuttgart-Bad Cannstatt: Frommann-Holzboog 1992.

Ertl, Wolfgang. 2002. Hume's Antinomy and Kant's Critical Turn. *British Journal for the History of Philosophy* 10: 617–640.

Euler, Leonhard. 1748. Reflexions sur l'espace et le temps. In: *Memoires de l'academie des sciences de Berlin* 4, 1750, 324–333. Engl. transl by M. Saclolo and P. Wake: Reflections on Space and Time. http://eulerarchive.maa.org/docs/translations/E149tr.pdf

Euler, Leonhard. 1763. *Vernünftige Gedanken von dem Raume, dem Orth, der Dauer und der Zeit*. Quedlinburg: Gottfried Heinrich Schwans Wittwe.

Falkenburg, Brigitte. 2000. *Kants Kosmologie*. Frankfurt a. M.: Klostermann.

Fischer, Kuno. 1860. *Immanuel Kant. Entwicklungsgeschichte und System der kritischen Philosophie Bd. I: Entstehung und Begründung der kritischen Philosophie. Die Kritik der reinen Vernunft*. Mannheim 1860 (= *Geschichte der neueren Philosophie* Bd. III).

Fischer, Kuno. 1909. *Geschichte der neueren Philosophie. Vierter Band. Immanuel Kant und seine Lehre*. I. Teil. 5. Auflage, betreut von Windelband. Heidelberg: Carl Winter.

Friedman, Michael. 2004. *Immanuel Kant: Metaphysical Foundations of Natural Science*. Translated and edited by Michael Friedman. Cambridge: Cambridge University Press.

Gerlach, Burkhard. 1998. Wer war der "große Mann", der die Raumtheorie des transzendentalen Idealismus vorbereitet hat? *Kant-Studien* 89: 1–34.

Grier, Michelle. 2001. *Kant's Doctrine of Transcendental Illusion*. Cambridge University Press.

Guyer, Paul. 1987. *Kant and the Claims of Knowledge*. Cambridge University Press.

Hinske, Norbert. 1970. *Kants Weg zur Transzendentalphilosophie. Der dreißigjährige Kant*. Stuttgart: Kohlhammer.

Kaulbach, Friedrich. 1982. *Immanuel Kant*. Zweite, durchgesehene Auflage. Berlin: de Gruyter.

Klemme, Heiner. 1996. *Kants Philosophie des Subjekts*. Meiner.

Kreimendahl, Lothar. 1990. *Kant — der Durchbruch von 1769*. Köln: Dinter.

Paulsen, Friedrich. 1875. *Versuch einer Entwicklungsgeschichte der Kantischen Erkenntnistheorie*. Leipzig: Fues.

Reich, Klaus. 1958. Einleitung zu: *Kant, Immanuel: Über die Form und die Prinzipien der Sinnen- und Geisteswelt*. Lat.-dt. Hamburg: Meiner, VII–XVI.

Riehl, Alois. 1924. *Der philosophische Kritizismus. Geschichte und System. Vol. 1: Geschichte des philosophischen Kritizismus*, 3rd ed. Leipzig: Kröner.

Thöle, Bernhard. 1991. *Kant und das Problem der Gesetzmäßigkeit der Natur*. Berlin: de Gruyter.

Part III
Critical Cosmology

This part of the book investigates the cosmological antinomy and its relevance for Kant's critical philosophy. Both have often been misunderstood, something which is hardly surprising, perhaps, given that Kant constructed proofs which he himself considered as fallacious and untenable. The clue for understanding the structure of the antinomy is its resolution; its philosophical significance, however, becomes more comprehensible against the background of the pre-critical project and Kant's lifelong efforts to give metaphysics the "secure course of a science" (*CPR* B vii–xv).

Chapter 5 analyzes the general structure of the antinomy of pure reason and the specific structures of its four versions, including a detailed logical reconstruction of all thesis and antithesis proofs. To understand the logical structure and epistemological significance of the antinomy, it is crucial to distinguish carefully between the pre-critical views which correspond to the standpoint of transcendental realism on which the proofs are based, and the critical point of view, which is the key to Kant's resolution of the antinomy. Kant was well aware that the proofs are defective. His critical diagnosis is that they seem conclusive from the point of view of transcendental realism, whereas transcendental idealism reveals that they derive from a self-contradictory cosmological concept. Our reconstruction shows that the proofs employ rationalist, empiricist, or verificationist arguments, including Kant's own pre-critical conception of the infinite, but do not depend fatally on claims of transcendental idealism; and that the proof results are due to the logical fallacy of an ambiguous middle term in the proofs derive from semantic equivocations inherent to the cosmological concept of the spatio-temporal world, which Kant considered to be inevitable in particular in the case of the "mathematical" antinomy.

Chapter 6 discusses the many facets of the relationship between the cosmological antinomy and Kant's transcendental idealism, from the question of whether the antinomy is a self-generated problem of Kant's critical turn, to his critical attempt to give metaphysics secure foundations, to his architectonics of metaphysics, the teleology of human reason, and his critique of naturalism. The "experiment of pure reason" explained in the preface to the second edition of the first *Critique* sheds light on the philosophical significance of the antinomy. It is a thought experiment that presents the cosmological antinomy as an argument in favour of transcendental idealism, and which may be understood as a transcendental argument against transcendental realism. Kant supports his thought experiment by once again drawing an analogy between the method of metaphysics and the analytic-synthetic method of Newtonian science. To this analogy we may add further common features of the pre-critical and the critical cosmology, which have been neglected in Kant research. For Kant, the antinomy is a new case against naturalism, as can also be seen from remarks in his later writings. Ultimately, its significance for Kant's critical project is to be a critical follower of the pre-critical physico-theology.

Chapter 5
The Antinomy of Pure Reason

> *The second class of sophistical inference is applied in general to the transcendental concept of absolute totality in the series of conditions for a given appearance; and from the fact that I always have a self-contradictory concept of the unconditioned synthetic unity in the series on one side, I infer the correctness of the opposite unity, even though I also have no concept of it.*
> (*CPR*, A 340/B 398)

The 1781 doctrine of the antinomy is a keystone of transcendental idealism. Kant took it over unrevised into the 2nd edition of the *CPR*, presented its central arguments in the *Prolegomena* of 1783 as the strongest impetus "to awaken philosophy from its dogmatic slumber" (4:338), and in a letter to Christian Garve written in 1798 still maintained that the antinomy of pure reason had aroused him from dogmatic slumber and drove him to the critique of reason (12:257–258). He himself thus retrospectively ascribes great significance to the antinomy for the systematic justification of his critical philosophy, which stands in considerable contrast to my reconstruction in Chap. 4 of the way in which it emerged through the course of the 1770s.

Yet without recourse to his later statements, the results of Chap. 4 also raise serious problems. He discovered the cosmological antinomy only on the basis of his critical theory of space and time. And he did not do so immediately, as the contradiction-avoidance strategy of the *Dissertation* of 1770 demonstrates. One might suspect that the cosmological antinomy is a self-generated problem that only arises when certain crucial assumptions of Kant's transcendental idealism are already taken for granted. Does he presuppose what needs to be proven, namely that cosmological cognition cannot be completed without falling into contradiction? Or are the proofs of the antinomy at least tenable from the perspective of the philosophical and scientific background of his time?

In order to shed more light on these questions, this chapter attempts a detailed reconstruction of the structure of the cosmological antinomy of the *CPR* in general, as well as its four versions. For Kant, the antinomy is due to an unavoidable fallacy

B. Falkenburg, *Kant's Cosmology: From the Pre-Critical System to the Antinomy of Pure Reason*, European Studies in Philosophy of Science 12,
https://doi.org/10.1007/978-3-030-52290-2_5

of pure reason which derives from a self-contradictory cosmological concept. In order to understand the doctrine of the antinomy, one has to take into account that he himself did not consider the proofs tenable. He claimed only that they look convincing as long as one maintains any dogmatic metaphysical point of view whatsoever, instead of transcendental idealism. Kant's claim that the proofs are inevitable certainly calls for careful investigation.[1] But let us first have a look at Kant's critics.

5.1 Objections and Refutations

From the beginnings of twentieth-century logic and mathematics to recent Kant scholarship, a range of objections have been raised against Kant's doctrine of the antinomy. A frequent reproach has been that the proofs of the antinomy are logically unsound. In the following discussion I set this objection aside, given that it misses the crucial point that Kant himself wanted to show that the proofs are defective, because their results are due to logical fallacies. A more specific version of the objection is deserving of consideration, however, namely that Kant's proofs depend on an untenable account of the mathematical infinite. In addition, suspicions that the doctrine of the antinomy is incoherent or circular deserve attention.

5.1.1 The Objection from Mathematics

In view of Cantor's transfinite set theory, Russell (1903, 458–461) accused Kant's proofs of being based on an inadequate concept of the mathematical infinite. More recently, van Benthem (1983, 33) emphasized that the proof of the thesis of the first antinomy of the *CPR*, which refers to the paradox of an infinite world age (see Sect. 5.4.1), derives from a conflation of the cardinal and the ordinal infinite. Van Bentham is indeed absolutely right. However, Kant himself deliberately constructed the proof in this way, in order to deconstruct its fallacy by means of the resolution of the first antinomy (see Sects. 5.4.1 and 5.3.3). Hence, the objection that Kant made logical errors misses the point.

All such assessments trace back to Cantor himself, who passes judgment on Kant's antinomy as follows in a letter dating from 1885:

> [...] it is unlikely that more has ever happened to discredit human reason and its abilities than with this section of 'critical transcendental philosophy'. I will show in passing that this author has only succeeded in asserting his antinomies through a vague, indiscriminate

[1]The analysis of Grier (2001, 172–229) is complementary to my approach, which I first presented in Falkenburg (2000, 177–254). Grier focuses on the origin of the four antinomies in Kant's theory of transcendental illusion; I focus on disentangling their logical, semantic, and epistemic aspects.

use of the concept of infinity [...], and that only for those who like to avoid a thorough mathematical treatment of such questions, as he does.[2] (Cantor 1932, 375)

The contradiction in the cosmological concept, which is the subject of the mathematical antinomy of 1781 (see Sect. 5.2.3), refers to the infinity of the world in space and time. The proof Kant gives for the conflicting claims about the finiteness or actual infinity of the world as a given totality is anything but stringent from a modern formal point of view (see Sect. 5.4.1). However, in a comment on the above passage, Zermelo correctly points out that Cantor's reproach of a lack of clarity in his conception of infinity does no justice to Kant's doctrine of the antinomy:

> Kant was not concerned here with a refutation or rejection of the *concept* of infinity, but with its application to the *world as a whole*, with the fact that human reason is equally urged by its inner nature to accept the world as limited as well as unlimited, as finite as well as infinite—a fact that can neither be eliminated by mathematical theories such as Cantor's set theory nor by his not very profound polemics.[3] (Cantor 1932, 377)

The antinomy of pure reason indeed concerns the real use of reason in cosmology. The way in which the doctrine of the antinomy emerged in the 1770s also supports Zermelo's defence of Kant against Cantor. In the *Theory of the Heavens* of 1755 (1:306–310), as well as in the symbolic cosmology of 1770 (2:387–2:388), Kant had still assumed that the world consists of an actually infinite number of substances. In the course of the critical turn, he changed only his view of the cosmological use of the concept of the infinite, but not his mathematical conception of it. It was only in the mid 1770s that he realized what Zermelo states in his comment on Cantor: around 1773–1775, Kant abandoned Leibniz's concept of the actual infinite for the cosmological use of reason, and this marked the decisive turning point from the contradiction-free cosmology of 1770 to the 1781 diagnosis of a self-contradictory cosmological concept (see Sect. 4.3.3). Concerning the mathematical structure of the world, in the *CPR* Kant rejects Leibniz's notion of the actual infinite in favour of a merely potentially infinite, adding a third option to the alternative of finiteness and actual infinity, as exposed in the resolution of the mathematical antinomy.

Zermelo's response to Cantor's harsh criticism is therefore entirely justified. The cosmological antinomy of 1781 does not arise from the concept of the mathematical

---

[2]"Es dürfte kaum jemals [...] mehr zur Diskreditierung der menschlichen Vernunft und ihrer Fähigkeiten geschehen sein, als mit diesem Abschnitt der 'kritischen Transzendentalphilosophie'. Ich werde gelegentlich zeigen, daß es diesem Autor nur durch einen *vagen, distinktionslosen* Gebrauch des Unendlichkeitsbegriffs [...] gelungen ist, seinen *Antinomien* Geltung zu verschaffen, und dies auch nur bei denen, die gleich ihm einer gründlichen mathematischen Behandlung solcher Fragen gern ausweichen." My translation.

[3]"Nicht um eine Widerlegung oder Ablehnung des Unendlichkeits*begriffes* handelte es sich hier bei Kant, sondern um seine Anwendung auf das *Weltganze*, um die Tatsache, daß die menschliche Vernunft sich durch ihre innere Natur ebenso gedrängt findet, die Welt als begrenzt wie als unbegrenzt, als endlich wie als unendlich anzunehmen—eine Tatsache, die weder durch mathematische Theorien wie die Cantorsche Mengenlehre noch durch seine wohl nicht sehr tiefgreifende Polemik aus der Welt geschafft werden kann." Zermelo's note [[1]] to Cantor's text; my translation.

infinite *per se*, but from its application to the spatio-temporal world as a whole. A mathematical theory of the actual infinite, as later developed by Cantor, would have had a similar status for the critical Kant as non-Euclidean geometry, with just one difference. According to the Transcendental Aesthetic, space is an infinitely given quantity which is actually infinite, albeit that it does not represent a given totality of mathematical objects. Apart from this, he would have considered a theory of the actual infinite (similarly to non-Euclidean geometry) as a logically possible symbolic cognition. From Kant's critical point of view, however, such a symbolic theory cannot apply to physical cosmology, given that it does not satisfy the conditions of the possibility of objective knowledge (this is the proviso that Kant did not take into account in his symbolic cosmology of 1770). These conditions force an epistemological definition of infinity, according to which 'infinite' and 'completely given in all parts or elements' are mutually exclusive concepts, in the real use of reason. Merging this epistemic concept of infinity with the cosmological concept of the world in space and time gives rise to the cosmological antinomy, according to the *CPR*.

5.1.2 *The Objection of Incoherence*

Zermelo's response to Cantor indicates that Kant was *not* dealing with a self-generated problem. In fact, Kant was concerned with a philosophical problem which is topical up to the present day, even though we no longer share his scientific and philosophical background. Kant wants to show that any attempt to complete cosmological cognition gives rise to some kind of naturalism, which he considers untenable. His diagnosis is that the cosmological antinomy in general is caused by a contradiction in the cosmological concept, and he analyzes four specific versions of it. The two versions of the "mathematical" antinomy are based on the part–whole relation. The two versions of the "dynamical" antinomy are based on relations of ground and consequence.

Some interpreters argue that all these problems are completely disparate. In particular, Bennett (1974, 114) claims that the antinomy chapter of the *CPR* "is in fact a medley, and the several sorts of unity claimed for it are all spurious." Grier (2001, 174–181) emphasizes that Kant's general account of the antinomy is due to his theory of transcendental illusion, and that the particular cosmological arguments of the four versions of the antinomy are grounded in this theory. As I will show in the following, the four versions of the antinomy do indeed form a coherent system of logical, semantic, and epistemological problems of cosmological cognition, to which Kant attributes the same origin—for good reasons.

5.1.3 *The Objection of Circularity*

Finally, it has been argued that Kant's doctrine of the antinomy is circular (Strawson 1966, 196), given that it only emerged on the basis of the 1770 theory of space and time as subjective forms of intuition. Does Kant hence presuppose in his proofs the very position of transcendental idealism which he finally employs to resolve the antinomy? To put this in other words: Does Kant's resolution of the antinomy just unravel a self-generated problem that only emerges in the context of transcendental idealism, but nowhere else? This objection is more substantial and harder to refute than the other ones. A preliminary reply to it is suggested in Section 5.2.2, but a more detailed refutation can only be attempted on the basis of the detailed logical and semantic analysis given in Section 5.4, and in the final chapter of this book (Sect. 6.1).

5.2 General Structure

Kant's general diagnosis is that the cosmological antinomy consists in an epistemic conflict of reason with itself (see Sect. 5.2.1), which can only be resolved by criticizing its tacit epistemic presuppositions. Hence, it is an *epistemic* antinomy, which for him is due to the structure of human reason, arising specifically from its demand that our knowledge be completed (see Sect. 5.2.2). But it also gives rise to a genuine *logical* antinomy, in his view; and in addition it has crucial *semantic* aspects. So, the cosmological antinomy is very complex. In order to understand its structure and to assess its tenability, we have to unravel its epistemic, logical, and semantic aspects.

Logically, Kant characterizes the antinomy as a contradiction inherent to our concept of the world, which results from conflating an intellectual cosmological concept with the concept of the sensible spatio-temporal world (see Sect. 5.2.3). The conflation of both cosmological concepts results from the quest to complete our cosmological cognition in four different regards, giving rise to four versions of the antinomy of pure reason: two "mathematical" and two "dynamical" antinomies. The antinomy chapter of the *CPR* then aims at demonstrating for each of the four variants of the antinomy that the corresponding specific cosmological concept is based on incompatible predicates. In order to prove this claim, Kant employs an "apagogic" proof procedure, making use of the skeptical method (see Sect. 5.3.3). The antithetic thesis and antithesis claims correspond to the philosophical positions of rationalism and empiricism, respectively. The arguments employed in the proofs stem from the philosophical tradition and the debates of eighteenth-century metaphysics, which he adopts specifically in order to give credibility to the proofs. Kant resolves the antinomy by pointing out that it is illegitimate to consider the spatio-temporal world as an object of cosmological cognition, and accordingly by limiting the cosmological use of reason (see Sect. 5.2.4).

Logical and semantic analysis shows that for all four versions of the antinomy, the proof results depend on equivocations in the use of cosmological concepts such as 'infinite', 'simple substance', 'cause', or 'ground of existence' (see Sect. 5.4), which have different meanings for the cosmological concepts of the intelligible and the sensible worlds. This result corresponds to Kant's own way of resolving the antinomy by stating a *sophisma figurae dictionis* or paralogism, i.e., the semantic fallacy of the ambiguous middle term of a logical syllogism. According to this diagnosis, the fallacies from which the contradictory conclusions in the antithetic proofs stem only become manifest from the critical point of view (see Sect. 5.2.4). Kant assumed that under the conditions of traditional cosmology, i.e., from a point of view of transcendental realism, the equivocations remain hidden and the proofs on both sides of each antinomy seem conclusive. Given that the proofs rely on arguments from the never-ending metaphysical debates to which he repeatedly referred, he considered them to be convincing. Hegel commented on the cosmological antinomy and Kant's resolution that it shows

> an excessive tenderness for the world to keep contradiction away from it, to transfer it to spirit instead, to reason, and to leave it there unresolved. (Hegel 1832, 21.232)

From a modern point of view one might agree with Hegel and ask, how it is possible to make an inconsistent cosmological theory consistent simply by limiting its scope. Does a theory of the spatio-temporal world in Kant's sense indeed become inconsistent by extrapolating it to a pre-critical cosmological concept? And can it be made consistent by restricting its scope to the empirical conditions of possible cosmological cognition?

5.2.1 Definition of the Cosmological Antinomy

The cosmological antinomy of the *CPR* is a "contradiction in the laws (antinomy) of pure reason" (A 407/B 434). Kant defines an antinomy as a conflict between two laws that can be claimed against each other on good grounds. The term 'antinomy' originates from legal theory and can in particular be traced back to Baumgarten.[4] In the Ethics lectures Kant uses the term 'antinomy' at the latest from 1775–1777,[5] hence around the time when the term also appeared in his notes on metaphysics (see Sect. 4.3.3).

According to Kant, the cosmological antinomy of the *CPR* derives from the laws, or principles, of pure reason. They give rise to antithetic claims about the cosmological concept of the world as a totality of spatio-temporal phenomena and their nomological relations. The background to this understanding is the notion of the world explained in Baumgarten's *Metaphysica* (Baumgarten 1757, 103–104; Guyer and Wood 1998, n. 46), which also underlies Kant's pre-critical cosmology from 1755 to 1770. In contrast to the pre-critical writings, in 1781 Kant explains this cosmological concept in terms of his speculative ideas of reason and the way in which they relate to the principles of the pure understanding. This concerns the

[4]See Hinske 1971. In contrast, the term 'antithetic' originates from Protestant controversial theology; see Hinske 1972.

[5]See the Ethics lectures Brauer (Menzer 1924, 59), Collins (27:280), and Mongrovius (27:1431) from the mid 1770s.

"absolute totality in the synthesis of appearances" (*CPR*, A 407–408/B 434). For Kant, this "transcendental idea" or "world-concept" (ibid.) emerges from the way in which we think about the world as a whole of spatio-temporal and nomological relations, i.e., the idea of the *empirical* world as an extensionally and nomologically *complete* totality.

According to the Transcendental Dialectic, "an idea or a concept of reason" is "a concept made up of notions which goes beyond the possibility of experience"; a concept is "made up of notions" if it is obtained from pure concepts of the understanding (A 320–321/B 377). In order to clarify his account of the ideas of reason, Kant distinguishes the following species of objective perceptions or cognitions: intuitions as singular concepts; concepts as general concepts; empirical and pure concepts; pure concepts that originate in the understanding (notions) or in the "pure image of sensibility" (ibid.; such as the concepts of mathematical objects); and ideas or concepts of reason as concepts from notions. Given that 'notion' and 'concept of the pure understanding' are synonymous for Kant, an idea or "concept made up of notions" is a second-order concept.[6]

The ideas or concepts from notions of the *CPR* do not correspond to the intellectual concepts of the *Dissertation* of 1770, which for the Kant of 1781 are concepts of the pure understanding. In the *CPR*, he emphasizes that the transcendental idea of the world as an absolute totality

> concerns nothing other than the exposition of **appearances**, it does not concern the understanding's pure concept of a whole of things in general. (A 416/B 443)

This idea differs in two regards from the intellectual concepts of 1770. On the one hand, it refers to the "exposition of appearances". On the other hand, it extrapolates

[6]According to Malzkorn (1999, 38–53), Kant's ideas of reason are defined as follows (my translations): they are concepts (I) "of the form of cognition" (ibid., 38), (II) "of the form of the relation between cognition of the understanding (judgments)" (ibid., 39), (III) "of the unity of all cognition of the understanding" (ibid., 45), and (IV) "of 'unconditioned conditions' of a uniform system of all empirical cognition" (ibid., 47). According to this determination, ideas are not concepts of empirical objects, but "meta-concepts of empirical science" (p. 45); the world as the "sum total of appearances [. . .] is not an object that could fall under such a meta-concept" (ibid., 76). Accordingly, he states that the world as the sum total of appearances for Kant is no possible object of an idea of reason (i.e., it lacks real possibility): "It only appears as such if one conflates the realm of appearances with the realm of cognition about them" (ibid., 77). However, he disregards that Kant locates the origin of the cosmological antinomy exactly in this problem. For Kant, the sensible world lacks not only real possibility, but also logical possibility as an object of an idea of reason. Malzkorn (1999, 77), however, criticizes Kant as follows: "A description of the world should not be conflated with the world itself. Such mistaken identity would make the world appear to be the object of a concept of reason, but it represented an illegitimate conflation of two distinct realms. Indeed, traces of such conflation can be found [. . .] in Kant's doctrine of the antinomy"; and he concludes: "Kant's theory of reason may [. . .] be coherently reconstructable and be an important part of his epistemology; however, it cannot convincingly present the world as an object of natural and inevitable speculation of reason". Indeed Kant himself attributes the inevitability of the antinomy precisely to an immanent tendency of reason to commit those logical fallacies of which today's interpreters would accuse him. For a comprehensive interpretation of Kant's theory of reason and the rationality of its natural disposition toward metaphysics, see Willaschek (2018).

the conceptual synthesis of the appearances to an absolute totality. By doing so, it attempts to complete the synthesis of the concepts of the pure understanding, which accomplishes the synthesis of sensory perceptions to appearances, to an absolute totality. Therefore, this idea is a second-order concept which relates to the concepts of pure understanding as the latter relate to the sensory perceptions.

For this transcendental idea or world-concept, the separation of the sensitive and the intellectual cosmological concepts of 1770 no longer works, nor does the antinomy-avoidance strategy of 1770 (see Sect. 4.3.1). The transcendental world-concept confounds the sensible and the intellectual cosmological concepts of 1770. It amalgamates the concept of nature as the sum total of concrete empirical phenomena under laws with the concept of the world as an abstract whole, i.e., a merelogical maximum. In terms of Kant's critical epistemology, this transcendental idea conceives of the world at once as a phenomenon and as a thing-in-itself, giving rise to a self-contradictory world-concept (see Sect. 5.2.3).

The synthesis at which reason aims according to this self-contradictory world-concept follows the principle of inferring from conditioned appearances to "the whole sum of conditions, and hence the absolutely unconditioned" (A 406/B 436). It comes in four versions, giving rise to the four kinds of cosmological antinomy, which emerge from a single kind of "sophistical inference", as Kant puts it at the beginning of Book Two of the Transcendental Dialectic, in the section on the dialectical inferences of pure reason:

> [...] and from the fact that I always have a self-contradictory concept of the unconditioned synthetic unity in the series on one side, I infer the correctness of the opposite unity, even though I also have no concept of it. (A 340/B 398)

Here, Kant refers to the apagogic proofs of the thesis and antithesis claims of the antinomy (see Sect. 5.3.3), i.e., the method of proving a proposition by a *reductio ad absurdum* of the opposite proposition. This is the skeptical method, which he started to employ in his notes on metaphysics from 1769 onwards. To apply it to the self-contradictory world-concept seems to give rise to compelling results, in his view. In this way, due to its very nature of asking for an unconditioned synthetic unity of the appearances, reason becomes entangled in incompatible conclusions about the totality of appearances. The result is the cosmological antinomy, as a conflict of reason with itself concerning contradictory laws:

> I will call the condition of reason with regard to these dialectical inferences the antinomy of pure reason. (A 340/B 398)

It should be noted that for Kant the antinomy of pure reason is primarily epistemological, even though it also has crucial logical and semantic aspects. The epistemic aspect lies in the persuasive power of the proofs of the thesis and antithesis claims. The logical aspect is that Kant considers the proof results of the four versions of the antinomy as based on sound logical deductions. The semantic aspect is the semantic presupposition that the thesis and antithesis propositions rely on meaningful concepts that apply to the world. According to this semantic presupposition, it is justified to assume that the proof results hold for the world, i.e., that they are not only logically but also semantically valid. This semantic aspect establishes the connection to the sphere of legal theory, from which the concept of an antinomy originates. In the legal sphere, 'validity' means nothing other than 'legitimacy'.

Hence, Kant's conception of the cosmological antinomy is quite complex. It implies an antinomy in the sense of modern logic, i.e., a contradiction that can be deduced from a system of propositions. In addition, it implies the corresponding semantic consequence that the underling cosmological theory is not satisfiable, i.e., that its concepts are empty, in contrast to the semantic presupposition mentioned above. Kant's account of the antinomy, however, is not exhausted by these logical and semantic aspects. For him, the epistemic justification of the antinomy is decisive. His claim that reason inevitably becomes entangled in the cosmological antinomy crucially depends on the epistemic aspects of the antinomy.

5.2.2 The Unity of Reason

Kant's conviction that the cosmological antinomy is inevitable is due to his account of reason, as the faculty of principles, in contrast to the understanding, as the faculty of concepts (A 299/B 356). This account of reason is closely related to the claim that reason is our cognitive faculty of establishing systems, and by its very nature strives for the completeness of cognition (see Willaschek 2018, 53–54; and Appendix A.1.4):

> Under the government of reason our cognitions cannot at all constitute a rhapsody but must constitute a system [...]. I understand by a system, however, the unity of the manifold cognitions under one idea. This is the rational concept of the form of a whole, insofar as through this the domain of the manifold as well as the position of the parts with respect to each other is determined *a priori*. The scientific rational concept thus contains the end and the form of the whole that is congruent with it. (*CPR*, A 832/B 860)

The connection between reason as our faculty of principles and reason as our systematizing faculty is closely connected to the issue that an idea or "concept made up of notions" is a second-order concept. Kant emphasizes the connection as follows:

> If the understanding may be a faculty of unity of appearances by means of rules, then reason is the faculty of the unity of the rules of understanding under principles. Thus it never applies directly to experience or to any object, but instead applies to the understanding, in order to give unity *a priori* through concepts to the understanding's manifold cognitions, which may be called "the unity of reason," and is of an altogether different kind than any unity that can be achieved by the understanding. (A 302/B 359)

Kant's claim about the systematic unity of reason provides a preliminary reply to the objection of circularity (see Sect. 5.1.3), according to which the cosmological antinomy may be a self-generated problem of transcendental idealism. The suspicion of circularity may be rebutted if the assumption that reason strives for systematic unity and completeness does not depend on transcendental idealism. In order to avoid circularity, the unity of reason as our systematizing faculty must not fatally depend on the key epistemic presupposition of transcendental idealism, the distinction between appearances (phenomena) and things-in-themselves (noumena). The logical reconstruction of the proofs below, as well as the final chapter of the book, will take up this issue in more detail (see Sects. 5.4, 6.2.2, and 6.3.3). Beyond the circularity objection, the question remains whether the proofs of the antinomy look valid from a pre-critical point of view (in Kant's terms, a position of transcendental realism).

5.2.3 A Self-Contradictory Cosmological Concept

Let us now look in more detail at the self-contradictory cosmological concept from which Kant's cosmological antinomy derives. It results from conflating an intellectual cosmological concept and the concept of the spatio-temporal world, hence conceiving of the world as both an appearance and a thing in itself. According to Kant, this leads to a contradiction inherent in the cosmological concept, to the self-contradictory concept of the sensible world as an intellectual totality of empirical parts and wholes, or of empirical grounds and consequences. This notion refers to the sensible world, as if it was an object of possible experience, even though as an intellectual totality of parts and wholes, or grounds and consequences, it is not an object of possible experience. Identifying the sensible world with an object gives rise to the epistemic category mistake of conceiving of the world in terms of an intellectual concept as an abstract *noumenon* and alike in terms of a sensitive concept as a concrete *phenomenon*, to take up the terminology of the 1770 *Dissertation*. In the *CPR*, this issue becomes less clear than in the *Prolegomena*. In §52b,c, Kant applies the example of the self-contradictory concept of a square circle to the mathematical antinomy:

> Now underlying the first two antinomies [. . .] is a contradictory concept of this type [. . .]. If I speak of objects in time and space, I am not speaking of things in themselves (since I know nothing of them), but only of things in appearance [. . .] and it is patently contradictory to say of a mere way of representing that it also exists outside our representation. (4:341–342)

> [. . .] the magnitude of the world with respect to space and time [. . .] must lie in itself, apart from all experience. But this contradicts the concept of a sensible world, which is merely a sum total of appearance, whose existence and connection [. . .] is itself nothing but a kind of representation. From this it follows that, since the concept of a sensible world existing for itself is self-contradictory, any solution to this problem as to its magnitude will always be false [. . .] The same holds for the second antinomy. (4:342)

The claim that "the concept of a sensible world existing for itself is self-contradictory" is the clue for understanding the source as well as the resolution of the cosmological antinomy. From a modern logical point of view, there are only contradictions between propositions. According to traditional (Aristotelian) logic, however, a contradiction can also arise within a concept. Accordingly, Kant explains the principle of contradiction in terms of non-contradictory predication:

> Now the proposition that no predicate pertains to a thing that contradicts it is called the principle of contradiction [...]. (*CPR*, A 151/B 190)

A concept composed of contradictory predicates is self-contradictory. Such a concept, like Kant's example of a "square circle" (*Prolegomena* §52b, 4:341), arises from attributing to an object defined by a predicate such as 'round' another predicate such as 'square' that is incompatible with its definition. In terms of Kant's distinction between analytic and synthetic judgments, a definition gives rise to analytic judgments, and predicates incompatible with it must stem from (illegitimate) synthetic judgments. In Kant's view, a self-contradictory world-concept arises from the transcendental illusion by attributing to the concept of the sensible world, predicates which are incompatible with its definition. Their attribution to the sensible world is induced by synthetic judgments which derive from the attempt to complete our cosmological cognition according the principles of reason; or, vice versa, the self-contradictory predication may arise from attributing to the concept of the intelligible world predicates which are incompatible with its definition, based on synthetic judgments which derive from applying the principles of the pure understanding to the appearances.

As noted above (in Sect. 5.2.1), for Kant the antinomy of pure reason has epistemic, logical, and semantic aspects. They lie in the persuasiveness of the thesis and antithesis proofs, the apparently sound logical deduction of the proof results, and the tacit semantic presupposition that the antithetic claims rely on a meaningful concept of the world. The self-contradictory predication takes place at three levels:

i. *Epistemic contradiction*: The world-concept inevitably conflates the sensible and the intelligible worlds of the 1770 *Dissertation*. It refers to the spatio-temporal world as a sum total of appearances, or as an aggregate of empirical phenomena, and at the same time to their totality as a thing in itself, that is, a noumenon.

ii. *Logical contradiction*: The world-concept conceives of the sum total of appearances in relational and alike in non-relational terms, in a kind of amphiboly of "concepts of reflection"[7] (A 260/B 316):

> In an object of the pure understanding only that is internal that has no relation (as far as the existence is concerned) to anything that is different from it. The inner determinations of a *substantia phaenomenon* in space, on the contrary, are nothing but relations, and it is itself entirely a sum total of mere relations. (A 265/B 321)

iii. *Semantic contradiction*: The world-concept has an ambiguous status as a first-order concept of the pure understanding and a second-order concept "made of notions". This is a type-theoretical mistake.[8] To refer to the empirical world as a sum total of appearances means to refer to a concrete aggregate and alike to an

[7]For the precise relationship between the amphiboly chapter and the antinomy, see Grier (2001).

[8]The parallels and differences between Kant's antinomy and the antinomies of set theory have been considered in particular by Zermelo (in Cantor 1932, 177) and Martin (1955, 55); for a detailed, critical discussion see Hallett (1984, 223–239).

abstract totality of things. As Kant explains in the *MFNS* for the idea of absolute space, this means to conflate "logical" and "physical universality":

> To make this into an actual thing is to transform the logical universality of any space with which I can compare any empirical space, as included therein, into a physical universality of actual extent, and to misunderstand reason in its idea. (4:482)

In order to understand the complex structure of the cosmological antinomy and its resolution, one has to unravel these three levels. Kant did not make this task easy, given that the crucial clues to his account of the self-contradictory world-concept are not found in the *CPR* but only in the *Prolegomena* and the *MFNS*.

5.2.4 The Resolution

In the *CPR*, Kant explains the origin of the antinomy starting from the intellectual cosmological concept of the world as an unconditioned totality that completes the series of all finite parts and conditions of the objects within the world. The concept of the sensible world of appearances comes into play by identifying the members of this series with the empirically given conditions of the objects of our cognition. To ask for the completion of such an empirical series gives rise to the transcendental idea of something unconditioned that makes the series of empirical objects and conditions complete. The antinomy then arises by inferring from the empirical partial series of objects and/or conditions to their ultimate non-empirical conditions,

> in accordance with the principle: **If the conditioned is given, then the whole sum of conditions, and hence the absolutely unconditioned, is also given,** through which alone the conditioned was possible. (A 409/B 436)

In the section *Critical decision of the cosmological conflict of reason with itself*, Kant emphasizes that this inference is "dialectical":

> The entire antinomy of pure reason rests on this dialectical argument: If the conditioned is given, then the whole series of all conditions for it is also given; now objects of the senses are given as conditioned; consequently, etc. (A 497/B 525)

The first (major) premise of the inference is a hypothetical judgment (9:120 and 9:129). Kant argues that it is possible to complete it in such a way that one obtains an analytic judgment, since the regress from the conditioned to its condition is demanded by the very concept of something conditioned:

> If the conditioned is given, then through it a regress in the series of all conditions for it is **given** to us **as a problem** [. . .]. (*CPR*, A 497–498/B 526)

Then he argues that for things in themselves the original non-completed major premise is true, whereas for appearances it is not. Finally, he claims that in the above "cosmological syllogism" the major premise takes the conditioned as a thing in itself, whereas the minor premise takes it as an appearance; their conflation gives rise to the logical fallacy of the ambiguous middle term, taking the middle term, the "conditioned", in different meanings (*Logic* §90, 9:134–135):

> From this it is clear that the major premise of the cosmological syllogism takes the conditioned in the transcendental signification of a pure category, while the minor premise takes it in the empirical signification of a concept of the understanding applied to mere appearances; consequently there is present in it that dialectical deception that is called a *sophisma figurae dictionis*. (A 499/B 527–528)

Hence, Kant locates the source of the cosmological antinomy in an equivocation concerning the conditioned objects of the series and the inference to their unconditioned ultimate condition. This equivocation emerges from the epistemic contradiction (i) in the self-contradictory cosmological concept sketched above (see Sect. 5.2.3)

5.3 Specific Structure

Kant specifies four kinds of incompatible predications of the world as an unconditioned totality of conditioned entities, and four pairs of corresponding antithetic judgments. Two of them concern the "mathematical" (spatio-temporal) aspects of the world, i.e., the extension of the world in space and time and the part–whole relations of the substances in the world. The other two concern the "dynamical" (nomological) aspects of the world, i.e., the causal relations between events and the necessity of the beings in nature.[9] The mathematical antinomy concerns the completion of the world, i.e., the spatio-temporal world as a totality. The dynamical antinomy concerns the completion of nature, i.e., the totality of the nomological conditions of the world.

5.3.1 The System of Cosmological Ideas

The four cosmological ideas or transcendental "world-concepts" derive from reason's quest for completeness. At first sight, they do not appear self-contradictory. To Kant, they were well known from the seventeenth- and eighteenth-century metaphysical debates:

1. "The **absolute completeness** of the **composition** of a given whole of all appearances" (A 415/B 443) gives rise to the idea of the universe as a spatio-temporal whole, as suggested in Kant's *Theory of the Heavens*, the precursor of modern physical cosmology.

[9]Here, Kant differentiates as follows between 'world' and 'nature'. The world is "the mathematical whole of all appearances and the totality of their synthesis in the great as well as in the small", whereas nature is the world "insofar as it is considered as a dynamic whole and one does not look at the aggregation in space or time" (*CPR*, A 418–419/B 446–447). In a corresponding footnote, Kant explains that here he means the concept of nature "taken substantively *(materialiter)*", in contradistinction to the concept of nature "taken adjectivally *(formaliter)*" (ibid.).

2. "The **absolute completeness** of the **division** of a given whole in appearance" (ibid.) gives rise to the idea of ultimate substances in the world, as suggested by the ancient atomists, Newton (1730, Query 31), and the dynamic atomism of Kant's *Physical Monadology* or Boscovich (1758), but also (in quite another sense) by Leibniz's *Monadology*.
3. "The **absolute completeness** of the **arising** of an appearance in general" (ibid.) gives rise to the idea of a spontaneous causality of freedom, which is crucial for morality, in contrast to the deterministic causality of nature, and
4. "The **absolute completeness** of the **dependence** of the **existence** of the alterable in appearance" (ibid.) gives rise to the idea of an absolutely necessary ultimate ground of the existence of all beings, which is closely related to the traditional proofs of the existence of God.

5.3.2 The Logical Oppositions

The cosmological ideas of the completeness of composition and division give rise to the mathematical antinomy, which consists in antithetic claims about the boundaries of the world in space and time and the existence of ultimate substances:

1. *Thesis*: "The world has a beginning in time, and in space it is also enclosed in boundaries." (A 426/B 454)
   Antithesis: "The world has no beginning and no bounds in space, but is infinite with regard to both time and space." (A 427/B 455)
2. *Thesis*: "Every composite substance in the world consists of simple parts, and nothing exists anywhere except the simple or what is composed of simples." (A 434/B 462)
   Antithesis: "No composite thing in the world consists of simple parts, and nowhere in it does there exist anything simple." (A 435/B 463)

The cosmological ideas of the completeness of the arising of appearances and the dependence of the existence of the beings in the world give rise to the dynamical antinomy, which consists in antithetic claims about the ideas of spontaneous causality or "self-activity (freedom)" (A 418/B 446) and the necessity of the beings in nature:

3. *Thesis*: "Causality in accordance with laws of nature is not the only one from which all the appearances of the world can be derived. It is also necessary to assume another causality through freedom in order to explain them." (A 444/B 472)
   Antithesis: "There is no freedom, but everything in the world happens solely in accordance with laws of nature." (A 445/B 473)

4. *Thesis*: "To the world there belongs something that, either as a part of it or as its cause, is an absolutely necessary being." (A 452/B 480)
   Antithesis: "There is no absolutely necessary being existing anywhere, either in the world or outside the world as its cause." (A 453/B 481)

It is instructive to analyze the antithetic claims in terms of the traditional square of logical oppositions, which derives from Aristotelian logic (Strawson 1952, 152), and was well known to Kant (*Logic* §§47–50, 9:116–117). Thesis and antithesis of the "mathematical" antinomy are pairs of *contrary* categorical judgments. In the second antinomy, a judgment of type A: 'All S are P' is opposed to a judgment of type E: 'No S is P' (where S is the subject, P a predicate). Singular judgments such as the claims about "the world" in the first antinomy also belong to these types of judgments, according to Aristotle's and Kant's logic. In contrast, thesis and antithesis of the "dynamical" antinomy are pairs of *contradictory* categorical judgments. In the third antinomy, a judgment of type I: 'Some S are P' is opposed to a judgment of type E: 'No S is P'. The thesis and antithesis claims of the fourth antinomy concerning the existence or non-existence of an absolutely necessary being are also contradictory.

This distinction between contrary and contradictory judgments is crucial for Kant's resolution of the cosmological antinomy. Kant locates the origin of the antinomy in the semantic aspects of traditional logic, that is, in the existential presuppositions of the antithetic claims. According to Aristotelian logic, a categorical judgment 'S is P' only has a truth value if its subject S exists. The existential presupposition of the antinomy of pure reason is that there is an object that corresponds to the cosmological idea. The opposite claims of the mathematical antinomy presuppose that the spatio-temporal world as a whole and its ultimate parts exist, while the opposite claims of the dynamical antinomy presuppose that there are ultimate causes and absolutely necessary beings in nature. Kant's resolution of both kinds of antinomy is that the corresponding existential presupposition is not satisfied.

For the contrary judgments of the mathematical antinomy, in Kant's view both opposing claims are false, given that they attribute properties to the non-existent object of a self-contradictory concept such as a squared circle, to take up his example of the *Prolegomena* (§52b, 4:341). For the contradictory judgments of the dynamical antinomy, in Kant's view both opposing claims may be correct, if one takes into account that they attribute contradictory properties to two different entities, namely to the intelligible and the sensible world, respectively. Then, they turn out to be subcontrary rather than contradictory (see Wolff 2017, 74–81). That is, in his view the self-contradictory cosmological concept of the spatio-temporal world as a totality is inevitable for the subjects of the mathematical antinomy, whereas for subjects of the dynamical antinomy it is not. For the latter, in 1781 he is still convinced that the solution of the 1770 *Dissertation* holds, whereas for the former he is not (see Sect. 4.3.3). Hence, according to the *CPR* and the *Prolegomena*, the mathematical antinomy is the genuine scandal of pure reason.

Kant considers the different logical structures of the mathematical and the dynamical antinomies as a clue to their different possible resolutions. Given his view that the antinomy emerges from a self-contradictory world-concept, there are two logical possibilities for resolving it: first, to reject the assumption that its subject exists as a given object and to refute any conclusion about its properties; second, to disentangle the self-contradictory cosmological idea into two distinct subjects to which the contradictory predicates may be attributed without getting into any conflict with each other. Kant argues that the path to either one or the other solution is indicated by the kind of opposition of the antithetic propositions. That is, if they are contrary, both judgments are false; if they are contradictory, both may be considered true, if understood in different respects.

5.3.3 The Apagogic Proof Structure

Kant thinks that the cosmological antinomy is inevitable, given that the equivocations in the cosmological syllogism (see Sect. 5.2.4) are an "entirely natural mistake of common reason" (*CPR*, A 500/B 528). As long as the difference between appearances and things in itself is not revealed from the point of view of transcendental idealism, the equivocations and the corresponding self-contradictory features of the cosmological idea remain undiscovered. From a non-critical point of view, the thesis claims can be substantiated with apparently good reasons by objections to the antithesis claims, and vice versa. In this way, pure reason becomes entangled in an "apagogic" proof procedure, which alternately makes it defend the thesis or the antithesis by refuting the respective opposite assumption on pain of contradiction (see *CPR*, A 340/B 398, quoted above in Sect. 5.2.1).

The "apagogic" proof procedure is based on the skeptical method (see Sect. 4.3.2), the only methodological tool available to Kant around 1770 after the collapse of his pre-critical system and his analytic method of metaphysics. His use of the skeptical method in the 1770s crucially contributed to his conviction that cosmological cognition gives rise to contradictions. In the *CPR*, he finally called the antithetic procedure itself the "skeptical method". It is worth quoting the passage at length, given that it again shows his distance from Hume's skepticism:

> This method of watching or even occasioning a contest between assertions, not in order to decide it to the advantage of one party or the other, but to investigate whether the object of the dispute is not perhaps a mere mirage at which each would snatch in vain without being able to gain anything [...], can be called the **skeptical method**. It is entirely different from **skepticism**, a principle of artful and scientific ignorance that undermines the foundations of all cognition [...]. For the skeptical method aims at certainty, seeking to discover the point of misunderstanding in disputes that are honestly intended and conducted with intelligence by both sides [...]. (A 423–424/B 451–452)

Unlike skepticism, the skeptical method does not aim at proving a lack of knowledge, but at certainty about the viability of our theories. In particular, it is intended to reveal internal contradictions in the formation of theories. The skeptical method does not prove anything in the positive, it only destroys. As long as it is not deliberately used as a critical tool, however, it gives the illusory impression of confirming the opposite of what has been destroyed.

Kant's "apagogic" proof procedure serves to show how prototypical philosophical positions inevitably come into conflict with their own presuppositions; and this is then seized upon by their opponents. The arguments employed in the proofs stem from the philosophical tradition; or, as Kant puts it, he takes them from the "battlefield" of metaphysical arguments that in his view prevent metaphysics from taking "the secure course of a science", according to the famous preface to the second edition of the *CPR* (B vii, xv). Kant himself associates the thesis claims with "dogmatism" or "Platonism", whereas he attributes the antithesis claims to "empiricism" or Epicureanism (A 465–471/B 493–499). The proofs are to show that the thesis party argues in rationalist terms, whereas the antithesis party argues in terms of empiricism, materialism, or naturalism (Kant had already associated Epicureanism with naturalism and materialism in the *Theory of the Heavens*, see 1:221–228). Both parties then play the concepts of the intelligible and the sensible worlds off against each other, without, however, being able to do so in a consistent way.

I will analyze the specific arguments in detail below (see Sect. 5.4). The thesis proofs mainly employ concepts and principles of the pure understanding, whereas the antithesis proofs also appeal to the empirical conditions of appearances in space and time, and, in the first antinomy, in particular to verificationist arguments.[10] However, it should already be noted that *both* parties are also compelled to make use of certain naturalistic arguments, given that their common subject of dispute is the spatio-temporal world as a sum total of appearances and their conditions.

In the proofs, Kant takes up traditional philosophical arguments for and against the existence of a beginning of the world in space and time, simple substances as the constituents of matter, spontaneous effects in nature, as well as an absolutely necessary ground for existence of the beings in the world. It needs to be examined whether these arguments are indeed based on a position of "transcendental" or metaphysical realism and are thus conclusive independently of Kant's transcendental idealism. In all four kinds of antinomy, the antithetic claims correspond to conflicting philosophical positions that have repeatedly been advocated in the history of metaphysics, from ancient atomism and Aristotle to Descartes and Newton, the Leibniz-Clarke debate, the French materialists, Christian Wolff and Euler to Kant's own pre-critical views.[11] Characteristic of the cosmological debates

[10]See Sect. 5.4.1 and my discussion of Allison (1983) vs. Allison (2004), Grier (2001) below in n. 21.

[11]Al Azm (1972) correctly observes that the arguments do not exactly correspond to rationalist and empiricist positions. Against the background of the Leibniz–Clarke debate, he then associates the thesis positions with Newton's views and the antithesis positions with Leibniz's views, overlooking that both Newton (or Clarke) as well as Leibniz themselves put forth rationalist as well as empiricist (or phenomenological) arguments, despite all their differences. See also the references to Bayle and Crusius in Heimsoeth (1960, 260), and the detailed discussion in Grier (2001, 182–229).

of the eighteenth century, in particular in the correspondence of Leibniz and Clarke, is the consolidation of one's own position by objections against the opposing line of argument. In doing so, each of the combatants tries to invalidate the opposite position under his own metaphysical conditions. In this way, no objective clarification of philosophical problems is achieved, and at best arguments deriving from incompatible premises are put against each other.

From Kant's critical point of view, these traditional arguments, as presented in the proofs, are fallacious, given that they are based on a self-contradictory cosmological concept and give rise to equivocations about the sensible or intelligible world to which this concept is supposed to refer. The "dialectic" character of such arguments is not obvious, in his view: Kant himself only discovered it step by step by using the skeptical method after 1769/1770, as indicated in his notes on metaphysics from the late 1770s (in particular N. 5037, and N. 5116; see Sects. 4.3.2 and 4.3.3). The *CPR* finally presents the arguments of the traditional cosmological disputes, or prototypical versions of them, as due to an inevitable internal conflict of reason with itself, which results from extending the cognition of the world of appearances in an unrestricted way, in accordance with the principles of reason.

For the rigour of the proofs the use of the skeptical method is obviously irrelevant. Instead of the indirect demonstrations via *reductio ad absurdum*, all the proofs could also have been carried out directly, as Hegel already noted in his *Logic* (Hegel 1832, 21.180–189), at least to the extent that the *reductio* does not merely consist in refuting the opponent's position from one's own presuppositions. However, for Kant's own goal of demonstrating self-contradictory predications, the indirect proofs are by no means irrelevant. By using the "apagogic" proof procedure, Kant wants to prove the traditional arguments supporting the thesis and antithesis claims to be convincing. At the same time he wants to expose the cosmological syllogism as a self-induced deception of reason, i.e., as transcendental illusion.

The apagogic proof structure of the antinomy is thus due partly to Kant's own long route to establishing the cosmological antinomy, partly to the cosmological debates of the eighteenth century, and partly to the concern to show that it is precisely the most powerful pros and cons of these debates that indicate their self-contradictory foundations. However, this approach raises two problems.

1. The first is the circularity objection already mentioned (see Sect. 5.1.3). The pair of proofs, which is meant to underpin each version of the antinomy, is intended to demonstrate a provable contradiction. In order to do so, it will not suffice to make the opponent's point of view absurd from his own critical position alone; it is necessary to do so on the basis of common pre-critical assumptions. Kant has to show that the proofs on each side of the antinomy seem to have true premises, given that for him syllogisms are only valid if the premises are true (*Logic* §59, 9:121). In addition, they should result partly from analytic and partly from synthetic judgments *a priori*, given that Kant wants to demonstrate that they are based on self-contradictory concepts. In order to look conclusive, the proofs ought to be based on the principles of the pure understanding. Nevertheless, they must also be due to a transcendental illusion, because otherwise the antinomy could not be resolved. The illusion of a contradiction can only arise by attributing to an object,

through illegitimate synthetic judgments, *a priori* predicates which contradict its very concept (if not each other). Kant can only achieve the intended proof results, however, if the proofs do not fatally depend on his own epistemology, but are also persuasive independently of it. In particular, they have to appear valid from the point of view of "transcendental realism" (i.e., metaphysical realism concerning the cognition of things in themselves). Otherwise, the cosmological antinomy is not an inevitable conflict of reason with itself, but a self-generated problem of Kant's transcendental philosophy.

2. The other problem is the fundamental difference between the structure and the proof results of the mathematical and the dynamical antinomy (see Sect. 5.3.2). Kant resolves the antinomy from the point of view of transcendental idealism by demonstrating that the proofs use equivocal concepts. For the mathematical antinomy, he argues that the thesis and antithesis are both false, given that they consist in contrary judgments about a subject that does not satisfy the semantic presuppositions of the predicates attributed to it. Correspondingly, he considers the mathematical antinomy to derive from the illegitimate presupposition that the spatio-temporal world as a sum total of appearances exists. For the dynamical antinomy, in contrast, he argues that the thesis and antithesis are both legitimate and may both be true. Here, the antithetic judgments are contradictory, but the contradiction may disappear by reference to different subjects. Correspondingly, he suggests that the thesis and antithesis claims of the dynamical antinomy refer to the distinct domains of appearances and things in themselves. Hence, according to the resolutions, only the mathematical antinomy is inevitable, whereas the dynamical is not. As he puts it in the *Prolegomena*, the mathematical antinomy presupposes that

> something self-contradictory (namely, appearance as a thing in itself) would be represented as being unifiable [...], (§53, 4:343)

whereas the dynamical antinomy presupposes that

> something that is unifiable is represented as contradictory [...]. (§53, 4:343)

Why does Kant assume that the antinomy-avoidance strategy of 1770 does not work for the mathematical antinomy of 1781, whereas he re-establishes it for the dynamical antinomy? And why does he nevertheless consider the latter, too, as a genuine cosmological antinomy?

5.3.4 Remarks Concerning the Logical Reconstruction

The claims and proofs of Kant's cosmological antinomy need to be analyzed at various levels. How persuasive are Kant's proofs in the context of traditional logic and eighteenth-century metaphysics, on the one hand, or from a modern point of view, on the other? Concerning his own views, Kant argues on two levels, both of which have to be taken into account. In particular, one has to distinguish the pre-critical position of transcendental realism from which he derives the antinomy, and the critical position of transcendental idealism from which he resolves it. Concerning the validity of his arguments, it is not only the relation between traditional and modern logic that deserves attention, but also his account of theory, experience, and appearances employed in the antinomy. In addition one has to take into account that Kant's views about the theories and experiments of natural science are only partially comparable with that of current science and philosophy of science. The question of whether the antinomy is still relevant today is beyond the topic of the present book and is therefore set aside in the following logical reconstruction.[12]

5.3.4.1 Experience and the Appearances

A consideration of Kant's use of the term 'appearances' in the antinomy doctrine is also important for a correct understanding of his proofs. Here, the difference between the pre-critical and the critical points of view is decisive.

From a pre-critical point of view, Kant did not differentiate between experience and the appearances. He understood them in a contemporary empiricist and sensualistic sense, that is, as sensory perception. In his pre-critical writings, he used the term 'appearance' either for natural phenomena, as in the *Theory of the Heavens* (1:248, 1:254), or for the ideas of inner experience, as in the *Prize Essay* (2:286). Indeed, the seventeenth/eighteenth-century rationalists and empiricists, from Descartes and Leibniz, to Locke, Berkeley, or Hume, shared this sensualistic concept of experience as sensory perception. In Kant's proofs of the cosmological antinomy, it is a common precondition of both prototypical pre-critical positions.

From a critical point of view, however, Kant differentiates between sensory perception and experience. Whereas the former corresponds to the sensualistic concept, the latter is based on the synthetic functions and principles of the pure understanding. It is only with these principles that the assumptions of full-fledged transcendental idealism about the spatio-temporal world could enter into the proofs. It should be noted that Kant could even use his 1770 distinction between appearances (*phaenomena*) and things in themselves (*noumena*) in the proofs, because this distinction on its own does not make the proofs dependent on transcendental idealism. The "appearances" of the 1770 *Dissertation* are sensory perceptions, whereas the "appearances" to which the cosmological concept of 1781 refers are the objects of experience. The "appearances" of 1781 presuppose the principles of the pure understanding, which are decisive for his critical account of objective cognition; the "appearances" of 1770 do not. However, in the 1770 *Dissertation* Kant had already differentiated sensory perception from the type of experience that depends on the logical use of the understanding (2:394), before his critical theory of objective cognition and the cosmological antinomy took shape.

[12] I attempted to give some hints as to its relevance for current science and its metaphysical generalization in Falkenburg (2000, 341–350), and Falkenburg (2004, 2005). See also Wind (1934).

Hence, the question of whether the proofs *fatally* depend on this key distinction of transcendental idealism is delicate and needs careful investigation.

In the following, I assume that Kant may indeed employ his views of 1770, including a sensualistic concept of appearances and verificationist arguments, as arguments in favour of transcendental realism. The proofs can thus make use of the premise that all things in space and time are objects of possible sensory experience, or appearances, without thereby abandoning the pre-critical point of view. This issue is crucial with regard to the circularity objection (see Sect. 5.1.3). Several interpreters fail to see that the arguments in the antithesis proofs rely on a sensualistic concept of appearances rather than on Kant's critical theory, and accuse Kant of offering proofs that depend on the preconditions of transcendental idealism.[13]

5.3.4.2 Traditional and Modern Logic

For Kant, a metaphysical theory is a doctrine made up of concepts, judgments, and syllogisms in the sense of traditional logic. Nowadays, the logical stringency of such a theory would be examined by reconstructing it in terms of first-order logic. The reconstruction has to take into account that in traditional logic, syntactic and semantic claims are interwoven. A traditional categorical judgment attributes or denies a predicate to a subject-matter, the existence of which it presupposes. If and only if the existential presuppositions of categorical judgments are satisfied, traditional categorical judgments and the square of logical oppositions can be unambiguously translated into modern first-order logic (Strawson 1952, 173–178).

The proofs of the cosmological antinomy, which are based on a pre-critical point of view, make precisely such existential presuppositions concerning the spatio-temporal world. Therefore, it is justified to reconstruct them in terms of modern first-order logic. However, from a critical point of view their existential presuppositions turn out to be fallacious. Kant presents the *proofs* of the antinomy on the *pre-critical condition* that the existential presuppositions concerning the respective cosmological concept *are met*. Then, he *resolves* the antinomy from his *critical point of view* as follows. In the mathematical antinomy the existential presuppositions are *not met*, whereas in the dynamical antinomy they *do not refer to one-and-the-same domain*. Discrepancies between Kant's way of presenting the

[13]See for example Kreimendahl (1990, 424–430). Schmucker (1990, 116–117), too, claims that Kant already takes the critical position when he relates the cosmological ideas to sensory perception. In my opinion, he conflates Kant's critical conception of the phenomena with the common sensualistic concept of experience of the seventeenth/eighteenth-century schools. Guyer (1987, 386–387) accuses Kant of either reducing truth to empirical confirmability, or rendering the argument in favour of transcendental idealism circular. His analysis of the proofs (ibid., 405–412) suffers from not distinguishing between the point of view of transcendental realism, which underlies the thesis and antithesis proofs, and the critical point of view, which gives rise to the resolution of the antinomy.

antinomy and a modern logical reconstruction may therefore be expected for this resolution above all.

Concerning the question of whether Kant's proofs are conclusive, a modern reconstruction should not obscure their crucial semantic features by the logical form of the reconstruction. A purely syntactic reconstruction makes the "transcendental illusion" of the above-mentioned existential presuppositions unrecognizable.[14] According to Kant's resolution, the antinomy stems from a semantic fallacy, or to be more precise: from the logical fallacy of an equivocation in the middle term of a syllogism (see Sect. 5.2.4). This logical fallacy is in his view inevitable from the point of view of transcendental realism; and he resolves it by taking into account the difference between appearances and things in themselves. In order to capture this crucial issue, I will use formal tools as moderately as possible.

5.3.4.3 Logical, Semantic, and Epistemic Facets of the Antinomy

Let me summarize the logical, semantic, and epistemic aspects of the cosmological antinomy, as understood from the point of view of Kant's resolution. They are closely interwoven, and to unravel them is not easy.

1. *Epistemic aspects*: The cosmological antinomy arises from conflating the intellectual cosmological concept of the spatio-temporal world as an extensional and nomological totality with the sensitive concept of the world as a sum total of appearances. The result is a transcendental conflation of the world as an abstract logical conception with a concrete object of experience. This in turn leads to the "dialectic" cosmological syllogism, which infers from something empirically conditioned to something unconditioned and gives rise to a self-contradictory cosmological concept. In all four versions of the antinomy, the "apagogic" proofs for thesis and antithesis draw their persuasive power from refuting the opposite claim via *reductio ad absurdum*.
2. *Semantic aspects*: The contradictory predication always results from proving the claims of the thesis and the antithesis under the presupposition that both refer to one-and-the-same existing object, the spatio-temporal world. The proofs, however, are based on equivocal predicates attributed to the sensible or the intelligible world, respectively. The resulting antinomy is resolved by demonstrating that the equivocal predication gives rise to the logical fallacy of a *sophisma figurae dictionis*, i.e., an equivocation in the middle term of a logical syllogism. Kant's conclusion is that therefore either the existential presupposition or the claim of unambiguous reference is illegitimate and must be false.

[14]Malzkorn (1999) reconstructs the semantic aspects of the antinomy as far as possible by syntactic tools, including temporal logic and existential statements. Guided by a "principle of charity" (ibid., 118–119) he overlooks the fact that each kind of antinomy emerges from a crucial *semantic* equivocation which Kant wants to reveal.

3. *Logical aspects*: Each specific kind of antinomy appears as strictly provable (i.e., logically derivable from analytic judgments and synthetic judgments *a priori*), as long as the logical fallacy is not revealed. The contradictory predication is always due to the attempt to infer from conditioned entities to an unconditioned entity. The conditioned is conceived as relational (with regard to our cognition), the unconditioned as non-relational (with regard to our cognition). The conflicting claims will thus result partly from monadic predication via non-relational predicates, and partly from non-monadic predication via relational predicates. From a logical point of view, the only source of attributing contradictory non-relational or relational predicates to the world is that the proof premises are either not satisfied or ambiguous.

5.4 Logical Analysis of the Proofs

The logical reconstruction is based on Kant's resolution of the antinomy. Otherwise, it would be hard to make transparent what Kant wants to demonstrate and how he proceeds. The following reconstruction confirms Kant's claim that the proofs in all four specific versions of the antinomy crucially depend on equivocal predicates attributed to the sensible or the intelligible world, respectively. To make it clear that Kant does indeed present the same general antinomy in two pairings of specific versions, I attempt to separate the syntactic and semantic claims of the mathematical and the dynamical antinomy in analogous ways.[15] The reconstruction makes use of the following logical tools:

i. first-order predicate logic,
ii. a transitive, asymmetric, and irreflexive relation $<$, which applies to appearances and which has a different interpretation in each version of the antinomy;
iii. a transitive, asymmetric, and irreflexive relation $\prec$, which denotes the metaphysical relation between an object and its sufficient reason; and
iv. the operator $\Diamond$ ('*it is possible that*') of modal logic.

The domain of the logical predicates is a logical universe W that corresponds to the world as a whole. For the intelligible world, W is considered as a formal or symbolic domain of logical individuals $x, y, z \ldots$. In contrast, for the sensible world the logical individuals of W are considered as spatio-temporal appearances. In the four versions of the antinomy, the relation $<$ is interpreted as follows:

[15] A thoroughly syntactic reconstruction of the semantic claim cannot capture these features. It has to violate the proof principles of consistency. Not all of Kant's proof premises can be consistent with one another. As regards Kant's concerns for a "dialectical" logic of transcendental illusion it is sufficient that the premises *seem to be* satisfied from the point of view of transcendental realism. Uncovering the transcendental illusion shows that indeed they *are not*. The syntactic reconstructions of Malzkorn admit of this insight only for the dynamical antinomy; see Malzkorn (1999, 199–200 and 221).

1. '*beyond*' (spatial interpretation) and '*earlier than*' (temporal interpretation)
2. '*proper part of*' (mereological interpretation)
3. '*cause of*' (causal interpretation)
4. '*ground of the existence of*' (ontological interpretation)

W is an abstract set or class of non-relational individuals. For Kant, relations belong to the sensible world, given that for him the "*substantia phaenomenon* is a sum total of mere relations" (*CPR* A 265/B 321). From a logical point of view, this does not preclude defining relations on W also for the *intelligible* world, considered as a logical universe. Reason's quest for the extensional and nomological completion of the *sensible* world then gives rise to the concepts of a least upper bound and a greatest lower bound for the individuals of W, with regard to the relation $<$.

The specific versions of the cosmological antinomy then derive from asking for the respective supremum or infimum of W, considered as the spatio-temporal world, with regard to $<$, in the respective interpretation. The thesis side claims that they belong to W, as its maximum or minimum. In contrast, the antithesis side claims that a maximum or minimum does not exist within W. The proofs result from confounding the sensitive and the intellectual cosmological concepts, by tacitly identifying the sensible world as a concrete multitude of appearances with an abstract multiplicity, i.e., the class of appearances $\{x \mid x \in W\}$. This is the type-theoretical mistake in the self-contradictory cosmological concept mentioned above (see Sect. 5.2.3). The four antinomies then derive from considering the world (or nature) as a concrete spatio-temporal object as well as an abstract class of such objects. The reasons to do so lie in the tacit semantic assumptions of the proofs, which give rise to equivocations that make the proof results plausible.

The structural difference between the mathematical and the dynamical antinomy is also based on the semantic aspects of the antinomy. Compared to the antinomy-avoidance strategy of 1770, Kant may justify the different resolutions as follows. For the mathematical antinomy, it is compelling to interpret the part–whole relation in spatio-temporal terms, including its mereological maximum (or supremum) 'world' and its mereological minimum (or infimum) 'simple substance'. Here, the "subreptic axiom" of 1770 concerning the conflation of the sensitive and intellectual cosmological concepts now seems inevitable to Kant, in contrast to his expectations of 1769–1770 (see Sect. 4.3.1). For the dynamical antinomy, however, he still considers the thesis and antithesis claims as untenable "subreptic axioms" in the sense of his 1770 cosmology. Here, he considers it possible to interpret the logical relation of ground and consequence in different ways for the sensible and the intelligible world, and hence to separate the respective domains of the thesis and the antithesis.

5.4.1 The Antinomy of Composition

The conflicting claims of the first antinomy refer to the beginning of the world in time and the spatial boundaries of the world. In his presentation, Kant seems to equate the predicates 'finite' (*endlich*), 'limited' (*beschränkt*), and 'bounded' (*begrenzt*), as well as their opposites 'infinite' (*unendlich*), 'unlimited' (*unbeschränkt*), and 'unbounded' (*unbegrenzt*). From a modern point of view, these predicates should be differentiated and mathematically specified. Kant already saw differences between (in)finiteness and having (no) boundaries, but he apparently regarded them as irrelevant for the formulation of the first antinomy.

According to the Transcendental Aesthetic of the *CPR*, space, as well as time, is "an infinite given magnitude" (B 39), and "every determinate magnitude of time is only possible through limitations of a single time grounding it" (B 47–48). Any limited part of space or time represents an infinite but limited magnitude. From Kant's critical point of view, such magnitudes are given as concrete wholes by intuition, but cannot be understood as a concrete multitude of points. According to the *CPR* neither space nor time is a real composite (*compositum reale*). Their composition is ideal and points are not components of space, but mere boundaries in space. In the doctrine of the antinomy, Kant also emphasizes that "mathematical points [. . .] are simple but are boundaries rather than parts of space" (A 439/B 467), in his remark to the antithesis proof of the second antinomy. One may argue that here he employs a precondition of transcendental idealism. He obviously does.

However, he had already made the same point in the 1770 *Dissertation*, in the corollary to his exposition of space and time as pure intuitions. There, he claimed for space and time,

> following the paradigm of sensitive intuition, it is rather the case that *the infinite contains the ground* of each *part* which can be thought, and, ultimately, the ground of the simple, or, rather, of the *limit*. For it is only when both infinite space and infinite time are given that any definite space and time can be specified by *limiting*. (2:405)

He also knew about the possibility of non-Euclidean geometry, in the sense of a non-contradictory "concept of a figure that is enclosed between two straight lines" (*CPR*, A 220/B 668), but claimed that such a figure has no objective reality (Friedman 1992, 81, n. 42, and 92–93). Speculations about other forms of space beyond three-dimensional Euclidean geometry can already be found in the *True Estimation* (1:23–24). He mentions both issues in the 1770 *Dissertation*, where he takes the properties of Euclidean 3-space as evidence for the nature of space as a pure intuition:

> That space does not have more than three dimensions, that between two points there is only one straight line, that from a given point on a plane surface a circle can be described with a given straight line, etc.—none of these things can be derived from some universal concept of space; they can only be *apprehended* concretely, so to speak, in space itself. (2:402–403)

He always considered such concepts of space as logically possible, but his 1770 theory of space and time made him think that they were irrelevant for the cosmological use of mathematics, given that they have no real possibility (as he put it in his pre-critical terms). In twentieth-century physics, the restriction of physical geometry to Euclidean geometry turned out to be obsolete. Kant obviously could not know that models of a finite geometry without (internal) boundaries would one day become relevant for physics.

For the logical structure of his first antinomy, however, the possibility of such models is indeed significant (see Mittelstaedt and Strohmeyer 1990 and the literature discussed there). Due to the different meaning of the predicates '(in)finite' and '(un)bounded', from a modern point of view the thesis and antithesis claims regarding space and time do not have exactly the same meaning. In addition, a model according to which physical spacetime is finite, but has no internal boundaries, escapes Kant's antinomy and his resolution, *without* any need to employ the epistemological principles of transcendental idealism (as already pointed out in Wind 1934).

Here, I set aside the difference between these predicates and reconstruct the antinomy for the pair of predicates 'bounded'/'unbounded', also setting aside the problem that the additional antithesis claim of an infinite spatio-temporal world may introduce an additional meaning. Thesis and antithesis then attribute contradictory predicates to the spatio-temporal world as a whole. The respective thesis (antithesis) claims correspond to the following propositions (*CPR*, A 426–427, B 454–455):

$P1_T$: The world has a (no) beginning or boundary in time.
$P1_S$: The world has (no) boundaries in space.

Thesis and antithesis are singular judgments. From a modern logical point of view, their claims relate as contradictory propositions Fa and $\neg Fa$, in which the individual constant a denotes the world as a logical individual. In the context of traditional logic, however, singular judgments are considered as universal (9:598), that is, to be reconstructed as $\forall x Fx$. For Kant, the thesis and antithesis claims of the first antinomy therefore represent a universal affirmative judgment and a universal negative judgment (of type A and type E), which according to the traditional logical square of oppositions are not contradictory, but contrary (Strawson 1952, Ch. 6).[16] Logically, contrary predicates may both be false. Kant's resolution indeed finally proposes to escape the antinomy in this way.

However, if one takes the thesis and antithesis claims as judgments that refer to the sensible world, as a logical universe of W of appearances $x, y, \ldots$ for which spatio-temporal relations hold, they become contradictory propositions of the logical form $\forall x Fx$ and $\neg\forall x Fx$ (see below). For the difference between singular and universal propositions, the reconstruction in modern first-order logic hence deviates from the logical form of Kant's first antinomy.

The beginning of the world in time or a boundary in space are an *infimum* or a *supremum* of W with regard to the relation $<$, interpreted in temporal or spatial terms. In the temporal version of the thesis ($T1_T$) or antithesis ($A1_T$), $<$ is interpreted as '*earlier than*'. In the spatial version of the thesis ($T1_S$) or antithesis

[16]Malzkorn (1999, 128) uses a predicate 'non-F' in order to preserve the contrary relation of thesis and antithesis.

($A1_S$), $<$ is interpreted as '*beyond*'. The different interpretations can obviously make no difference for the logical reconstruction of the antithetic claims and their proofs. Following van Benthem (1983, 32–33), we note that the formal antithetic propositions of the thesis ($T1_{T,S}$) and the antithesis ($A1_{T,S}$) should be reconstructed as follows, and turn out to be contradictory:

($T1_{T,S}$) $$\exists x \neg \exists y \ (y < x)$$

($A1_{T,S}$) $$\forall x \ \exists y \ (y < x)$$

According to traditional logic, the outer quantifiers give rise to an I-judgment (particular and affirmative) and an E-judgment (universal and negative), which are contradictory, too. With the one-place predicate B

(D_0) $\quad Bx \leftrightarrow \neg \exists y \ y < x$ $\quad$ ('*is a beginning*' or '*is a boundary*')

thesis ($T1_{T,S}$) and antithesis ($A1_{T,S}$) become:

($T1_{T,S}$) $$\exists x \ Bx$$

($A1_{T,S}$) $$\neg \exists x Bx \ \text{ (or: } \forall x \neg Bx)$$

For Kant, the thesis and the antithesis are contrary. Formally, his view is due to his considering the world as an individual that is subject to two contradictory predicates, which give rise to contrary singular instead of particular judgments.

Kant presents the proofs for the spatial and the temporal interpretation of the relation $<$ separately, but in a completely analogous way. On the thesis side, the core of both proofs is an argument based on the concept of the infinite (*argument of infinity*). On the antithesis side, an argument based on the ideality of space and time is crucial (*argument of ideality*). Kant employs both arguments apagogically. The infinity argument serves to support the assumption that the world has boundaries in space and time by refuting the antithesis; the ideality argument serves to support the infinity of the world in space and time by refuting the thesis.

5.4.1.1 Thesis Proof: The Argument of Infinity

The argument of infinity aims at supporting the thesis by refuting the antithesis. For the temporal version of the antinomy, Kant presents the argument in the form of the traditional paradox of an infinite world age. His argument runs as follows. The concept of an infinite world age means that an infinite sequence of temporal states has elapsed until the present day. This in turn means that the present moment could not have been reached in the greatest time period; while on the other hand, the present is the very time that has been reached right now. Through a kind of measurement consideration, Kant transfers this argument to the infinity of the world in space.

As one can show, in both cases the paradox only arises from a specific concept of infinity which is incompatible with the antithesis. According to it, 'infinite' means 'never completely given by successive synthesis', or, as Kant expresses it in the proof of the thesis: the "infinity of a series consists precisely in the fact that it can never be completed through a successive synthesis" (*CPR*, A 426/B 454). According to this concept of infinity, which corresponds to the *ordinal* concept of the infinite, the present cannot be reached by the passing of an infinite amount of time. Analogously, it is impossible to measure the extension of an infinite space, because "we can think of the magnitude of a quantum [...] only through the completed synthesis, or through the repeated addition of units to each other" (A 426–428/ B 454–456).

In reconstructing the formal propositions of the proofs step by step, I follow van Benthem (1983, 32–33), without, however, sharing his view that Kant *naively* conflated the ordinal and the cardinal concepts of infinity:

(1) Let us assume against the thesis,

  ($A1_T$) the world has no beginning in time, and
  ($A1_S$) no boundaries in space:

$$\neg\exists x \neg\exists y \; y < x$$

(2) Then, up to a given point in time an infinite series of states of things in the world has passed (from ($A1_T$)), and the world is an infinite given whole of simultaneously existing things (from ($A1_S$)):

$$\forall x : \{y \mid y < x\} \text{ is infinite}$$

(3) [However, the concepts of a series and of a magnitude imply the following meaning (I) of 'infinite' and (U) of 'unbound', in contrast to what is given as a finite or bounded entity within intuition:]

  (I) For a series, 'infinite' means that it cannot be completed through successive synthesis.
  (U) For a magnitude, 'unbound' means that it cannot be completed through successive synthesis.

  Hence:

$$\forall x : \{y \mid y < x\} \text{ has no least upper bound}$$

(4) Therefore, no infinite time series can have elapsed up to the present (from (I)), and "the successive synthesis of the parts of an infinite world would have to be regarded as completed", but "in the enumeration of all coexisting things" by

this successive synthesis "an infinite time would have to be regarded as elapsed, which is impossible" (from (U) and (I)):

$$\exists x\{y \mid y < x\} \text{ is not infinite/unbound}$$

(5) Therefore the world must have a beginning in time and boundaries in space (*modus tollens* results in $\neg(\mathrm{A}1_T)$ und $\neg(\mathrm{A}1_S)$):

$$(\mathrm{T}1_{T,S}) \qquad \exists x \neg \exists y\ (y < x) \qquad \text{or:} \qquad \exists x\ Bx$$

Step (1) consists in the introduction of premises, while step (2) is a logical consequence of (1). In the decisive step (3), the ordinal concept of the infinite comes into play, which Kant, in another crucial step of his critical turn, had seen since 1773–1775 as in conflict with a cardinal infinite conception of the universe. The *Dissertation* of 1770 still understood the ordinal concept of infinity in the sense of a mathematical limiting size which is larger than any given number, and Kant did not consider it as incompatible with the assumption that the intelligible world is actually infinite (2:388–389; see Sect. 4.4). In 1775, on the other hand, he defined the ordinal concept of infinity no longer as a mathematical notion, but as an epistemic limiting concept, which depends on the synthetic faculties of the understanding (N. 4780, 17:725; see Sect. 4.4). From now on, he understood infinity as the absolute impossibility of a complete synthesis of the composition or decomposition of a given object. Hence we have to ask whether the thesis proof presupposes a crucial element of Kant's critical epistemology, which goes substantially beyond his 1770 theory of space and time as pure forms of intuition. The issue is not made any easier by the fact that Kant refers to (I) as the "true (transcendental) concept of infinity", in the remark on the thesis proof, which he then explains in the corresponding footnote as a "multiplicity (of given units) that is greater than any number", i.e., "the mathematical concept of the infinite" (A 432/B 460), as defined in the *Dissertation* (2:388 n.).

In order to clarify the question which infinity concept the proof actually employs, let us first consider the time argument. Here, step (3) indeed makes use neither of the mathematical concept of infinity of 1770 nor of the epistemic concept of infinity found in the later notes on metaphysics. Kant rather appeals to the way in which his readers conceive of an infinite series which is successively generated in time. According to him, the mathematical concept of an infinite sequence corresponds to the ordinal infinite and not to the cardinal concept of an infinite given set or number, which the dogmatic metaphysician assumes when he claims that "up to every given point in time an eternity has elapsed" (A 426/B 454).[17] Accordingly, I interpret step (3) as an analytic argument, i.e., as a conclusion that derives from the very concept

[17] According to Allison (2004, 367–371), this is an assumption of transcendental realism, which presumes that the spatio-temporal world is a *totum syntheticum* and contradicts the definition (I) of infinity. My reconstruction of the thesis proof essentially agrees with Allison's interpretation; see also his refutation of the objections raised by Russell (1903), Moore (1953), Strawson (1966), and Bennett (1974).

of a series. The ordinal concept of infinity is contained in the concept of an infinite time series, because any infinite series or sequence is ordinally and not cardinal infinite, according to this concept.[18]

The space argument is also analytic. Like the concept 'age of the world', the concept 'magnitude of the world' refers to a given whole or *totum syntheticum* (Allison 2004, 369–370). In complete analogy to the time series argument, Kant argues against this concept that 'infinite' for a spatial whole means 'cannot be completed by the repeated composition of units'. This argument, too, may be suspected of having a fatal dependence on the assumptions of transcendental idealism. A closer look, however, shows that the proof employs Kant's 1770 concept of the infinite, which is not yet suspected of depending fatally on transcendental idealism. The space argument may be derived from three different premises:

(a) 'infinite' means 'cannot be completed by successive synthesis' (critical concept of infinity);
(b) the concept of an (infinite) spatial magnitude corresponds to an (unfinishable) composition of units, i.e., to a measurement (verificationist/empiricist concept);
(c) 'infinite' means a 'maximum magnitude that we can conceive of' (rationalist concept).

Of these options, (a) corresponds to the critical concept of infinity, whereas (b) and (c) represent two opposed positions of transcendental realism. In his remark on the thesis, Kant indeed criticizes (c) as a "defective concept" of the infinite as a maximum magnitude and he emphasizes that he does not want to employ it in his proof (A 430/B 458), as he did in the *Dissertation* of 1770 (2:388 n.; see Sect. 4.4). Instead, the proof relies on his own pre-critical concept of infinity of 1770, according to which the *mathematical* infinite is "a magnitude which, when related to a measure treated as a unit, constitutes *a multiplicity larger than any number*" (2:388 n.), to which he indeed refers in the note to the remark on the thesis (A 432/B 460).

In his 1770 cosmology, Kant still strictly separated the concepts of the intelligible world (which is subject to the purely intellectual or logical use of reason) and the sensible world (which is subject to the real use of reason), and he assumed that the intelligible world (conceived as a symbolic logical universe) may be actually infinite. After 1773–1775, he no longer considered this separation tenable. Now he was convinced that beyond mathematics, in the cosmological use of reason, this concept of the infinite gives rise to identifying 'infinite' with 'immeasurable', and

[18]Malzkorn (1999, 257) concedes this interpretation as admittable, in contrast to Strawson (1966, 176), Mittelstaedt and Strohmeyer (1990, 156), and Schmucker (1990, 115–116). However, he thinks that the proof of the thesis in this case fatally depends on a transcendental philosophical premise: namely on (4a) "When a series has elapsed, then it can be completely synthesized successively" (Malzkorn 1999, 257; my translation). According to my interpretation, this premise is not needed, given that "elapsed" for an infinite time series would mean that this time series was actually or cardinally infinite, in contradiction to the very concept of a series. The thesis proof confounds two concepts of the infinite which are incompatible regarding their real possibility; see below.

thus to (b). Hence, the ordinal and the cardinal concepts of the infinite differ from a semantic rather than a logical point of view. The main difference between Kant's verificationist concept (b) of 1770–1775 and his critical concept (a) of the infinite seems then to be that the former applies to the concepts of space and time, whereas the latter is a general epistemic account of the infinite.

The appearance of fatal dependence on a premise of transcendental idealism thus arises for three reasons: first, because Kant does not distinguish between the critical or "transcendental" concept and the "mathematical" concept of the infinite (which indeed stems from 1770), in the remark on the thesis; second, because he employs in the time argument an analytical argument based on the meaning of an infinite series, which is identical with the "true (transcendental) concept", but simply corresponds to the ordinal concept of an infinite series; and third, because after having rejected the paradoxical metaphysical concept of the infinite as a maximum magnitude, in 1770, when he still attributed actual infinity to the intelligible world, all that was left to him was the mathematical concept of a "multiplicity larger than any number". Therefore in the thesis proof of 1781 he sees himself committed to the "true (transcendental) concept of infinity", in the time argument, and to a verificationist concept of infinity or unboundedness with regard to the magnitude of the world, in the space argument. In sharp contrast to 1770, however, he now uses these arguments to argue *against* the possibility of an actually infinite world. His formal cosmology of 1770 indeed had served to avoid the later antinomy (see Sects. 4.2.3 and 4.3.3). In the end, it is little wonder that the Kant scholars and critics have been confused about the thesis proof.

Applied to the world in space and time, the ordinal concept of infinity introduced in (3) by (I) and (U) is semantically incompatible with the cardinal concept, on the basis of which (2) derives from ($A1_T$) and ($A1_S$).[19] It is brought into play by introducing a hidden additional premise, which is obscured by the fact that the conclusion (2) from ($A1_T$) and ($A1_S$) already implies claims about an infinite series of temporal states and about the world as a totality. In this way, the proof confounds two concepts of infinity, which are incompatible with regard to the real use of reason. To think of the world as a given spatio-temporal totality corresponds to the cardinal concept of infinity, whereas to think of it as resulting from successive synthesis of its temporal states and the composition of its spatial parts corresponds to the ordinal concept. To employ both at once gives rise to a self-contradictory concept of the spatio-temporal world. The cosmological theory from which the proof derives is inconsistent, as the contradictions between (4) and (2) indicate, as well as the formal derivations of (5) from (1), i.e., $(A1_T) \vdash \neg (A1_T)$ and $(A1_S) \vdash \neg (A1_S)$.

The logical and semantic reconstruction reveals the inconsistent theory that underlies the proof, but it does not capture Kant's main point. In the end, Kant

[19] In contrast, van Benthem (1983, 33) assumes that both concepts are *logically* rather than semantically incompatible. In Kant's view, in the face of the resolution of the antinomy, this is just a consequence of the logical fallacy of taking the middle term of the cosmological syllogism in different meanings, which is what gives rise to the antinomy.

wants to demonstrate that the arguments of transcendental realism in favour of a world of finite age and magnitude rely on unsatisfiable premises. From his critical point of view, the 'apagogic' argument against the logical possibility of an infinite world age or magnitude is inevitably based on contradictory presuppositions, which remain invisible from a pre-critical point of view. Kant certainly does not confound the two incompatible concepts of infinity in naively flawed arguments. He does so consciously. He argues at two levels. Critically, he constructs an argument which contains a logical fallacy due to a semantic mistake. Pre-critically, he takes up a traditional argument in favour of a finite world age and magnitude that the defenders of a certain variant of transcendental realism consider to be valid. He then can criticize the transcendental realist for not being clear about the presuppositions of his argument. In his view, the concepts of a cardinal infinite temporal series and of an infinite given spatial whole give rise to a self-contradictory cosmological concept of the world as a spatio-temporal totality. Of course, the underlying conflation of an ordinal and a cardinal concept of the world is *not* based on analytic judgments, in contrast to the other steps of the proof. From Kant's critical point of view, it is enforced by the synthetic faculties of our understanding and reason's quest for the completion of our cognition. Hence, the conflation results from synthetic judgments *a priori* which aim at extending the categories into the unconditioned (A 409/B 436). They give rise to an equivocal use of the middle term 'infinite' in the cosmological syllogism, disguising the fact that the proof result is subrepted by a logical fallacy.[20]

5.4.1.2 Antithesis Proof: The Argument of Ideality

The situation is different when it comes to proving the antithesis. The argument of ideality is based on the following premise: Space, like time, is not a real, but an ideal composite (*compositum ideale*). That is, the components of space and time are not real entities, but only ideal magnitudes, which are obtained from the composition or partition of concrete parts of space given in intuition. With this premise Kant appeals to Leibniz's view of space and time and the corresponding arguments from the Leibniz–Clarke debate. According to Leibniz, the beginning of the world in space and time is undetermined (i.e., there is no sufficient reason) with regard to the question of where or when such a beginning of the world should be. There is nothing 'outside' the world in relation to which space and time could have a beginning (or an end). In Kant's view, the contrary thesis claim is absurd because the concepts of the beginning of the world in time or of a boundary of space and time have neither meaning nor reference—or, as Kant emphasizes in the section on the *Critical decision of the cosmological conflict of reason with itself*: all positions are only in the universe, but the universe itself has no position (A 501/B 530–531). The ideality argument is due to Leibniz; but the pre-critical Kant gave him the

[20]Malzkorn (1999, 118–119 and 130–141) neglects this crucial point. To avoid the diagnosis of a *non sequitur*, he employs a syntactic "principle of charity" and reconstructs a logically valid proof.

verificationist interpretation (see Sect. 3.3), which fits in well with the empiricist antithesis position. The apagogic proof may be roughly reconstructed as follows (see A 427–429/B 455–457):

(1) We assume, against the antithesis, that

($T1_T$) the world has a beginning in time
($T1_S$) and also boundaries in space:

$$\exists x \neg \exists y \ (y < x)$$

(2) Then the time before the beginning of the world was empty (from ($T1_T$) and the meaning of 'beginning'). Likewise, space beyond the boundaries of the world is empty (from ($T1_S$) and the meaning of 'boundary').

$$\exists x \{y \mid y < x\} \text{ is empty}$$

(3) However, nothing can happen in an empty time (temporal interpretation of (2) and Leibniz's principle of indiscernibles). Likewise, an empty space beyond the world does not contain objects of intuition, "and hence no correlate of the world to which the world could stand in relation" (A 429/B 457), i.e., it is nothing that can have a spatial relation to the world (spatial interpretation of (2) and Leibniz's principle of indiscernibles).
(4) Therefore the world cannot arise in empty time (from (3)) and "the relation of the world to empty space would be a relation of the world to **no object**" (ibid.; from (3)). Such a relation is nothing, and therefore nothing in the world can exist:

$$\neg \exists x \ x \in W$$

(5) Therefore, "the world itself cannot have any beginning, and so in past time it is infinite" (A 427/B 455), and "the world is not bounded at all in space, i.e., in its extension it is infinite" (A 429/B 457; modus tollens):

($A1_{T,S}$) $\forall x \exists y \ (y < x)$ or: $\neg \exists x \ Bx$

Step (1) is again the introduction of premises, while step (2) is logically equivalent to (1). In the proof, Kant puts the relationality of space and time forward against the assumption that an empty space or an empty time could be real entities. The claim that space and time are relational is obviously a synthetic judgment. Step (3) is decisive, which consists in applying Leibniz's principle of indiscernibles to the concepts of space and time, in a verificationist interpretation.[21] Then, Kant goes

[21] Allison (1983, 49) also emphasizes that the proof result depends decisively on Leibniz's principle of indiscernibles, and interprets this in the sense of a verificationist position. In the revised edition, Allison (2004, 373) no longer accepts this view, but argues that "the argument takes an

back and forth between formal claims and their temporal or spatial interpretation, which in turn conforms to his diagnosis of the conflation between the intellectual and the sensitive world-concepts. Formally, the proof result differs slightly from the result of the thesis proof. Kant does not claim a logical contradiction to the established premise, but rather a semantic inconsistency. The theory on which the proof is based has an empty domain and thus cannot correspond to the sensible world. But, most importantly, Kant does not consider this proof result to be surreptitiously obtained by logical fallacy, in contradistinction to the (rationalist) thesis claim. In fact, a critically limited version of the antithesis claim serves him as the key to the resolution of the first antinomy.

5.4.1.3 Conclusions on the First Antinomy

In view of this reconstruction of the proofs, we can now refute the objection of circularity (see Sect. 5.1.3) for the first antinomy. The only ingredient of transcendental idealism proper that enters into the proofs of the thesis and the antithesis is the 1770 theory of space and time as forms of pure intuition. In the 1770 *Dissertation*, however, Kant did not yet consider this theory to be in conflict with his formal cosmology. In addition to his 1770 theory of space and time, in the 1781 arguments he only employs his 1770 concept of infinity, according to which the *mathematical* infinite is "a magnitude which, when related to a measure treated as a unit, constitutes *a multiplicity larger than any number*" (2:388 n.). This concept of the infinite gives rise to a verificationist account of the concept of an infinite world age, which Kant adopts for the "apagogic" thesis proof because he considers the rationalist conception of the infinite as a maximum number to be defective. The antithesis proof, in turn, also makes use of Leibniz's principle of indiscernibles in a verificationist interpretation. Hence, Kant proves the First Antinomy on the grounds of verificationism rather than transcendental idealism.

epistemological turn, which is not to be confused with verificationism"; in the corresponding note he claims that it "is a confusion because it (falsely) implies that the assumption of an absolute beginning is meaningless rather than simply false" (ibid., 504, n. 33). I am puzzled about this claim; it fits in with the critical resolution of the antinomy, but not with the point of view of transcendental realism taken in the proof. Against the background of Kant's 1755 and 1758 arguments (see Sect. 3.3.1), the verificationist interpretation seems plausible to me; whereas I do *not* share the view that Kant employs verificationist arguments in *all four* antinomies (Allison 1983, 61 and 312): he only does so in the first antinomy, as far as I can see. In his revision, Allison (2004) seems to take into account the criticism of Grier (2001, 190); she suggests an interpretation of the proof in terms of real relations, according to which "an empty time would lack any 'distinguishing condition of existence' (A 428/B 456)", i.e., any real (ontic) ground of the world. However, the claim in step (4) that "the relation of the world to empty space would be a relation of the world to **no object**" may well be understood as a claim about a meaningless concept (given that relations are usually understood as two-place predicates), and hence as a verificationist argument.

5.4.2 *The Antinomy of Division*

The second antinomy deals with the composition of matter from simple parts. Its subject is the cosmological idea of the "The **absolute completeness** of the **division** of a given whole in appearance" (A 415/B 443).[22] It consists in opposing claims about the result of completely dividing a given whole into its ultimate components, and the corresponding proofs. The thesis claims atomism, the antithesis anti-atomism. The thesis (antithesis) propositions of the antinomy break down into two partial claims:

$P2_1$ All (no) compound substances or things in the world consist of simple parts.
$P2_2$ In the world, only the simple or what is composed of simples (nothing simple) exists.

The partial claims of the thesis and the antithesis are formalized by means of the part–whole relation $<$ ('is a proper part of') of mereology.[23] To simplify the notation, let us define three one-place predicates by means of the relation $<$: [24]

$$(D_1) \qquad Cx \leftrightarrow \exists y\,(y < x) \qquad (\text{'}\textit{is composite}\text{)'}$$

$$(D_2) \qquad Sx \leftrightarrow \neg Cx \qquad (\text{'}\textit{is simple}\text{'})$$

$$(D_3) \qquad Px \leftrightarrow \exists y\,(y < x \wedge Sy) \qquad (\text{'}\textit{has simple proper parts}\text{'})$$

According to (D_1) and (D_3), everything that has simple proper parts is composite:

$$\forall x\,(Px \rightarrow Cx)$$

Then, the thesis and antithesis claims are:

$$(T2_1) \qquad \forall x\,(Cx \rightarrow Px)$$

$$(T2_2) \qquad \forall x\,(Sx \vee Px)$$

---

[22] For a much more comprehensive analysis of the second antinomy, its historical background, its genesis, and its relation to Kant's theory of cognition, see Engelhard (2005).

[23] For the axioms which the relation $<$ of mereology obeys, see e.g., Simons (1987). The following reconstruction does not depend on specific mereological axioms.

[24] The following reconstruction is basically identical to the one suggested in Falkenburg (2000), but differs from the one in Falkenburg (1995). My current definition of the predicate P agrees with that of Malzkorn (1999, 172).

$$(A2_1) \qquad \forall x\,(Cx \rightarrow \neg Px)$$

$$(A2_2) \qquad \neg\exists x\, Sx$$

The partial claims ($T2_1$), ($T2_2$) of the thesis are logically equivalent. Therefore, I suppress the indices, referring to both as (T2) in the following. In contrast, ($A2_2$) is stronger than ($A2_1$). That is, ($A2_1$) precludes compound substances that have simple parts, but admits of simple substances that do *not* form the parts of a whole. This reading of the antithesis would even be compatible with the thesis. The conjunction of (T2) and ($A2_1$) results in a theory according to which there are only simple substances:

$$(T2) \wedge (A_1) \qquad \forall x\,[(Cx \rightarrow Px) \wedge (Cx \rightarrow \neg Px)] \quad \leftrightarrow \quad \forall x\, \neg Cx$$

This case corresponds to Leibniz's *Monadology*, or a reinterpretation of it in terms of real or physical monads.[25] It is a genuine logical alternative to the weak version ($A2_1$) of the antithesis that does not run into the antinomy. In order to refute monads, the antithesis proof has to demonstrate that the stronger version ($A2_2$) of the antithesis holds. Accordingly, in his remark on the thesis with respect to the difference between ($A2_1$) and ($A2_2$), Kant emphasizes that:

> This second proposition of the antithesis goes much further than the first, since the first banishes the simple only from the intuition of the composite, while the second, on the other hand, does away with the simple in the whole of nature. (A 437/B 465)

This objective of the proof corresponds to the logical relationships of the antithetical claims. (T2) and ($A2_{1,2}$) are not contradictory, but just *contrary*. (T2) and ($A2_1$) have the traditional form of contrary categorical A- and E-judgments. (T2) and the stronger partial antithesis claim ($A2_2$) are also not contradictory. The negation of (T2) ['There are substances that are neither simple nor have simple parts'] is logically weaker than ($A2_2$) ['There are no simple substances']. (T2) and ($A2_2$) therefore do not constitute an exhaustive alternative. Hence, a formal model of composite substances is possible in which neither claim will hold. In addition to (T2) and ($A2_2$), a mereological hybrid of atomism and anti-atomism is logically possible, i.e., a theory according to which some composite substances consist of atoms and others do not:[26]

$$(H) \qquad \exists x\, Px \wedge \exists x \neg Px$$

[25] According to (A 440–442/B 468–470), the thesis does not directly refer to Leibniz's monads, but to real monads, see Engelhard (2005, 176–177). Kant's own explanation of the thesis proposition in terms of "transcendental **atomistic**" or "the dialectical principle of **monadology**" (A 442/B 470) does not really contribute to clarifying this question. However, see also below.

[26] The axiom SF5 of Simons (1987, 42) is equivalent to (H), his SF3 to (T2), and his SF4 to ($A2_2$).

(T2) and (H) are genuine logical alternatives to ($A2_2$); that is, both claims are incompatible with ($A2_2$). From (T2) as well as (H) we can derive an existential claim concerning simple substances which is contradictory to ($A2_2$):

(T2*) $\exists x\, Sx$

With this logical consequence (T2*), a proof of ($A2_2$) via *modus tollens* would result in refuting (T2) as well as (H). To justify ($A2_2$) or refute (T2*) is absolutely crucial for Kant's objective.

In order to preclude the real possibility of simple substances, Kant employs an epistemological argument that again seems to make use of some crucial claims of his transcendental philosophy. It runs counter to the real possibility of monads, i.e., the option that non-relational substances may exist in the spatio-temporal world. Here, he does not argue apagogically, but directly. This proof is easily (mis-)understood as a transcendental argument by which Kant diminishes his proof results through having recourse to his position of transcendental idealism.[27] He argues that it is impossible to experience simple, non-relational substances, given that all sensory phenomena are relational:

> The second proposition of the antithesis, that in the world nothing at all exists that is simple, is here supposed to signify only this: The existence of the absolutely simple cannot be established by any experience or perception, whether external or internal, and the absolutely simple is thus a mere idea [...] and hence in the exposition of appearances it has no application or object. (A 435–437/B 463–655)

As he emphasizes in the note to the thesis proof (A 440–442/B 468–470), Leibniz's monads are, unlike atoms, simple substances that cannot be understood as elements of a composite and cannot be identified with the spatial components of material things. For Leibniz, the monads are *noumena*, whereas material things are *phaenomena*. Kant's choice of the words "substance" for the subject of the thesis instead of "thing" in the antithesis seems to point to this distinction.[28] But the argument can also be substantiated from an empiricist or rationalist point of view. According to Leibniz, too, the phenomena are thoroughly relational. And also according to Locke it is impossible to experience simple, non-relational substances in the world.

Let us now look in more detail at the 'apagogic' proofs of ($T2_2$) and ($A2_2$). They take up arguments from the seventeenth/eighteenth-century discussions on atomism, as found in Descartes, the English empiricists, the Leibniz–Clarke debate, Wolff, Euler, and in Kant's own *Physical Monadology* of 1756. Yet the problem is much older. It goes back to Zeno's paradoxes, ancient atomism, and Aristotle's conception of the continuum. To the latter Kant indeed refers when he emphasizes in the remark to the thesis (A 438/B 466) that points are not elements of space, but boundaries in space.

[27] In Falkenburg (1995, 15), I still did so.

[28] However, the issue is not completely clear; see n. 25 above.

5.4.2.1 Thesis Proof: The Mereological Argument

In the proof of the thesis, however, the concept of space plays no role at all. The proof is provided in the terminology of Wolff's and Baumgarten's ontology, using the concept of a composite substance. In fact, Kant had already given more or less the same proof in the *Physical Monadology* (1:477; see Sect. 2.2.3.2). Kant employs in it formal assumptions about the relation between a composite and its parts, about substances as something persisting by themselves, and about the world as the whole of all substances, i.e., the mereological maximum or supremum of the part–whole relation. The concept of substance in the proof corresponds to the substance metaphysics of rationalism from Descartes to Baumgarten, in 1756 as well as 1781/1787. According to it, a substance is a non-relational entity that exists for itself, an *ens per se* of which one can imagine that "all composition is removed in thought"; for such substances, "composition is only a contingent relation, apart from which, as beings persisting by themselves, they must subsist" (A 434–436/B 462–464).

The modern formal correlate of this (intellectual) concept of a substance is not the concept of a mereological atom in the sense of an infimum of the part–whole relation, but the concept of a logical individual considered without any relations. According to it, a composite formed from substances consists by definition of simple elements.[29] The composite is then merely an aggregate in which the whole relates to its parts like a concrete class to its elements, or as a multiplicity to its units. Accordingly, a substance is then nothing but "something [. . .] that subsists without any composition, i.e., the simple" (A 434/B 462).

Directly based on this logical concept of a substance, the proof of the thesis would become trivial. The thesis claim would then merely be an analytic judgment about composite substances, where a substance only means an abstract "thought-entity", i.e., an "empty concept without object" or *ens rationis* (A 292/B 348). This criticism has been raised by Russell and others.[30]' Then it would lie in the concept of a composite substance that it has simple parts, following rationalist substance metaphysics and in contrast to current formal mereology.

A more detailed analysis of the proof, however, shows that it is not that trivial. It is crucial that the proof is based on an equivocation of the predicate "simple",

---

[29]Malzkorn (1999, 279, n. 96), in contrast, neglects the traditional metaphysical concept of a substance and defends the view that Kant does *not* employ an analytical argument.

[30]See Russell (1903, 460): "It is indeed obvious that the proposition, true or false, is concerned purely with whole and part, and has no special relation to space and time. Instead of a complex substance, we might consider the numbers between 1 and 2, or any other definable collection. And with this extension, the proof of the proposition must, I think, be admitted; only that *terms* or *concepts* should be substituted for *substances*, and that, instead of the argument that relations between substances are accidental (*zufällig*), we should content ourselves with saying that relations imply terms, and complexity implies relations." Vogel (1975, 299; my translation) charges the proof with circularity: "That the argument of the removal of all composition can indeed only apply to an object which already presupposes what is to be proved, we have already commented upon as against the proof in the *Monadologia physica* [. . .].'

resulting in the conflation of a non-relational and a relational concept (A 434–436/B 462–464).

(1) We assume against the thesis that a composite substance does not consist of simple parts:

$$\exists x\,(Cx \wedge \neg Px)$$

(2) If one then considers all composition of this substance to be removed, i.e., one conceives of the multiplicity of parts $\{y \mid y < x\}$ disregarding their part–whole relations (abstracting from their relation $<$),
(3) neither (trivially) composite nor (according to (1)) simple parts remain. Hence, the considered multiplicity of parts is empty, i.e., there are no (parts that may form) composite substances:

$$\neg\exists x Cx$$

(3′) [This result contradicts the existential presupposition of the thesis according to which there are composite substances, i.e,. (3) must be wrong:]

$$\exists x Cx$$

(4) Therefore it must be either impossible to remove all composition in thought (*modus tollens* for the conclusion from (2) to (3)),
(5) or [the conclusion from (2) to (3) is flawed and] after removing the composition, something remains that exists regardless of all composition, i.e., that is simple.

Due to the generality of the assumption (1), this conclusion is correct for all composite substances. However, the predicate "simple" attributed to a substance is ambiguous, and the conclusion may be understood in two ways:

(5a) [The composition is thought to be removed; i.e., one abstracts from the part–whole relation, and "simple substance" just means "logical individual":]

$$\forall x\,[Cx \rightarrow \exists y\,(y < x)]$$

(5b) [The thesis is about substances "in the world". The world is the supremum with regard to the part–whole relation, therefore "simple substance" means "mereological atom":]

$$\forall x\,[Cx \rightarrow \exists y\,(y < x \wedge Sy)]$$

(5a) is identical with the mereological definition (D_2) of the predicate "composite", while (5b) is based on the mereological definition (D_1) $Sx \leftrightarrow \neg Cx$ of the predicate S and is logically equivalent to the formal claim (T2) of the thesis. The proof does not distinguish between these two possible formal options. But the next proof steps obviously employ the rationalist concept of a substance, committing to option (5a):

(6) Since composition is a contingent relation of the substances, i.e., a relation which is accidentally satisfied in the multiplicity W,
(7) according to (4) [and the traditional definition of the concept of substance] the composite would not consist of substances:

$$\neg\exists x\, Cx$$

(8) However, this result contradicts the [existential] presupposition [concerning the composite substances] (*modus tollens* for the conclusion from (4) to (7)).
(9) This result only allows for (5), i.e., after removal of all composition something simple remains.

To interpret this conclusion in terms of logical individuals (5a) obviously gives a much weaker proof result (9a) than the intended result (9b) which refutes (1), and which is only obtained by interpreting the "simple substances" in (9) as mereological atoms (5b):

(9a) All composite substances consist of simple parts that have to be understood as logical individuals:

$$\forall x\, [Cx \rightarrow \exists y\, (y < x)]$$

(9b) (1) is false, i.e., all composite substances consist of simple parts:

$$\text{(T2)} \qquad \forall x\, (Cx \rightarrow Px)$$

Reconstructed in this way, the proof no longer looks trivial but indeed appears rather intricate. Kant achieves the intended result by conflating the concepts of a logical individual (as a substance of the intelligible world) and a mereological atom (as a substance in the sensible world). The conflation is based on the equivocal use of the term 'simple substance', which is the middle term of the cosmological syllogism (see Sects. 5.2 and 5.2.4). For logical individuals, to form composites is merely a "contingent relation" of intelligible substances. On the other hand, the substances are supposed to exist "in the world", i.e., as real composites within the sensible world, the part–whole relations of which are *not* to be neglected.

We should note, however, that Kant did not employ the abstract concept of a set. He identified multiplicities with concrete classes of objects and he also conceived of the logical universe W in this way.[31] Therefore, he was not yet able to distinguish

[31]Engelhard (2005, 175) challenges this claim, pointing to Kant's third and fourth arguments against the discursive character of space in the Transcendental Aesthetic (B 39–40). However, Kant's theory of mathematics does not admit of the abstract concept of a set, or a logical

logical individuals from mereological atoms. From his point of view, a composite is an aggregate, and the simple parts of a composite are likewise conceived as mereological atoms and as the elements of a class of objects. For Kant therefore the conflation of both concepts of simple substances must have been a compelling logical consequence of forming the concept of a 'simple substance in the world'. In his view, the proof is only based on analytic judgments about the substances in the world and their composition.[32] Due to the equivocation of the predicate 'simple' attributed to a substance, the proof has nevertheless been obtained from a logical fallacy and from a self-contradictory concept, according to Kant's critical point of view.

5.4.2.2 Antithesis Proof: The Argument of Spatial Extension

At the end of the thesis proof, Kant adds that "composition is only an external state of these beings" (A 436/B 464), preparing the proof of the antithesis. The proof of the antithesis interprets the part–whole relation $<$ as a spatial relation. Kant again starts from the concept of a "composite thing (as substance)", but then claims that composition, as an "external relation between substances, [...] is only possible in space" (A 435/B 463). If the "contingent" relation of simple substances is an external relation, it must be constructed in space, according to Kant's view of mathematics. The construction of a relation in pure intuition is synthetic. Hence, on the side of the antithesis Kant attempts to prove a synthetic judgment *a priori* about composite spatial substances. The spatial interpretation of the part–whole relation in the antithesis proof adds to the logical concept of a composite substance far-reaching assumptions about the structure of the logical universe W on which the relation operates. According to the traditional concept of substance, the parts of a substance are also substances. For a spatial part–whole relation, the property of being spatially extended is transferred from a whole to all its parts; "every part of the composite must occupy a space [...] as a real composite" (A435/B 463). Under this condition, it is easy to prove the claim that a substance in space has no simple parts (that is, in the spatial interpretation, point-like constituents). The basic idea behind the antithesis proof also originates from the *Physical Monadology* (1:479). The apagogic proof starts from $\neg(A2_1)$, in order to arrive at the proof result $(A2_2)$, in the spatial interpretation:

multiplicity, given that from 1770 on he considers the objects of mathematics to be generated *in concreto* in pure intuition.

[32]The extent of the multiplicity, or the number of individuals that form a concrete composite, remains open. The proof does not claim that a composite has finitely or countably infinitely many simple parts. The thesis here does not imply a position of finitism; in contrast to the *Prolegomena* (§52c, 4:342), where Kant states the thesis and antithesis claims as follows: "[...] that bodies in themselves consist of infinitely many parts or of a finite number of simple parts."

(1) We assume against the [first partial claim of the] antithesis that a composite thing (as substance) consists of simple parts:

$$\exists x\,(Cx \rightarrow Px)$$

(2) Since composition, as an external relation, is only possible in space (spatial interpretation of $<$),
(3) a composite thing has as many parts as spatial parts (from (1) and (2)).
(4) Now the spatial parts of a space are spaces (continuum properties of space).
(5) Hence each simple part of the composite must be spatial [i.e., extended].

$$\forall x\,[Ex \rightarrow (x \text{ is spatial})]$$

(6) Since a real spatial thing is extended, i.e., encompasses a multitude of different parts, it must (as a substance) in turn be composed of substances (spatial interpretation of the traditional concept of substance):

$$\forall x\,[(x \text{ is spatial}) \rightarrow Cx]$$

(7) So the simple parts of space are composed of substances (from (5) and (6)):

$$\forall x\,(Sx \rightarrow Cx)$$

(8) Given that this implies a contradiction [for composite substances in space], (1) must be false:

$$\forall x\,(Cx \wedge \neg Px)$$

The proof result is equivalent to $\forall x\, Cx$ and thus to ($2A_2$), but obviously only applies to spatial substances. The proof again switches from formal statements to their interpretations, that is (from Kant's critical point of view) from an intellectual to a sensitive cosmological concept. The synthetic steps of the proof are (3) and (4). In (3) Kant identifies the parts of substances extensionally, i.e., according to the scope of their class, with the parts of space; and in (4) he brings the continuum properties of space as a form of intuition into play. For a spatial part–whole relation $<$, the domain 'world' on which the relation $<$ operates from a modern point of view has the formal properties of the power set of the mathematical continuum. The class of all composite substances has the same scope as the class of all space regions; and the class of all simple substances corresponds to the set of all points (which has the cardinality of the continuum, as we know today). According to the traditional concept of substance and Kant's concept of space as a form of pure intuition, however, the simple parts of a spatial continuum (that is, points in space) do not have the character of spatial substances. For Kant, they are not parts of space, but mere boundaries in space, as he emphasizes in his remarks to the thesis and the antithesis of the second antinomy (A 438, 439/B 466, 467). The assumption that a composite has simple parts, which is drawn from the traditional concept of a substance, is therefore incompatible with the assumption that the spatial extension of a substance in space is transferred to all its parts, which the antithesis proof employs. Thus the (rationalist!) premise introduced at the beginning of the proof is refuted. As in the first antinomy, Kant does *not* consider this proof result of the antithesis as due to a logical fallacy. It is in accordance with the resolution of the antinomy and with *Proposition 4* of the *Dynamics* part of the *MFNS* (4:503).

5.4.2.3 Conclusions on the Second Antinomy

The arguments in the proofs of the thesis and the antithesis merge several logical and semantic levels. To my understanding, to a large extent Kant does so consciously. To be more precise, he does so in the *CPR*, whereas in his earlier, pre-critical versions of the proofs he obviously did not. Indeed the proofs of the thesis and the antithesis merge the following concepts:

(α) in the thesis proof: logical individuals and mereological atoms;
(β) in his addition to the thesis proof, making the transition from "contingent" to "external" and hence "spatial" relations: a formal and a spatial part–whole relation; and, finally,
(γ) in the antithesis proof: the structural properties of a spatial continuum and the logical properties of a multiplicity or class of substances.

According to the proof of the thesis, to be composite is a formal relation in a class or multiplicity of concrete substances. Due to his lack of the abstract concept of a set, and his concrete account of a multiplicity or class of objects, Kant probably did not see that one should not conflate "simple" substances in the sense of logical individuals with mereological atoms. Unlike in 1770, however, in 1781 he realized that the formal concept of an intelligible substance from Wolff's or Baumgarten's ontology should not be conflated with the cosmological concept of a substance "in the world" of spatio-temporal appearances.

From Kant's critical point of view, as well as from a modern perspective, the defect of the thesis proof lies precisely in this conflation. Kant's own diagnosis is that the arguments of the traditional debate on atoms and monads confound a formal domain of intelligible substances with the spatio-temporal world as the sum total of all sensible substances. His diagnosis perfectly conforms to the modern diagnosis that the proofs of the second antinomy conflate a logical universe of individuals without relations (logical individuals) and a relational structure (mereological atoms). According to the view of mathematics which Kant advanced from the mid 1770s, only spatial substances, i.e., appearances, may stand in "external" part–whole relations. From his critical point of view, the thesis proof is defective in postulating external or mathematical relations at all for the formal substances of traditional ontology. The antithesis proof, however, only becomes defective by applying the proof result to formal substances in the sense of traditional ontology instead of material substances in the sense of a *substantia phaenomenon* in space.

Why did Kant consider the antinomy of division to be inescapable? Apparently in 1781, as well as in 1770, he was convinced that cosmology, if it is possible to establish such a discipline at all, has to employ a concept of substance; and that the cosmological concept of the world as a totality as well as the concept of their ultimate parts refer to intelligible substances. On the other hand, he had meanwhile come to two new conclusions: first, that the merely logical use of the part–whole relation, i.e., its application to intelligible substances, is obtained only surreptitiously; and second, that the concept of a substance becomes non-monadic in real use, i.e., when applied to the spatio-temporal world. To talk of the "simple parts" of a composite sensible substance is thus surreptitious, too. It makes no sense to attribute or deny simple parts to sensible substances. The concept of a simple, non-composite substance in the spatio-temporal world is self-contradictory for Kant in 1781, and the proof results are obtained from equivocal use of the predicates 'simple' (logical vs. mereological) and 'composite' (non-spatial *vs* spatial, or logical vs. real). Therefore, Kant considers both proofs as persuasive, as long as the underlying logical fallacy is not revealed.

5.4.3 The Antinomy of Causality

The third antinomy concerns the opposition of the causality of nature and causality through freedom. Its subject is the transcendental concept of freedom, as an idea of theoretical reason. The thesis claims that there are appearances which are not due to the causality of nature but to a causality through freedom. They are assumed to be spontaneous with regard to the relation $<$, which now has a causal interpretation. Spontaneous events are considered to have no cause in the sense of $<$, i.e., to be logical minima with regard to $<$. The antithesis makes the opposite claim that there are no appearances due to a causality through freedom, but everything in the world happens due to the causality of nature (A 444–446/B 472–474):[33]

P3 There are (no) appearances in nature which are not due to the causality of nature.

The antithetic assertions are contradictory:

(T3) $$\exists x \neg\exists y\,(y < x)$$

(A3) $$\forall x\,\exists y\,(y < x)$$

The thesis and antithesis claim that the principle of causality is, or respectively is not, universally valid. According to the *principle of causality* (P_C), every event x

[33]My reconstruction sets aside all details concerning the cosmological concept of freedom.

underlies some law of nature, according to which it is due to a cause y:[34]

$$(\mathrm{P_C}) \qquad \forall x \exists y\, (y < x)$$

The antithesis is equivalent to the principle of causality, the thesis to its negation:

$$(\mathrm{A3}) \equiv (\mathrm{P_C}), \qquad (\mathrm{T3}) \equiv \neg(\mathrm{P_C})$$

The formal claims of thesis and antithesis are identical to those of the first antinomy. The interpretation of $<$ is what makes the difference, given that here $<$ is understood as a causal relation. This time the proofs merge the principle of causality and the principle of sufficient reason.

According to the principle of causality, an event x occurs under two necessary conditions, i.e., it is governed by a law of nature and it has a cause y. Both necessary conditions together result in a sufficient condition for the occurrence of x. Here, the cause y is a *real* ground for the occurrence of x, and it has the status of a necessary condition. According to the principle of causality, however, it can be regarded as a contingent condition for the application of a law of nature. In the terminology of the fourth antinomy, or rather its resolution, it has only *relative*, but not *absolute* necessity (see Sect. 5.4.4). In contrast to the principle of causality, the principle of sufficient reason claims that there is a sufficient condition y for the occurrence of x. Logically, it calls for the existence of a complete condition z, i.e., a complete series of necessary conditions $(z_1, z_2, \ldots, z_n)$ which together are sufficient for the event x to occur. This is a very strong logical condition. In Kant's view, such a complete condition z is subject to the principle of the determining ground of 1755, i.e., it means "to posit a predicate while excluding its opposite" (1:391), or to give a necessary and sufficient *logical* ground for the occurrence of x (see Sects. 2.2.2.1 and 3.1).

Interpreted in this way, the relation '*is sufficient reason of*' is just a *logical* relation between the antecedent G and the consequence F of a hypothetical judgment 'If G, then F'. The thesis, however, is a rationalist claim about the world. Hence, it presumes that a relation $\prec$ holds which links *real* grounds g and consequences f in the world in such a way that the existence of g is the necessary and sufficient condition for f to occur. $\prec$ is a transitive, asymmetric, and irreflexive relation (so far, it has the same formal properties as $<$), which however in addition satisfies a uniqueness condition (UN):

[34]One could also express this claim by a predicate N with the meaning 'obeys a law of nature', or within second-order logic by a proposition of the form $\forall x \exists N\; Nx$, where N stands for the law of nature. Malzkorn (1999, 194) defines two predicates 'causality according to the laws of nature' and 'causality through freedom', which he formalizes as three-place relations of two events x, y at a time t. This approach makes it possible to distinguish the absence of a cause (indeterminism or contingency) explicitly from causality out of freedom. In his approach, contradictory claims are also only obtained if the middle term 'world' of the cosmological syllogism is used equivocally (Malzkorn 1999, 199–200).

(UN) $$z \prec x \rightarrow \forall y\,(y \prec x \rightarrow y = z)$$

The *principle of sufficient reason* (P_{SR}) has the same logical form as the principle of causality, but it employs the stronger relation $\prec$:

(P_{SR}) $$\forall x \exists z\,(z \prec x)$$

5.4.3.1 Thesis Proof: The Argument of Sufficient Reason

The apagogic proof of the thesis may then be reconstructed as follows (A 444–446/B 472–474):

(1) Suppose there is no other causality than the causality of the laws of nature,

$$\neg\exists x\, \neg\exists y\,(y < x)$$

(2) then everything that happens presupposes something previous that caused it (from (1)):

(A3) ≡ (P_C) $$\forall x \exists y\,(y < x)$$

(3) This is valid for the preceding cause and for the one that preceded it, etc.; thus there is an infinite regress of causes:

$$\forall x \forall n \exists y_1 \ldots \exists y_{n-1} \exists y_n\,(y_n < y_{n-1} \wedge \ldots \wedge y_1 < x)$$

(4) This means that there is no first cause, and hence no complete series of causes:

> If, therefore, everything happens according to mere laws of nature, then at every time there is only a subordinate but never a first beginning, and thus no completeness of the series on the side of the causes descending one from another. (A 446/B 474)

$$\neg\exists y_n \forall x\,(y_n < x)$$

(4′) [This means, however, that there is no sufficient reason for anything to happen:]

$$\forall x \neg\exists y\,(y \prec x)$$

(5) But nothing happens without sufficient reason:

(P_{SR}) $$\forall x \exists y\,(y \prec x)$$

(6) This principle contradicts the "unlimited universality" of (3), making (1) absurd. Hence there must exist causes that do not obey a law of nature, against the principle of causality:

(T3) $$\exists x \neg\exists y\,(y < x)$$

In (5), Kant explicitly introduces the principle of sufficient reason (P_{SR}) as an additional premise, implicitly assuming that it contradicts (4). (P_{SR}) reinterprets the logical relation of ground and consequence as a relation $\prec$ between real events, and this reinterpretation seems to make the introduction of this new premise legitimate. The reinterpretation seems plausible to the rationalist defender of the thesis, who identifies the complete sequence of all necessary conditions with a sufficient condition, given that only the *totality* of all necessary conditions yields a condition that is "*a priori*" sufficient. The argument runs in parallel to the thesis proof of the first antinomy; note that (4) is equivalent to the claim made there in step (2) (see Sect. 5.4.1.2):

(2) of ($T1_{T,S}$) $\qquad \forall x : \{y \mid y < x\}$ is infinite

From Kant's critical point of view, the sufficient condition *a priori* that exists according to (P_{SR}) does not belong to the sensible world. It is nothing but the abstract class of all conditions of the appearances, i.e., it is a merely intellectual, logical concept rather than a real entity. To interpret the relations of these conditions in the sense of $\prec$ leads to assuming that their abstract totality belongs to the sensible world, in contradiction to the presupposition made at the beginning. Hence, the proof result is due to the logical fallacy of confounding logical and real grounds.

5.4.3.2 Antithesis Proof: The Argument of Spontaneous Actions

The apagogic proof of the antithesis, on the other hand, employs the principle of causality against the assumption that there is causality through freedom. Here, 'causality through freedom' is formally reconstructed as 'has no cause and initiates a causal chain'. The proof runs as follows (A 445–447/B 473–475):

(1) Suppose there is causality through freedom as a faculty of initiating an occurrence in the world and a series of its consequences,

$$\exists x \neg \exists y \, [(y < x) \; \wedge \; \exists z \, (x < z)]$$

(2) then there are spontaneous occurrences in nature which in turn have no cause (from (1)):

(T3) $\qquad \exists x \neg \exists y \, (y < x)$

(3) Every beginning of action [with consequences], however, presupposes a state of it before the action:

$$\forall x \, [\exists y \, (x < y) \; \rightarrow \; \exists z \, (z < x)]$$

(4) This consequence of the principle of causality contradicts the premise that there is causality through freedom, which is thus absurd:

$$\neg\exists x\,[(\neg\exists y\,(y < x) \wedge \exists z\,(x < z)]$$

The crucial point of the argument is that (4) is logically equivalent to the first of the following two statements, but not to the second:

(5a) Every event has a cause, or it cannot initiate a causal series:

$$\forall x\,[\exists y\,(y < x) \vee \neg\exists z\,(x < z)]$$

(5b) Every event has a cause, there are no spontaneous actions in nature:

$$\neg\exists x \neg\exists y\,(y < x)$$

In (3), Kant's proof introduces the principle of causality (P_C) as a hidden premise, which contradicts the presupposition (1) and is identical with the antithesis to be proven; this gives the stronger result (5b). So far the proof is circular and analytic. Kant, however, continues to argue:

(6) "Thus transcendental freedom is contrary to the causal law [...] and hence is an empty thought-entity" (A 445–447/B 473–475):

$$\{x \mid \neg\exists y\,(y < x)\} \text{ is empty}$$

From Kant's critical point of view, the result of the argument is illegitimately interpreted in such a way that the principle of causality (P_C) is understood as a sufficient condition. This interpretation is inevitable if one assumes that every event x that is causally related to other events exclusively obeys the laws of nature. Then, a deterministic understanding of Kant's principle of causality excludes the possibility of freedom. Only indeterministic law-like connections within nature would still allow it. According to Kant's critical point of view, the proof result is achieved by conflating a real cause, which is only a necessary condition, with a sufficient reason. Real causes are appearances under laws of nature, relative to which they have the status of contingent initial conditions of a causal process. The transition from (5b) to (6), however, reinterprets them as sufficient reasons, that is, as necessary *and* sufficient conditions.

5.4.3.3 Conclusions on the Third Antinomy

Both proofs introduce the contradictory opposite of the initial premise as an additional, new premise. In the proof of the thesis, the new premise is the principle of sufficient reason, which is justified by putting 'real cause' and 'logically sufficient reason' on a par, and which is played off against the principle of causality. The proof of the antithesis plays the principle of causality off against the assumption that there are spontaneous actions in nature, and then it identifies a real cause with a sufficient reason. In both proofs, these steps are synthetic. Both proofs invoke the principle of causality, which applies to all occurrences in the sensible world, against the principle of sufficient reason, without making a clear distinction between them.

From Kant's critical point of view, the antinomy is resolved by detecting that the proofs conflate a non-logical variant of the principle of sufficient reason, which contradicts the principle of causality, with the merely logical principle of the determining ground. In the thesis proof, Kant presumably considers this conflation to be inevitable. According to his view of mathematics, an infinite regress of causes is merely an abstract logical concept, which however suggests the equivocal use of the term 'cause' for 'real ground' (= cause in nature) and 'logical determining ground' (= sufficient condition). On the antithesis side, according to Kant's resolution of the antinomy, the proof is built on an equivocal use of the term 'causality', when the principle of causality (of nature) is employed to argue against the possibility of a causality through freedom, i.e., the existence of spontaneous causes in nature. The proof of the antithesis appears to be compelling if one assumes that the principle of causality is universally valid in the sensible world.

5.4.4 The Antinomy of Necessity

The fourth antinomy refers to the necessity of the beings in the world, or to the ground of existence of all that exists in the world. It asks for the nomological completion of our cognition of the world with respect to its ontic ground or *ratio essendi* in the sense of the *New Elucidation*. The thesis claims that there is an absolutely necessary being which belongs to the world, whereas the antithesis claims that such a being exists neither inside nor outside of the world (A 452, 453/B 480, 481). This asymmetry is due to the prototypical positions for which thesis and antithesis stand.

The rationalist starts from the intellectual concept of the world. Thus for him the concept 'outside the world' is meaningless. Either he considers God to be the basis of the world in an all-encompassing sense, or he follows Spinoza and regards God as being one with nature. The thesis proof, however, deals with the "world of sense, as the whole of all appearances" (A 452/B 480), in accordance with Kant's previous remark that the antinomy "concerns nothing other than the exposition of **appearances**" (A 416/B 443). Hence it starts from a naturalistic premise and its proof therefore supports Spinozism.

The empiricist, on the other hand, proceeds from the concept of the sensible world or nature, which offers him the following options: (i) From the naturalistic point of view, a necessary being would have to be part of the world. (ii) However, as Berkeley's idealism shows, empiricism is also compatible with a non-naturalistic position according to which a necessary being is assumed to exist outside the world, as its cause (see Kant's account of "empirical idealism", *CPR*, A 369–373).

The antithesis proof also argues against this position, i.e., the antithesis represents naturalism, too.

In a precise reformulation, thesis and antithesis are accordingly *not* contradictory. It turns out that the logical option of a necessary being, which is *neither* part *nor* cause of the world *nor* identical with the world as a whole, does not contradict the antithesis claim (Malzkorn 1999, 218). Under the naturalistic assumptions associated with the attempt at a nomological completion of the cognition of nature, however, this option (which points to Kant's resolution of the antinomy) is irrelevant from a rationalist as well as an empiricist point of view, which correspond to the thesis or antithesis claims respectively. Hence, a formal reconstruction of the further claims about the necessary being as being a part or the cause of the world is dispensable. Therefore I only reconstruct the following reformulation of the antinomy, according to which the thesis claims that an absolutely necessary being exists as part or cause of the world, and the antithesis claims the contradictory opposite:

P4 There is an (no) absolutely necessary being, either in the world as a part of it or outside the world as its cause.

Kant uses the term 'necessary' as a monadic predicate, which is assigned to a 'being' x inside or outside the world W. According to Kant this is not a *real* predicate, which attributes any characteristic mark to x, but a *modal* predicate $\mathcal{N}$. With it, the contradictory claims of the thesis and the antithesis become:

$$\exists x \mathcal{N} x \quad \text{and} \quad \neg \exists x \mathcal{N}$$

According to the second edition of the *CPR*, "*necessity* is nothing other than the existence that is given by possibility itself" (B 111). For the logical use of reason, this definition basically agrees with the relation between the operators $\Box$ ('*it is necessary that*') and $\Diamond$ ('*it is possible that*') of modern modal logic, according to which the necessity and the possibility of a proposition p relate as: $\Box p \leftrightarrow \neg\Diamond\neg p$.

The meaning of necessity and possibility employed in the fourth antinomy is much stronger. It is based on traditional metaphysics, and in particular on Leibniz's account of real possibility and his principle of the compossibility of all substances. According to it, the modal predicate $\mathcal{N}$ expresses the strange property of making a really possible thing or event x *exist*, for instance (according to the ontological proof) God as the sum total of all realities. The concept of necessity used in the fourth antinomy stands in Leibniz's tradition. For a description F, Kant's modal predicate $\mathcal{N}$ can be formally defined as follows:

$$\text{(D4)} \qquad \mathcal{N} x \leftrightarrow (\Diamond F x \rightarrow F x) \qquad \text{('}\textit{is absolutely necessary}\text{')}$$

Then, the contradictory modal claims of the thesis and the antithesis become:

$$\text{(T4)} \qquad \exists x (\Diamond P x \rightarrow P x)$$

(A4) $$\neg\exists x(\Diamond Px \rightarrow Px)$$

The antithesis claim (A4) is compatible with any modern system of modal logic, the thesis claim (T4) with none.

In addition to the modal concept of necessity, and in contrast to it, there is the *relational* concept of what is necessary, according to which all sensory appearances are relatively necessary or conditioned. The relative necessity or conditionality is equivalent to the relational concept of contingency of the second edition of the *CPR*:

> [...] but then the concept of the contingent is already taken in such a way that it contains, not the category of modality (as something, the non-existence of which can be thought), but that of relation (as something that can only exist as the consequence of something else). (B 290)

Relatively contingent things and events obey the principle of causality. As a claim about the relatively contingent, the principle of causality is an analytic judgment (ibid.). Based on it, we may define the predicate R as follows:[35]

(D5) $$Rx \leftrightarrow \exists y\, y < x$$ (‘*is relatively contingent*’)

5.4.4.1 Thesis Proof: The Argument of Absolute Necessity

Like the epistemological additional proof to the antithesis of the second antinomy, the proof of the thesis is not carried out in an apagogic form but directly. In accordance with the relational concept of necessity that applies to all appearances, the proof begins with a reflection that first returns to the line of thought of the thesis proof of the third antinomy and then transforms it into a modal claim (A 452–454/B 480–482):

(1) As a condition of the possibility of our empirical conception of time, the world of sense contains a series of alterations (temporal interpretation of $<$):

$$\exists x_1 \exists x_2 \ldots \exists x_{n-1} \exists x_n\, (x_1 < x_2 \wedge \ldots \wedge x_{n-1} < x_n)$$

(2) Every alteration, however, is subject to a necessary condition which precedes it according to a law of nature (re-interpretation of $<$ in terms of the relational concept of necessity and causality):

(P$_C$) $$\forall x\, \exists y\; y < x$$

(3) Now for every conditioned there is a complete series of conditions identical to an infinite regress of causes, as in the thesis proof of the third antinomy:

[35] In contrast, in his reconstruction Malzkorn (1999, 231–234) uses a modal operator of contingency, which is defined by $\Diamond Px \vee \Diamond\neg Px$.

$$\forall x \forall n \exists y_1 \ldots \exists y_{n-1} \exists y_n \, (y_n < y_{n-1} \wedge \ldots \wedge y_1 < x)$$

(4) [However, the totality of all necessary conditions is a sufficient reason for the existence of the whole series:]

(P_{SR}) $$\forall y \exists x \; x \prec y$$

(5) Hence, there exists something absolutely unconditioned that conditions the whole series of alterations and has absolute necessity (modal re-interpretation of (4)):

(T4) $$\exists x (\Diamond Px \rightarrow Px)$$

The formal objective of the proof has thereby been achieved. The proof basis is identical with that of the thesis of the third antinomy. The principle of sufficient reason is used in (4) as a hidden premise. In passing to (5), it is then re-interpreted as the modal claim to be proven (this is new), given that from the thesis point of view a sufficient reason is an absolutely necessary condition. However, the proof does not end here. The proof result stated at the very end of the thesis proof is in line with Spinozism and seems to go beyond the thesis claim:

> Thus in the world itself there is contained something absolutely necessary (whether as the whole world-series itself or as a part of it). (A 454/B 482)

Schmucker (1990, 159) objects to this result that it differs from the thesis claim. But the difference is simply due to the assumption that "the world" is the "whole of appearances", given that the antinomy "concerns nothing other than the exposition of **appearances**" (A 416/B 443), as already mentioned above. Kant indeed proves by means of the concept of a series of appearances in time that the absolutely necessary being, the existence of which has so far been stated in the proof, is not an external cause but part of the world. He argues that the beginning of a time-series cannot be outside the world of sense, but itself is subject to temporal conditions and hence belongs to time. Kant thus achieves a proof result which is stronger than his thesis claim, as also noted by Malzkorn (1999, 219 and 230). In fact it exactly corresponds to Spinozism, the rationalist version of naturalism.

5.4.4.2 Antithesis Proof: The Argument of Relative Necessity

The antithesis proof is carried out in apagogic form (A 453–455/B 481–483):

(1) Let us suppose that there is an absolutely necessary being as cause or part of the world:

(T4) $$\exists x (\Diamond Px \rightarrow Px)$$

(2) Then there is an absolutely necessary being *either* in the series of the alterations in the world (causal interpretation of the absolutely necessary being, confounding cause and sufficient reason):

(T3) $$\exists x \neg \exists y\ y < x$$

(3) But this contradicts the principle of causality (P_C) [according to which every event is conditioned by a law of nature], so it cannot be the case:

$$\neg \exists x \neg \exists y\ y < x$$

(4) *Or* the sequence of all changes has no beginning in the world (logical alternative to (2), equivalent to (3)):

($A3 \equiv P_C$) $$\forall x \exists y\ y < x$$

(5) Then each part of the series must be [relatively] contingent, even though the series as a whole should be an [absolutely] necessary being:

(A4) $$\neg \exists x (\Diamond Px \rightarrow Px)$$

(6) Hence, no absolutely necessary appearance can exist *within* the world. The assumption that there is an absolutely necessary cause of the world *outside* of the world, however, again contradicts the principle of causality.

Here again, in (2) a relatively necessary cause is mistaken for an absolutely necessary sufficient reason; and in (3) the principle of causality is introduced as an additional premise. Both proof steps derive from the naturalistic presuppositions of the proof. The modal re-interpretation of (4) in (5) then contradicts the initial presupposition and is identical with the antithesis to be proven.

5.4.4.3 Conclusions on the Fourth Antinomy

Both proofs are based on conflating two incompatible concepts of necessity: on the one hand the "relative" necessity in the sense of causal conditions; and on the other hand the "absolute" necessity of a being whose concept implies its existence, as in the case of the ontological proof, which identifies God with the sum total of all realities. The conflation is due to the equivocal use of the term 'condition' in a relative or in an absolute sense. According to Kant, the reason for confounding both concepts of necessity is due to reason's quest for completeness, which employs the principle of causality in an unrestricted use and is committed to naturalism on the part of both thesis and antithesis.

From Kant's critical viewpoint, the antinomy is resolved similarly to the third antinomy. This time, the solution is to detect that the proofs conflate a non-logical variant of the principle of sufficient reason with the principle of causality, mistaking relatively necessary conditions for an absolutely necessary ontic ground or *ratio essendi*. Kant considers this conflation to be inevitable as long as relative and absolute necessity are not distinguished. Again, the proof of the antithesis appears to be compelling if one assumes that the principle of causality is universally valid in the sensible world.

5.5 Conclusions: The Equivocations in the Cosmological Concept

It remains to summarize how the contradiction in the cosmological concept recurs in each of the four antinomies, according to Kant. To consider the world as the sum total of all appearances with a concrete spatio-temporal object gives rise to four kinds of equivocations, on which the proof results decisively depend and which consist in the conflating of concepts that characterize the intelligible or the sensible world, respectively.[36] In the "mathematical" antinomy, only the thesis proof is based on this equivocation, while the antithesis proofs are carried out by applying the concepts of space and time to the concept of the world of appearances. In the "dynamic" antinomy, on the other hand, the proofs of thesis and antithesis are each based on one and the same equivocation.

I *Antinomy of composition*:
Thesis argument:
Equivocation of 'infinite' (cardinal infinite ↔ ordinal infinite)
Antithesis argument: Ideality and relationality of space and time

II *Antinomy of division*:
Thesis argument:
Equivocation of 'simple substance' (logical individuum ↔ mereological atom)
Antithesis argument: Mereological structure of a spatial substance

III *Antinomy of causality*:
Thesis and antithesis arguments:
Equivocation of 'condition' (sufficient reason ↔ cause in nature)

[36]The cosmological antinomy has this aspect in common with the paralogism of pure reason. The paralogism is based on a fallacious categorical syllogism in which the middle term 'I' is used equivocally in an empirical or intelligible sense. The antinomy, on the other hand, is based on a fallacious hypothetical syllogism in which the complete series of empirical conditions is conflated with an intelligible unconditioned; see the distinction between paralogism and antinomy (A 407/B 433–434) as well as the diagnosis that the antinomy is based on a *sophisma figurae dictionis* (A 499/B 527–528, and *Logic* §90, 9:135). See also Malzkorn (1999, 110). Seifert (1989) concludes from this equivocation that Kant's argument is inconclusive, similar to Malzkorn (1999), who does so for the third and fourth antinomy. Kant, however, only has the burden of showing that the proof *seems* conclusive from a dogmatic metaphysical point of view. He himself was convinced that they are *not*.

IV *Antinomy of necessity*:
Thesis and antithesis arguments:
Equivocation of 'necessity' (absolute necessity ↔ relative necessity)

In accordance with Kant's system of the four versions of the cosmological antinomy, the terms used equivocally fall under the categories of quantity, quality, relation, and modality (A 411–415/B 438–443). Corresponding to the basis of the proofs, the respective equivocations in the "mathematical" or "dynamic" antinomy have different logical and semantic consequences.

All four versions of the cosmological antinomy derive from considering the world as a whole at once as a concrete spatio-temporal object and as an abstract class of such objects. The reasons for doing so lie in the tacit semantic assumptions of the proofs, in specific equivocations which make the proof results plausible. In Kant's terminology, the antinomy derives from the cosmological concept of the world as the sum total of appearances. This is the pre-critical cosmological concept used in Wolff's and Baumgarten's metaphysics, which also underlies Kant's own pre-critical cosmology from 1755 to 1770. The "mathematical" antinomy concerns the extension and the composition of the spatio-temporal world, the "dynamical" antinomy concerns the nomological relations within nature. Kant's development shows that he detected the mathematical antinomy in the 1770s, most probably between 1772 and 1775 (see Sects. 4.3.3 and 4.4). For him, as we have noted (see Sect. 5.3.2), the true scandal of reason is the mathematical antinomy. For both the thesis and the antithesis, it makes claims about the spatio-temporal properties of appearances, according to which the above cosmological concept is self-contradictory.

5.5.1 First Antinomy

Kant's own pre-critical cosmology advances the thesis position by assuming a finite world age, and the antithesis position by assuming that the world is infinite in space. In the *Theory of the Heavens* he did not employ the thesis or antithesis arguments in favour of these assumptions. My analysis of Kant's critical turn shows that in the mid 1770s he became aware of the antinomic structure of claims about the size of the world in space and time (see Sect. 4.3.3). According to my reconstruction, the proofs of the thesis and the antithesis are based only on Kant's 1770 theory of space and time and his 1770 concept of infinity, which give rise to verificationist arguments (Sect. 5.4.1.3), but *not* on the specific assumption of transcendental idealism that things in themselves are epistemically unaccessible. Indeed, Kant emphasizes in the *CPR* that the antinomy only concerns the "absolute totality in the synthesis of appearances" (A 407–408/B 434). As far as I can see, the proofs of the first antinomy proceed on the grounds of Kant's pre-critical point of view of 1770, including his 1770 theory of space and time as pure forms of intuition, but *not* on his later assumptions of transcendental idealism. From his pre-critical views up to 1770, the proofs seemed conclusive, given that they take up a traditional argument against an infinite world age as well as arguments from the Leibniz–Clarke debate which were well known to his readers. It is only from Kant's critical point of view that the thesis proof turns out to be defective, given that it conflates an ordinal and a cardinal concept of the infinite (Sect. 5.4.1). And Kant only realized that they do so by distinguishing the mathematical and the metaphysical conceptions of the infinite, which he did from the mid 1770s onwards (see Sect. 4.4).

After his critical turn, he considers the equivocal use of 'infinite' as an inevitable consequence of identifying the sum total of appearances with a concrete object; and hence of identifying a logical concept whose content is successively generated (i.e., ordinal infinite) with a given quantity (i.e., with a cardinal infinite). The critical Kant makes a sharp distinction between the corresponding concepts of the potential and the actual infinite. The thesis proof, which in the time argument is based on the traditional paradox of the infinite world age and in the space argument on Kant's own verificationist concept of infinity of 1770, is therefore defective from Kant's critical point of view. In contrast, the proof of the antithesis, Leibniz's argument against Newton's absolute concepts of space and time, remains correct from Kant's critical point of view, if correctly understood in terms of a potentially instead of an actually infinite world size.

5.5.2 *Second Antinomy*

The proofs of the second antinomy rely on Kant's own pre-critical proofs concerning the existence of physical monads and the infinite divisibility of a spatial continuum filled with matter, as given in the *Physical Monadology* (1:477, 1:479). The arguments are based on eighteenth-century views about material substances and the space they fill. These views were already subject to debate, to which Kant had contributed with his 1756 work. From a modern point of view, the proofs merge the following concepts: logical individuals and mereological atoms; formal (mereological) and spatial part–whole relations; and the formal domain of logical individuals and the spatio-temporal world (Sect. 5.4.2). In order to assess Kant's proofs and his view of their fallacies, we should be aware that neither Cantor's later account of the mathematical continuum nor the concepts of twentieth-century set theory were available to him. Against the background of Wolff's or Baumgarten's ontology, it seemed compelling to consider simple substances (or monads) to be parts of the spatio-temporal world; and hence to conflate the structure of a spatial continuum with the logical properties of a class of substances (see Sect. 5.4.2.3). This conflation, which is illegitimate from Kant's critical point of view, gives rise to the antinomy. It is based on an equivocal use of the term 'simple substance' for the element of a class and the part of a whole. To use the term 'simple' for logical and mereological simplicity and to apply the latter to spatio-temporal substances leads to the interpretation of simple substances as both non-relational and relational.

Kant himself considered the thesis proof as convincing as long as the underlying logical fallacy is not revealed. In 1781, Kant considers the apagogic refutation of the antithesis in favour of the thesis achieved in this way as defective, because it is based on contradictory premises. The proof of the antithesis is different. Kant regards the apagogic refutation of the claim that a spatial substance may have simple spatial parts as compatible with the claim that material substances are divisible to infinity. Analogously to the first antinomy, the proof result remains correct from Kant's critical point of view, if correctly understood as a potentially instead of an actually infinite number of parts. This resolution is in accordance with the matter theory of the *MFNS*. In contrast, the additional proof that refutes the existence of simple non-spatial substances, such as Leibniz's monads, is direct. It employs an epistemological argument concerning the reference to the objects in the world of sense.

5.5.3 Third and Fourth Antinomy

The case of the dynamic antinomies is different. For them, Kant continues to consider the dualism of *phaenomena* and *noumena* to be tenable. The arguments in both antinomies are largely parallel. The proofs confound logical and real grounds, or the principle of causality and the principle of sufficient reason, in similar ways. Against the background of rationalist metaphysics from Leibniz to Wolff and Baumgarten, including French materialism and the kind of naturalism or Epicureanism against which the preface of the *Theory of the Heavens* had argued, Kant presumably thought that this conflation is inevitable in any kind of dogmatic metaphysics.

Here, from Kant's critical point of view, the proofs of both thesis and antithesis are similarly defective, in particular as far as the conclusiveness of the "apagogic" proof procedure is concerned. According to the resolution of the antinomy, the assumption that thesis and antithesis refer to the same world (as a sum total of appearances) is illegitimate. In both variants of the antinomy, the antithesis position corresponds to naturalism. The thesis proofs, in turn, purports to trace conditions in the world of sense back to intelligible grounds. In the third antinomy, the thesis claim involves a commitment to a naturalistic interpretation of the principle of sufficient reason; and in the fourth antinomy, to the rationalist variant of naturalism, i.e., Spinozism. According to the resolution of the "dynamic" antinomy, thesis and antithesis do not form contradictory, but subcontrary pairs of judgments (see Wolff 2017, 74–81). The thesis refers in each case to the intelligible world, the antithesis to the world of sense. On the part of the antithesis, the arguments are still acceptable from a critical point of view. They make use of the principle of causality, which, however, again needs to be restricted to a potentially (rather than actually) infinite series of conditions. On the part of the thesis, however, from Kant's critical point of view the arguments lead only to subjectively plausible cognition of the intelligible world, which does not satisfy objectively sufficient truth conditions.

The structural difference between the mathematical and the dynamical antinomy is due to the mathematical vs. dynamic interpretation of the relation $<$ in the antithetic claims. The mathematical antinomy employs part–whole relations, the dynamical antinomy causal relations and logical vs. real grounds. For part–whole relations, the "subreptic axiom" of 1770 concerning the conflation of the sensible and intelligible world concepts now seems inevitable to Kant, in contrast to his expectations of 1769–1770 (see Sect. 4.3.1). From the defective proofs and the logical structure of the antithetic claims he concludes that thesis and the antithesis are both false, given that they rely on an unsatisfied existential presupposition concerning the spatio-temporal world as a sum total of appearances. In contrast, for the dynamical antinomy he still considers the thesis and antithesis proofs to be "subreptic axioms" in the sense of his 1770 cosmology. According to his resolution of the antinomy, here it is possible to give the logical relation of ground and consequence different interpretations for the sensible and the intelligible world, and hence to separate the respective realms of the thesis and the antithesis. To what extent this resolution is convincing is another question, and indeed a substantial philosophical problem that is beyond the scope of the present book.

References

Allison, Henry E. 1983. *Kant's Transcendental Idealism: An Interpretation and Defence*. New Haven: Yale University Press.
Allison, Henry E. 2004. *Kant's Transcendental Idealism: An Interpretation and Defence*. Revised and Enlarged Edition. New Haven: Yale University Press.
Al Azm, Sadik J. 1972. *The Origin of Kant's Arguments in the Antinomies*. Oxford: Clarendon Press.
Bennett, Jonathan. 1974. *Kant's Dialectic*. Cambridge: Cambridge University Press.
Baumgarten, Alexander Gottlieb. 1757. *Metaphysica*. Halle: Hemmerde. Quoted after *Kants gesammelte Schriften*, vol. 17.
Boscovich, Roger. 1758. *Theoria philosophiae naturalis. A Theory of Natural Philosophy*, ed. J.M. Child. Chicago and London: Open Court Publishing Company 1922. https://archive.org/details/theoryofnaturalp00boscrich.
Cantor, Georg. 1932. *Gesammelte Abhandlungen mathematischen und philosophischen Inhalts*, ed. Ernst Zermelo. Berlin: Springer.
Engelhard, Kristina. 2005. *Das Einfache und die Materie. Untersuchungen zu Kants Antinomie der Teilung*. Kantstudien-Ergänzungshefte 148. Berlin: de Gruyter.
Falkenburg, Brigitte. 1995. Kants zweite Antinomie und die Physik. *Kant-Studien* 86: 4–25.
Falkenburg, Brigitte. 2000. *Kants Kosmologie*. Frankfurt a. M.: Klostermann.
Falkenburg, Brigitte. 2004. Experience and Completeness in Physical Knowledge: Variations on a Kantian Theme. In *Philosophiegeschichte und Logische Analyse, Schwerpunkt: Geschichte der Naturphilosophie*, ed. U. Meixner and A. Newen, 153–176. Paderborn: Mentis.
Falkenburg, Brigitte. 2005. Some remarks on cosmology and scientific realism. *Kairos* 26: 229–246.
Friedman, Michael. 1992. *Kant and the Exact Sciences*. Cambridge: Harvard University Press.
Grier, Michelle. 2001. *Kant's Doctrine of Transcendental Illusion*. New York: Cambridge University Press.
Guyer, Paul. 1987. *Kant and the Claims of Knowledge*. Cambridge: Cambridge University Press.
Guyer, Paul, and Allen W. Wood. ed. 1998. *The Critique of Pure Reason*. Cambridge: Cambridge University Press.
Hallett, Michael. 1984. *Cantorian Set Theory and Limitation of Size*. Oxford: Clarendon Press.

Hegel, Georg Friedrich Wilhelm. 1832. *Wissenschaft der Logik*. Repr. in: *Gesammelte Werke*, Hamburg: Meiner (quoted here: Vol. 21, 1985). Engl.: *The Science of Logic*. Translated and edited by George Di Giovanni. Cambridge University Press 2010.

Heimsoeth, Heinz. 1960. *Atom, Seele, Monade*. Verlag der Akademie der Wissenschaften: Mainz und Franz Steiner Verlag: Wiesbaden.

Hinske, Norbert. 1971. "Antinomie". In *Historisches Wörterbuch der Philosophie*, ed. Joachim Ritter, vol. I, Darmstadt, col. 393–396.

Hinske, Norbert. 1972. Kants Begriff der Antithetik und seine Herkunft aus der protestantischen Kontroverstheologie des 17. und 18. Jahrhunderts. Über eine unbemerkt gebliebene Quelle der Kantischen Antinomienlehre. *Archiv für Begriffsgeschichte* 16: 48–59.

Kreimendahl, Lothar. 1990. *Kant—der Durchbruch von 1769*. Köln: Dinter.

Malzkorn, Wolfgang. 1999. *Kants Kosmologie-Kritik. Kant-Studien Erg.-Hefte 134*. Berlin: de Gruyter.

Martin, Gottfried. 1955. *Kant's Metaphysics and Theory of Science*. Manchester University Press. Transl. of: *Immanuel Kant: Ontologie und Wissenschaftstheorie* (Köln: Universitätsverlag 1951).

Menzer, Paul, ed. 1924. *Eine Vorlesung Kants über Ethik*. Im Auftrag der Kantgesellschaft herausgegeben von Paul Menzer. (Philosophia practica universalis). Second ed. 1925, Berlin-Charlottenburg: Pan Verlag Rolf Heise, Berlin-Charlottenburg.

Mittelstaedt, Peter, and Ingeborg Strohmeyer. 1990. Die kosmologischen Antinomien in der Kritik der reinen Vernunft und die moderne physikalische Kosmologie. *Kant-Studien* 81: 145–169.

Moore, Gerald E. 1953. *Some Main Problems of Philosophy*. London: Allen & Unwin.

Newton, Isaac. 1730. *Opticks*. With a preface by I.B. Cohen, based on the 4th ed., New York. 4th ed., London 1930. With a Preface by I.B. Cohen, Dover 1979.

Russell, Bertrand. 1903. *The Principles of Mathematics*. New York: W.W.Norton & Company Inc.

Schmucker, Josef. 1990. *Das Weltproblem in Kants Kritik der reinen Vernunft*. Bonn: Bouvier.

Seifert, Josef. 1989. Das Antinomienproblem als ein Grundproblem aller Metaphysik: Kritik der "Kritik der reinen Vernunft". *Prima Philosophia* 2: 143–168.

Simons, Peter. 1987. *Parts. A Study in Ontology*. Oxford: Clarendon Press.

Strawson, Peter F. 1952. *Introduction to Logical Theory*. London: Methuen.

Strawson, Peter F. 1966. *The Bounds of Sense. An Essay on Kant's Critique of Pure Reason*. London: Methuen.

van Benthem, Johan. 1983. *The Logic of Time: a modal-theoretic investigation into the varieties of temporal ontology and temporal discourse*. Dordrecht: Reidel.

Vogel, Karl. 1975. *Kant und die Paradoxien der Vielheit*. Meisenheim am Glan: Hain.

Willaschek, Marcus. 2018. *Kant on the Sources of Metaphysics. The Dialectic of Pure Reason*. New York: Cambridge University Press.

Wind, Edgar. 1934. *Das Experiment und die Metaphysik*. Tübingen: Mohr. Repr.: B. Buschendorf (ed.), with an introduction by B. Falkenburg, Frankfurt am Main: Suhrkamp 2000. Engl. transl. by C. Edwards: *Experiment and Metaphysics. Towards a Resolution of the Cosmological Antinomies*. Oxford: Legenda 2001.

Wolff, Michael. 2017. *Der Begriff des Widerspruchs. Eine Studie zur Dialektik Kants und Hegels*, 3rd ed. Berlin: Owl of Minerva Press.

Chapter 6
Cosmology and Transcendental Idealism

> *The questions whether the world has a beginning and its extension in space a boundary; whether there is anywhere, perhaps in my thinking self, an indivisible and indestructible unity [...]; whether my actions are free or, like those of other beings, controlled by the strings of nature and fate; whether, finally, there is a supreme cause of the world, or whether natural things and their order constitute the ultimate object [...]—these are questions for whose solution the mathematician would gladly give up his entire science; for that science cannot give him any satisfaction in regard to the highest and most important ends of humanity.* (*CPR, A 464/B 492*)

What remains of Kant's pre-critical cosmology, in view of the antinomy of pure reason? In Chap. 5, I discussed the relationship between the cosmological antinomy and transcendental idealism, mainly with regard to the circularity problem (see Sect. 5.1.3). In this chapter I examine this relationship from a more general point of view. First I will summarize my results regarding the circularity problem (see Sects. 5.3.3 and 5.4), in order to clarify whether, or to what extent, the antinomy is a self-generated problem of Kant's transcendental idealism (Sect. 6.1). I then proceed to aspects of the philosophical significance of the antinomy which have been neglected in Kant research over recent decades. Kant considered the antinomy as an independent argument in favour of transcendental idealism. The strength of this argument obviously depends on the conclusiveness of the doctrine of the antinomy, in particular on the persuasiveness of the proofs and Kant's claim that the antinomy can only be resolved on the grounds of transcendental idealism. In the preface B of the *CPR*, he illustrates the significance of the antinomy in terms of an "experiment of pure reason" (B xxi). The "experiment of pure reason" is a thought experiment which aims at proving that transcendental idealism is the only viable philosophical position (Sect. 6.2.2). Kant here again employs an analogy between Newtonian science and the method of metaphysics, in the context of his project to give metaphysics the secure course of a science, via transcendental idealism.

B. Falkenburg, *Kant's Cosmology: From the Pre-Critical System to the Antinomy of Pure Reason*, European Studies in Philosophy of Science 12,
https://doi.org/10.1007/978-3-030-52290-2_6

Finally, I will examine what remained of Kant's pre-critical unification project in his critical philosophy, discussing hitherto overlooked aspects concerning the continuity between Kant's pre-critical and critical cosmology (Sect. 6.3), not only with regard to the methods of metaphysics (Sect. 6.3.1) and the structure of the critical system (Sect. 6.3.2), but also concerning the role of the cosmological antinomy as a new case against naturalism (Sect. 6.3.3). The new shape which he gave to these issues in his critical system is closely related to his general topic of the teleology of reason, as expressed in terms of the "necessary and essential ends of humankind" (A 850/B 878) at the end of the Architectonics chapter of the *CPR* (Sect. 6.4).

6.1 The Antinomy: A Self-Generated Problem?

Let me first take up once more the unresolved questions of the last two chapters. Is Kant's diagnosis of the cosmological antinomy a self-generated problem of his critical turn, which results from proofs that fatally depend on crucial assumptions of transcendental idealism? Or are the proofs of the four antinomies compelling at least against the background of eighteenth-century philosophy? As we put it at the end of Section 5.3.3: Why does Kant assume that the antinomy-avoidance strategy of 1770 does not work for the mathematical antinomy of 1781, whereas he re-establishes it for the dynamical antinomy? And why does he nevertheless consider the latter, too, as a genuine cosmological antinomy, if he already knew the resolution (i.e., the distinction between noumena and phenomena) around 1770, before the exact formulation of the problem? To give a definite answer to the second question would require a much more detailed analysis of the topics of the dynamical antinomy, which is beyond the scope of the present book. Here, I focus on the mathematical antinomy, which is decisive for Kant's claim that the cosmological antinomy is inevitable.

It has to be emphasized again that Kant himself did not consider the proofs of the antinomy to be tenable. In the *CPR*, he claimed only that the proofs *seem* conclusive as long as one maintains a dogmatic metaphysical point of view, instead of transcendental idealism. According to my logical reconstruction in Section 5.4, the proofs of the mathematical antinomy rely on traditional metaphysical and epistemological arguments, as well as on Kant's own pre-critical views up to 1770, including verificationist principles. These arguments were complemented by astronomical problems concerning the options of a finite or infinite universe (De Bianchi 2013, 42–44), and the resolution of the mathematical antinomy also offered a way out of them. Against the background of the metaphysical debates of his time, Kant's own pre-critical contributions to them, and the parts of his pre-critical theory of the universe that still seemed tenable to him,[1] the proofs must have

[1] In 1791, Kant authorized the publication of an excerpt of the *Theory of the Heavens* as an appendix to the German translation of William Herschel's works (Herschel and Gensichen 1791). For details, see De Bianchi (2013); for the differences between the pre-critical and the critical cosmogonies see Ferrini (2004).

seemed conclusive. Indeed, the objections against Kant's proofs discussed above in Section 5.1 did not arise before the late nineteenth century.

With regard to the size of the world, the proof of the thesis depends decisively on restricting the conception of the infinite to the real use of reason in cosmology. By doing so, Kant relies on his own pre-critical concept of infinity as formulated in 1770. His dissertation of 1770 strictly distinguished the mathematical limiting concept of the infinite, which has the meaning 'larger than any given finite number' (corresponding to the ordinal infinite), on the one hand, and the metaphysical concept of the actual infinite, i.e., the infinite in the sense of a multiplicity the scope of which exceeds the comprehension of the human mind (corresponding to the cardinal infinite), on the other (2:389; see Sect. 4.4).

In his abstract or symbolic cosmology of 1770, however, he still applied the concept of the actually infinite to the world as a whole and its simple parts (see Sect. 4.2.3). There, he defended this cosmological use of the concept by differentiating the limits of the human mind from the "limits within which the very essence of things is contained":

> For [...] *the abstract ideas which the mind entertains when they have been received from the understanding very often cannot be followed up in the concrete and converted into intuitions.* But this *subjective* resistence often creates the false impression of an *objective* inconsistency. And the incautious are easily mislead by this false impression into taking the limits, by which the human mind is circumscribed, for the limits within which the very essence of things is contained. (2:389)

We have seen that Kant had rejected the actual infinite for cosmological cognition since 1773–1775. Indeed, he did so in the course of developing his theory of objective cognition. He had seen the need for such a theory from 1772 at the latest, as is shown by his famous letter to Marcus Herz dated February 21, 1772 (see Sect. 4.3.3). It was then that he started to recognize that the (ordinal) mathematical limiting concept of the infinite is needed for the real use of reason in cosmology, and he could no longer consider the strict separation between an abstract or symbolic cosmology and the spatio-temporal world as legitimate for the real use of reason in metaphysics. However, he only became aware of the antinomy by trying to retain dogmatic metaphysics based on his increasing insights into the subjective conditions of metaphysical cognition, if we believe his note N. 5116 on metaphysics from the late 1770s (see Sects. 4.3.3 and 4.5):

> It took a long time before in such a way I found the whole dogmatic theory to be dialectical. [...] But while I thereby certainly became careful, I was still not instructed. [...] for I still believed I could find a method for extended dogmatic cognition through pure reason. (18:95–96)

Around 1770, when Kant began "earnestly" to make use of the antithetic method given the lack of a better philosophical method, he himself saw "this doctrine" only "as if in twilight" (18:69), with the rudimentary epistemology based on his new theory of intuition (see Sect. 4.1). From his own later point of view, Kant probably thought that the epistemological insights of the years 1770–1775 had opened his eyes to the true grounds of the never-ending cosmological debates of his time. The "great light" of 1769 had apparently not yet carried so far as to illuminate these grounds, consisting at this stage mainly in the revision of the rationalist doctrine of ideas by identifying space and time with pure forms of intuition. Compared with the transcendental philosophy of 1781 (including the doctrine of the antinomy), this "great light" was indeed only a dawn.

After all, Kant did not want to resolve the so-called "antinomy problem" with his dissertation of 1770, as has for a century been claimed. Such a claim confuses ground and consequence. The antinomy took shape for him only on the grounds of the 1770 dissertation, after rejecting the concept of the actual infinite for cosmology, and after realizing that one should not conceive of a (relationless) "simple substance" as a (relational) "substance in the world". Hence, in as much as Kant's discovery of the cosmological antinomy emerged in the course of developing the foundations of transcendental idealism, to a certain extent it is indeed a self-generated problem of the critical turn. He might have continued doing formal ontology, as a theory of the world that may or may not be true, on the one hand, while on the other developing a new theory of cognition. However, given that his new theory of cognition was still directed at the spatio-temporal world as an object of metaphysics, this was not what he did. By rejecting the actual infinite for cosmology, Kant became convinced that the cosmological concept of the spatio-temporal world confuses the abstract concept of a sum total of appearances with a concrete multitude of things (see Sects. 4.3.3 and 4.4). As he put it in *N. 4525*: "The absolute-whole in appearance is a contradiction" (18:96).

But how are this reconstruction, and the related circularity objection, compatible with his own later statements on the crucial significance of the antinomy for his path towards transcendental idealism? According to the circularity objection raised by Strawson (1966, 195–196), Kant presupposes in the proofs of the antinomy what has to be proven, that is, transcendental idealism. My main point in defence of Kant is that he considered the proofs to be convincing from the point of view of transcendental realism. As shown in Section 5.4.1.3, the proofs for the thesis and antithesis of the first antinomy in the *CPR* can indeed be coherently interpreted in such a way that they do not crucially depend on the foundations of transcendental idealism, but rather on verificationist arguments based on different positions of transcendental realism. Similarly, the proofs of the second antinomy rely on rationalist assumptions about simple substances "in the world" on the thesis side, and on traditional assumptions about spatially extended substances, on the antithesis side (see Sect. 5.4.2.3).

Let me add a few remarks on the third and fourth antinomies. Their subjects connected rational cosmology to the other parts of Wolff's metaphysical system, that is, to rational psychology and theology. Their thesis and antithesis claims were also subject to never-ending philosophical debates, and their proofs rely on a confusion between logical and real grounds, or sufficient reasons and causes, which in Kant's view was a typical fallacy of dogmatic metaphysics, a fallacy from which he however had already escaped in his 1763 writings. The thesis proofs employ rationalist arguments, while the antithesis proofs depend on naturalist arguments current at the time. Here, too, the arguments may be defended without recourse to Kant's position of transcendental idealism.

Kant's later retrospective autobiographical remarks, up to his letter to Garve of September 20, 1798, only make sense if he himself considered the cosmological antinomy to be a discovery of far-reaching philosophical implications. Finding the mathematical antinomy was decisive for him, and the way he shaped the dynamic antinomy was a consequence of this finding. Kant's crucial new insight of 1773–1775 was that human reason is inescapably involved in conflicting cosmological claims, as long as it does not acknowledge its limitations. The claim that Kant considered the cosmological antinomy to be an internal problem of dogmatic metaphysics needs therefore to be qualified. The inevitability claim concerning the antinomy resulted from reformulating the traditional cosmological arguments by taking into account that they apply to objects in space and time. Given that, they inevitably make use of naturalistic premises. In this way the antinomy finally provides him with a new, critical argument against naturalism, after the collapse of his pre-critical system (see Sect. 6.3.3).

6.2 Against Transcendental Realism

After his critical turn, Kant considered transcendental idealism as the only viable alternative to the traditional philosophical positions of his day, in particular, to

(i) "transcendental realism", that is, any version of a metaphysical realism about the world or nature,
(ii) "empirical idealism", that is, Berkeley's dogmatic idealism, and
(iii) "skepticism", among which he ranked Descartes's radical doubt as well as Hume's skepticism.

With the doctrine of the antinomy he wanted to demonstrate that no version of transcendental realism is tenable, given that it leads to an inescapable conflict of cosmological claims. It has to be noted that the doctrine of the antinomy attacks neither Berkeley's idealism nor Descartes's skeptical doubt (see Sect. 6.2.1), but all versions of transcendental realism, in particular Leibniz's monadology, Spinoza's pantheistic doctrine of substances (as Kant's refutation of the thesis proofs of the second and fourth antinomy demonstrates), the metaphysical views of the Wolffians and Newtonians, and finally, naturalism, that is, the position which Kant calls "Epicureanism", e.g., French materialism (which corresponds to the antithesis claims of the third and fourth antinomy). It has to be taken into account, however, that Kant's arguments in the doctrine of the antinomy do not rule out Hume's skepticism. By attacking transcendental realism, indeed, they seem rather to stand in Hume's anti-metaphysical tradition. In the *Prolegomena*, Kant even repeats his rhetorical appeal concerning the interruption of his "dogmatic slumber" by Hume (4:260), but this time with regard to the cosmological ideas and their self-contradictory consequences:

> This product of pure reason in its transcendent use is its most remarkable phenomenon, and it works the most strongly of all to awaken philosophy from its dogmatic slumber, and to prompt it toward the difficult business of the critique of reason itself. (4:338)

As emphasized above (in Sects. 4.3.2 and 4.3.3), this rhetoric does not indicate that there was any direct impact of Hume's skepticism on the genesis of Kant's skeptical method and the discovery of the antinomy. By referring to Hume in the *Prolegomena*, and by emphasizing the significance of the antinomy in the preface B of the *CPR*, Kant was actually only seeking greater popular appeal for his doctrine of transcendental idealism. He wanted to demonstrate that the antithetic of pure reason provides an independent argument in favour of transcendental idealism, an argument which is easier to grasp than the difficult explanations of the Transcendental Analytic. He regarded the "apagogic" proofs of the cosmological antinomy, which were clothed in traditional metaphysical and epistemological arguments, as a means of converting to his "revolution in the way of thinking" (B xii), even those readers whom the Transcendental Aesthetic and Analytic did not convince. According to the preface B of the *CPR*, the cosmological antinomy represents the negative result of an "experiment of pure reason" (B xxi; see Sect. 6.2.2). Given the importance of this thought experiment, it is remarkable that Kant research has not paid more attention to it.[2] It is completely new in the second edition of the *CPR*, which, on the other hand, completely detaches Berkeley's idealism from the antinomy of pure reason. Before analyzing Kant's experiment of pure reason in more detail, let me briefly explain the relation between the antinomy and Berkeley's idealism.

6.2.1 The Refutation of Idealism

One of the few passages changed in the Transcendental Dialectic of 1787 concerns the refutation of idealism. In the 1781 edition of the *CPR*, the refutation of idealism is located at the end of the paralogism chapter of the Transcendental Dialectic. It is only here that Kant relates Berkeley's "empirical idealism" directly to the antinomy of pure reason. In particular, he distinguishes (Berkeley's) "dogmatic" idealism from (Descartes's) "skeptical" idealism and adds a remark concerning the relation of the former to the doctrine of antinomies:[3]

> The **dogmatic idealist** would be one who denies the existence of matter, the **skeptical idealist** one who doubts them because he holds them to be unprovable. The former can be so only because he believes he can find contradictions in the possibility of a matter in

[2]There are a few exceptions: Buzzoni (2011, 2017) takes up Kant's experiment of pure reason in the general context of thought experiments in science and philosophy; Fulkerson-Smith (2013a,b) focuses on interpreting it as an illuminating experiment in Bacon's sense; and Zuckert (2020) discusses Kant's *a priori* experimentation as a proof that the attempt to exit the human perspective fails. Kalin (1972) considers Kant's transcendental arguments in general to be thought experiments.

[3]Heidemann (1998, 79–80) relates the passage to A 381–395. Due to Kant's phrase "reason in its internal conflict" this is not plausible, however. See also Guyer and Wood (1998, 739, n. 34).

> general, and just now we are not yet dealing with that. The following section on dialectical inferences, which represents reason in its internal conflict regarding the concepts belonging to the possibility of the connection of experience, will also help us out of this difficulty. (A 377)

In the second edition of the *CPR*, Kant shifts the refutation of idealism to the end of the section on the principles of the pure understanding (B 274–279); this is in fact a reaction to Garve's and Feder's review of the *CRP* in the *Göttingische Gelehrte Anzeigen* and the resulting debate (Heidemann 1998, 87–93). His refutation of idealism of 1787 distinguishes Descartes's "problematic" or "skeptical idealism" from Berkeley's "dogmatic idealism" as follows:

> Idealism (I mean **material** idealism) is the theory that declares the existence of objects in space outside us to be either merely doubtful and **indemonstrable**, or else false and **impossible**; the **former** is the **problematic** idealism of Descartes, who declares only one empirical assertion (*assertio*), namely **I am**, to be indubitable; the **latter** is the **dogmatic** idealism of Berkeley [. . .]. (B 274)

In the "refutation of idealism" of 1787, Kant argues only against Descartes's form of doubt, and no longer against Berkeley's dogmatic idealism. He now claims that Berkeley

> declares space, together with all the things to which it is attached as an inseparable condition, to be something that is impossible in itself, and who therefore also declares things in space to be merely imaginary. (ibid.)

Thus here he no longer associates Berkeley's position with the topics of the cosmological antinomy, i.e., transcendental realism about the world of appearances; rather, he now associates it with metaphysical realism about space and time, which he already sees as refuted by the Transcendental Aesthetic:

> Dogmatic idealism is unavoidable if one regards space as a property that is to pertain to the things in themselves; for then it, along with everything for which it serves as a condition, is a non-entity. The ground for this idealism, however, has been undercut by us in the Transcendental Aesthetic. (ibid.)

Correspondingly, the Transcendental Aesthetic of 1787 contains an additional remark according to which attributing space and time to things in themselves makes Berkeley's idealism plausible:

> For if one regards space and time as properties that, as far as their possibility is concerned, must be encountered in things in themselves [. . .]; then one cannot well blame the good Berkeley if he demotes bodies to mere illusion [. . .]. (B 70–71)

Neither this passage nor the remark on Berkeley's dogmatic idealism quoted above (B 274) refers exclusively to Newton's theory of absolute space and time. In my view, Kant here is concerned rather to remind his readers of the problem (discovered by him in 1768) that neither Newton's nor Leibniz's theories of space and time are tenable as objective theories. Indeed, Kant now claims that space as a property of things in themselves is a "non-entity" (B 274). According to the amphiboly chapter of the *CPR*, a "non-entity" is an "**empty object without concept**, *nihil negativum*" (A 292/B 348), that is, something corresponding to a self-contradictory concept. The puzzle of 1768 resulted in a concept of space according to which space, considered as a thing in itself, is neither relational not absolute, i.e., non-relational. Kant escaped this with his 1770 theory of space and time as forms of intuition, long before stating and resolving a cosmological antinomy in the sense of the *CPR*. Hence, in 1787 Kant removes a self-contradictory concept of space comparable to his puzzle of 1768 from the context of his cosmological antinomy. This shift indicates that Kant did not want to merge his 1768 puzzle of incongruent counterparts with the antinomy of 1781, which is in accordance with my reconstruction of his critical turn (see Chap. 4).

6.2.2 The Experiment of Pure Reason

Kant's "experiment of pure reason" (*CPR*, B xxi) comes in the context of his famous remarks on the secure course of a science. Let me start with the well-known passage in which Kant compares metaphysics with natural science:

> When Galileo rolled balls of a weight chosen by himself down an inclined plane, or when Torricelli made the air bear a weight that he had previously thought to be equal to that of a known column of water, or when in a later time Stahl changed metals into calxc and then changed the latter back into metal by first removing something and then putting it back again,* a light dawned on all those who study nature. They comprehended that reason has insight only into what it itself produces according to its own design; [...] Reason, in order to be taught by nature, must approach nature with its principles in one hand, according to which alone the agreement among appearances can count as laws, and, in the other hand, the experiments thought out in accordance with these principles—yet in order to be instructed by nature not like a pupil, who has recited to him whatever the teacher wants to say, but like an appointed judge who compels witnesses to answer the questions he puts to them. (B xxii–xxiii)
>
> * Here I am not following exactly the thread of the history of the experimental method, whose first beginnings are also not precisely known. (B xxiii)

Here, Kant distinguishes his position from empiricism. An experiment is not merely an observation of natural phenomena; it is theory-laden, that is, it poses questions to nature on the basis of theoretical presuppositions. From the point of view of the philosophy of science, Kant here takes a partly constructivist, partly critically rationalist position. To be more precise, he combines the constructivist view, according to which the law-like connection of phenomena is only guaranteed by the principles of the pure understanding, with a hypothetico-deductive approach, according to which experimentation is theory-driven and serves to test specific theoretical assumptions.

Here another suspicion of circularity may arise. Does Kant interpret the experiments of natural science according to the principles of his transcendental philosophy in order to confirm that philosophy has to proceed in the same way if it is to become a science? He does not argue that Galileo, Torricelli, or Stahl agreed with him as regards the nature of experimentation. Instead, he stresses in the footnote to the above passage that he does not claim to give a precise historical outline. However, a similar conception of the nature of experiment is due to Francis Bacon, to whom Kant refers immediately before the passage just quoted (B xxii). Bacon shares the constructivist understanding of explanations in natural science proposed by Kant (Gloy 1996, 68–78; Kim 2008, 10). In the *Instauratio Magna*, from which Kant takes the motto for the second edition of the *CPR*, Bacon emphasizes that the "interpretation of nature" is based on defeating and capturing nature through action.

What consequences for metaphysics does Kant want to draw out from this approach to natural science? In the well-known passage on the Copernican Revolution, he compares the paradigm shift in astronomy with the results of his critical turn. The comparison demonstrates the importance which Kant attributes to his epistemological paradigm shift, but it does not specifically contribute to understanding the "secure course" of metaphysics which he seeks. Next, he characterizes the specific problem of metaphysics as understanding how cognition *a priori* of objects is possible. Then, he emphasizes that he wants to probe his "revolution in the way of thinking" through an experiment, the "experiment of pure reason" (B xxi), which deals with the speculative metaphysical objects which are given "merely through reason" but not by experience:

> As for objects insofar as they are thought merely through reason, and necessarily at that, but that (at least as reason thinks them) cannot be given in experience at all—the attempt to think them (for they must be capable of being thought) will provide a splendid touchstone of what we assume as the altered method of our way of thinking, namely that we can cognize of things *a priori* only what we ourselves have put into them.* (B xviii)

Kant's footnote to the passage makes two points. First, the experiment of pure reason must be a thought experiment, given that it aims at probing the validity of principles *a priori*. Second, its subject is the dualism of appearances and things in themselves. The experiment addresses the distinction between objects as *phaenomena* and *noumena*, as the "elements of pure reason" which are here under test:

> * This method, imitated from the method of those who study nature, thus consists in this: to seek the elements of pure reason in that **which admits of being confirmed or refuted through an experiment**. Now the propositions of pure reason [...] admit of no test by experiment with their **objects** (as in natural science): thus to experiment will be feasible only with **concepts** and **principles** that we assume *a priori* by arranging the latter so that the same objects can be considered from two different sides, **on the one side** as objects of [...] experience, and **on the other side** as objects that are merely thought [...] for isolated reason striving beyond the bounds of experience. (ibid.)

Kant's thought experiment aims at demonstrating that his new, transcendental foundations of metaphysics are significantly more stable than the metaphysical principles of his precursors. He wants to present his "experiment of reason" in a way that is as rigorous as the experiments of natural science. In analogy to the question posed to nature by an experiment of physics or chemistry, he understands his thought experiment as a question posed to pure reason. The outcome of the experiment settles the question of whether the underlying theory is tenable or not. In the case of natural science, the experiment tests whether its result agrees with the theoretical expectation; in the case of pure reason, the experiment tests whether reason conforms with itself. As expected in a thought experiment, here the decisive instance is whether the presupposed theory is consistent. Hence, Kant's key question is how reason can remain in agreement with itself: through a metaphysics that cares about the difference between objects of possible experience and objects of pure thought; or through a metaphysics that does not? Accordingly, the footnote continues:

> If we now find that there is agreement with the principle of pure reason when things are considered from this twofold standpoint, but that an unavoidable conflict of reason with itself arises with a single standpoint, then the experiment decides for the correctness of that distinction. (ibid.)

The experiment of pure reason is thus intended to examine under which conditions reason remains in agreement with itself, or entangles itself in contradictions. Kant's concern is to show that the latter occurs if reason assumes that it has access to things in themselves or "unconditioned" objects of knowledge. This concern raises the question of how a thought experiment can justify a philosophical theory. Of course we know, as did Kant, that proving the consistency of a theory does not prove its truth. So how is it he appears to be convinced that his experiment of reason promises to give metaphysics the "secure course" of a science? The text following the footnote takes up this question. There Kant remarks that his deduction of the possibility of cognition *a priori* in the Transcendental Analytic gives rise to

> a very strange result, and one that appears very disadvantageous to the whole purpose with which the second part of metaphysics concerns itself, namely that with this faculty we can never get beyond the boundaries of possible experience, which is nevertheless precisely the most essential occupation of this science. (B xix–xx)

This frustrating result corresponds to the famous dictum of the "all-destroying Kant" (*Alleszermalmer*), which Mendelssohn circulated in his lectures on the existence of God (Mendelssohn 1785, 3), establishing Kant's reputation as a destroyer of metaphysics. Kant counters by emphasizing that the "disadvantageous" result is balanced by offering a "checkup" (*Gegenprobe*) on truth, through his thought experiment:

> But herein lies just the experiment providing a checkup on the truth of the result of that first assessment of our rational cognition *a priori*, namely that such cognition reaches appearances only, leaving the thing in itself as something actual for itself but uncognized by us. (B xix–xx)

What is on trial here is the possibility or impossibility of metaphysical cognition of the unconditioned, that is, "transcendental realism" about things in themselves and the possibility of completing their cognition:

> For that which necessarily drives us to go beyond the boundaries of experience and all appearances is the **unconditioned**, which reason necessarily and with every right demands in things in themselves for everything that is conditioned, thereby demanding the series of conditions as something completed. (B xx)

The experiment of reason is supposed to prove that transcendental realism gives rises to contradictions and is therefore impossible, whereas the cognition of objects as phenomena does not and is therefore possible:

> Now if we find that on the assumption that our cognition from experience conforms to the objects as things in themselves, the unconditioned **cannot be thought at all without contradiction**, but that on the contrary, if we assume that our representation of things as they are given to us does not conform to these things as they are in themselves but rather that these objects as appearances conform to our way of representing, then **the contradiction disappears** [...] (ibid.)

What is more, this is supposed to give an independent proof of Kant's transcendental idealism that does not rely on the principles of the Transcendental Analytic:

> and consequently that the unconditioned must not be present in things insofar as we are acquainted with them (insofar as they are given to us), but rather in things insofar as we are not acquainted with them, as things in themselves: then this would show that what we initially assumed only as an experiment is well grounded.* (B xx–xxi)

The significance of this conclusion remains unclear here; for the refutation of a theory by proof of a contradiction does not yet prove that a contradiction-free alternative theory is true. Indeed Willaschek (2018, 245–249) discusses different versions of transcendental realism and shows that all of them are contrary but not contradictory to transcendental idealism. Hence, transcendental idealism is only sufficient, but not necessary for avoiding the antinomy, just as Kant's way out of the 1768 puzzle was (see Sect. 3.4.2.2). Kant, however, assumes that the options he discusses form a complete alternative, i.e., that we may either have cognition of things in themselves, or of phenomena. But how can he justify this claim?

6.2.2.1 Once Again: The Analytic-Synthetic Method

The footnote to the phrase just quoted indicates how Kant wants to support it. He appeals once again to the traditional analytic-synthetic method, which aims at establishing truth through a two-step procedure:

> * This experiment of pure reason has much in common with what the **chemists** sometimes call the experiment of **reduction**, or more generally the **synthetic procedure**. The **analysis of the metaphysician** separated pure *a priori* knowledge into two very heterogeneous elements, namely those of the things as appearances and the things in themselves. The **dialectic** once again combines them, in **unison** with the necessary rational idea of the **unconditioned**, and finds that the unison will never come about except through that distinction, which is therefore the true one. (B xx–xxi)

There is an important distinction, however, between the way in which Kant uses this method here, and his pre-critical use of the analytic method. In his pre-critical writings, he followed the philosophical *zeitgeist* by splitting the two parts of the traditional analytic-synthetic method and opposing the analytic method of philosophy to the synthetic method of mathematics (see Sect. 2.1). Nevertheless in the *Prize Essay* of 1764, he already understood the analytic method in analogy to *both* steps of Newton's method (see Sect. 2.3). Now he emphasizes again that only the *complete* analytic-synthetic method gives rise to "well grounded" philosophical conclusions. The crucial passage in the text quoted at the end of Section 6.2.2 is the addendum:

> [...] then this would show that what we initially assumed only as an experiment is well grounded.* (B xx–xxi)

This phrase recalls Pappus's original method, according to which analysis has to be followed by synthesis in order to justify the principles of a theory, and Newton's corresponding claim that in physics analysis comes first, "[...] assuming the Causes discover'd, and establish'd as Principles", followed by synthesis, "explaining the Phaenomena proceeding from them, and proving the Explanations" (Newton 1730, 405). In the associated footnote, Kant compares the "experiment of pure reason" with the second step of the analytic-synthetic method in chemistry. According to this comparison, the Transcendental Analytic corresponds to the first, analytic step of the analytic-synthetic method, and the Transcendental Dialectic to the second, synthetic step. Kant chooses chemistry as a benchmark here, given that in this way he can compare the distinction between *phenomena* and *noumena* of the Transcendental Analytic with the results of chemical analysis; and the "unison" of the appearances with the idea of the "unconditioned" in the Transcendental Dialectic with the results of chemical synthesis.

In his comparison between philosophical and chemical analysis and synthesis, Kant is obviously not referring to chemical elements in a modern sense. He had a chemical theory of the analysis (or decomposition and division) and synthesis (or dissolution and combination) of matter (McNulty 2017, 93–100), which was close to Stahl's theory (Carrier 1990, 194–198), mentioned before (B xii–xiii) in Kant's analogy between the experiments of natural science and the method of philosophy. According to Stahl, combustion is the loss of phlogiston, and metals are composed of phlogiston and the ash-like materials obtained by calcination or oxidation (Hudson 1992, 47). The main idea behind chemical analysis and synthesis in Stahl's chemistry was Newtonian, as it is still in chemistry today. It remained unchanged when Lavoisier, Dalton, and their followers replaced Stahl's phlogiston theory by the modern theories of oxidation, chemical atoms, and the synthesis of compound substances out of atoms.[4] Up to the present day, chemical analysis has meant the dissection of compound substances into elements, whereas chemical synthesis has meant the composition of compound substances from the elements.

Kant's account of analysis and synthesis fits perfectly with Newton's account of analysis and synthesis in the *Opticks*, according to which analysis proceeds "from Compounds to Ingredients" (Newton 1730, 404), and synthesis proceeds as "assumed in the Method of Composition for explaining the Phaenomena arising from them: An Instance of which Method I gave in the End of the first Book" (Newton 1730, 405). There, Newton referred to his optical experiments that showed

[4]Molecules are compound quantum systems in which the electrons belonging to the atomic nuclei can no longer be individuated, similar to the quantum states of solids and the constituent models of nuclear and particle physics. Nevertheless, sum rules for charge, mass, and other dynamic quantities hold, as a residue of classical atomism and the analytic-synthetic method. See Falkenburg (2007, 2015, 2019a).

how it is possible to recompose white light from the colored light spectra by means of two or more prisms (Newton 1730, 147–148 and 186–189).

But Kant's analogy between chemical synthesis, or the "experiment of reduction", and the experiment of pure reason, has yet another punch line. According to Stahl's chemical theory, reduction is the recovery of a metal from its ash-like calx by recombining the latter with phlogiston (Hudson 1992, 47). According to Kant's experiment of pure reason, the Transcendental Dialectic is the recovery of metaphysics from its ashes by uniting the elements of pure reason in the right way, namely such that the "rational idea of the **unconditioned**" takes into account the distinction between *phenomena* and *noumena*. By doing so, the "reduction" of metaphysics gives rise to the use of the traditional metaphysical ideas as regulative principles, in the domain of the appearances.

A further crucial point has to be made here. The way Kant employs the experiment of pure reason does not just recall Pappus's analytic-synthetic method, but in particular Pappus's account of *problematic analysis* (see Appendix A.2.1). According to Pappus, the "problematic" analysis and the corresponding proof by synthesis are a touchstone of the logical tenability of a given assumption:

> In the case of the problematic kind, we assume the proposition as something we know, then proceeding through its consequences, as if true, to something established, if the established thing is possible and obtainable, which is what mathematicians call 'given', the required thing will also be possible, and again the proof will be the reverse of the analysis; but should we meet with something established to be impossible, then the problem too will be impossible. (Pappus 1589, 84)

Kant's experiment of pure reason aims at showing that the metaphysical position of transcendental realism gives rise to contradiction, and hence is impossible. The preface to the second edition of the *CPR* gives an outline of the experiment which the doctrine of the antinomy of pure reason in the Transcendental Dialectic performs in detail. The resolution of the antinomy given in the Transcendental Dialectic, which is based on transcendental idealism, in turn demonstrates in Kant's view that the "thing" which is "established" by the Transcendental Analytic, i.e., the philosophical position of transcendental idealism, is "possible and obtainable". In contradistinction to transcendental idealism, the option of transcendental realism is ruled out, given that the cosmological antinomy demonstrates its impossibility, in his view.

Accordingly, the experiment of reason aims at proving that the problematic assumption of a distinction between appearances and things in themselves, which was initially assumed as given, is appropriate for justifying metaphysical cognition; while its omission leads to a self-contradictory (i.e., logically impossible) theory of metaphysical cognition. Now, then, either to make this distinction or to refrain from doing so is indeed a complete, exhaustive alternative. Hence, in view of Pappus's method it becomes clear that Kant, with this brief addendum (and the associated footnote), indeed wants to provide the missing justification of his claim that his thought experiment, by ruling out transcendental realism, supports transcendental idealism.

To what extent he succeeds is another question, as already emphasized (see end of Sect. 6.2.2 above). However, Kant's renewed reference to the analytic-synthetic method must be understood as a metaphysical justification for transcendental idealism, which he considers the only viable alternative to transcendental realism. He wants to prove the viability of his own "revolution in the way of thinking" and the impossibility of any metaphysical cognition that does not distinguish between *phaenomena* and *noumena*. In doing so, he no longer separates the two partial steps of the method, as he did in his pre-critical writings, but follows Pappus and Newton in considering them to belong together.

At the same time, his "experiment of reason" ought to show that his "altered method of thinking" can claim scientific precision. In this way, Kant's thought experiment and its comparison with chemical synthesis finally is to demonstrate that the aim of his critical project is far from leaving nothing at all of metaphysics, contrary to Mendelssohn's dictum of the "all-destroying Kant".[5]

The importance of Kant's "experiment of reason" has been underestimated in Kant research, which is perhaps understandable given that it is only explained in two footnotes to the rhetorical arguments of the preface B of the *CPR*. By referring to the traditional version of Pappus's analytic-synthetic method in the main text, however, Kant supports the explanatory power of this experiment. And this is not only rhetoric. On the one hand, he appeals to the undisputed success of the traditional analytic-synthetic method in eighteenth-century science; whereas on the other, he points out the crucial systematic function of the Transcendental Dialectic for the metaphysical foundations of his critical project.

6.2.2.2 Kant's Thought Experiment: A Transcendental Argument

But how does his argumentation look from the point of view of today's philosophy of science, and what does it actually prove? Kant's argument has the character of an inference to the only possible explanation, which is an extreme version of an inference to the best explanation. Indeed, a thought experiment is expected to have a stronger result than any experiment of natural science, given that it is a test of the consistency rather than the empirical tenability of a theory.

It is possible to reconstruct Kant's experiment of pure reason as a transcendental argument, as understood in the recent philosophical discussion (see Falkenburg 2018a). Kant's critical method aims at justifying objective knowledge via the necessary conditions *a priori* for the possibility of such knowledge. A transcendental argument takes up this line of reasoning (Stern 2017): It takes a premise X (here, the *a priori* conditions of cognition explained in the Transcendental Analytic) to be a necessary condition for the possibility of a given Y (here, objective knowledge).

[5]In this context it has to be mentioned that transcendental idealism aimed at providing the foundations of a new, critical metaphysics, with transcendental philosophy as its *metaphysica generalis* and the *MFNS* as the remaining part of a *metaphysica specialis* with objective reality.

The "experiment of pure reason" follows this line of thought. Here, the necessary premise X of the transcendental argument is the position of transcendental idealism; whereas the given Y, whose possibility depends on this premise, is cosmological cognition, that is, objective knowledge of the world as a sum total of appearances. According to the Transcendental Dialectic of the *CPR*, the attempts to complete such knowledge of the world give rise to the antinomy of pure reason.

The conclusiveness of Kant's thought experiment obviously depends on the strength of the arguments of the mathematical antinomy in the Transcendental Dialectic. In Kant's day the proofs of the mathematical antinomy must have seemed convincing, whereas from a modern mathematical and cosmological point of view the situation is different. Indeed on the basis of modern (i.e., post-Cantorian) conceptions of the infinite, the logical fallacies of the proofs may be revealed without Kant's critical distinction between *phenomena* and *noumena*, as van Benthem (1983) has clearly shown for the first antinomy (see Sect. 5.4.1). A definitive assessment is difficult, however, given that Kant's mathematical antinomy does not deal with an inner-mathematical problem but with the possibility of cosmological cognition. We have already seen that Zermelo insisted on this crucial difference in reaction to Cantor's harsh criticism of Kant's mathematical antinomy (see Sect. 5.1.1).

6.2.3 The Analogy Between Science and Metaphysics

Kant's thought experiment of pure reason employs again an analogy to Newtonian science in order to demonstrate that metaphysics has the "secure course of a science". In common with the pre-critical approach of the *Prize Essay*, it makes analogical use of Newton's method. The experiment of pure reason investigates the logical consequences of metaphysical theories, just as the experiments of natural science investigate the consequences of theories about nature. In this way, Kant continues the analogy between philosophical cognition and Galileo's or Stahl's experiments that require nature to take the witness stand (*CPR*, B xiii–xiv). His thought experiment aims at confirming or refuting the claims of pure reason, just as an experiment of physics or chemistry aims at confirming or refuting a specific scientific hypothesis.

Kant's comparison between Newtonian science and metaphysics in the preface B forms a multi-layered analogy (Falkenburg 2018b). Kant differentiates between mathematical and philosophical analogies; both conceptions trace back to Greek mathematics and philosophy. Traditionally, an analogy is a proportional relation based on the mathematical concept of a double ratio:

$$A : B = C : D$$

If A, B, and C are given, D may be inferred from the analogical equation. Accordingly, analogical reasoning infers from two given phenomena A and B, and a third given phenomenon C, to an unknown entity X, under the premise that A relates to B in a similar way as C relates to X. A well-known case of analogical reasoning in metaphysics dates back to Plato's *Timaeus*: God behaves to the world like a watchmaker to his clock, i.e., he has made the world and hence is the sufficient reason, or cause, of the world. This example, according to which God is the creator of the world, obviously falls prey to Kant's criticism of traditional metaphysics. Conclusions by analogy are the kind of inductive inferences on which demonstrations of the existence of God and other questionable metaphysical arguments are based.

According to Kant's critical philosophy, analogies in this traditional sense can only give rise to symbolic cognition. The *CPR* employs analogies in a constitutive as well as a regulative sense. Mathematical analogies give rise to constitutive principles of human cognition, philosophical analogies only to regulative principles (A 179–180/B 222). The analogies of experience provide regulative principles which support inferences from given to non-given appearances. In addition, human reason needs to make regulative use of analogies in order to establish the systematic unity of empirical cognition; that is, reason establishes the "analogue of a scheme" of the "systematic unity of all concepts of the understanding" (A 665/B 693; see Appendix A.1.4.2), which gives rise to regulative principles for extending our knowledge of nature.

The analogy between Newtonian science and metaphysics in the preface B of the *CPR*, however, is different. It obviously has neither the regulative function of the analogies of experience nor the function of symbolic cognition based on the "analogue of a scheme" of systematic cognition. It is a methodological analogy between scientific knowledge and metaphysical cognition. This analogy between Newtonian science and metaphysics has also the traditional structure of a proportion relation, and it comes on several levels:

(1) Kant compares human cognition to the insights of Galileo, Toricelli, or Stahl. He claims that human cognition relates to its objects as the experiments of natural science relate to their results, in "that reason has insight only into what it itself produces to its own design" (B xiii).

(2) Next, he compares the experiments of natural science to a trial. According to this part of the analogy, an experiment relates to nature as the judge relates to the witnesses, making them "answer the questions he puts to them" (B xiii).

(3) Then he makes his famous remarks about the Copernican turn, now establishing an analogy between epistemology and astronomy (B xvi). By changing the observer's point of view, astronomy replaced Ptolemy's geocentric system of the world with Copernicus' heliocentric system. Analogously, changing the observer's point of view in epistemology gives rise to the critical insight that the objects of experience depend on the constitution of our cognitive capacities, and not vice versa.

(4) Next, Kant extends levels (1) and (2) of the analogy from epistemology to metaphysics, that is, from empirical knowledge to cognition *a priori*. He states

that his *a priori* approach to human cognition, according to which "we can cognize of things *a priori* only what we ourselves have put into them", is an experiment with a positive outcome, insofar as it "promises to metaphysics the secure course of a science in its first part", however, only as far as the objects of possible experience are concerned (B xviii).

(5) Concerning the second, speculative part of metaphysics "beyond the boundaries of possible experience", the experiment of pure reason provides an additional touchstone for the falsity or truth of the opposing metaphysical options, i.e., transcendental realism vs. transcendental idealism. According to this part of the analogy, this metaphysical thought experiment relates to consistency as the experiments of science relate to falsifiability (note on B xviii–xix).

(6) Finally, in the footnote Kant equates the method of metaphysics to the method of chemistry, by claiming that the Transcendental Analytic of the *CPR* relates to the Transcendental Dialectic (in particular, to the antinomy of pure reason) as chemical analysis relates to chemical synthesis (note on B xxi).

What, in the end, does Kant achieve with this multi-layered analogy between Newtonian science and a scientific metaphysics, and what does he not? To answer the latter question, Gloy (1996) compares Kant's transcendental philosophical method with the experimental method of physics. Her distinctions between the physical experiment, a "transcendental" and a "metaphysical" experiment correspond to the above levels (1–2), (3–4), and (5). Gloy emphasizes that even the physical experiment can only partially realize the "secure" course of a science that Kant was aiming at, given that no experiment can give a theory the absolute certainty that Kant attributed to Newton's physics and hoped for metaphysics. Kant was wrong on one crucial point: all human cognition is fallible, and so is scientific knowledge. His epistemic ideal of apodictic certainty can at most be realized in mathematics and logic, but not in the foundations of physics nor in metaphysics. But in addressing his contemporaries Kant was referring to the success of the analytic-synthetic method in eighteenth-century physics; the twentieth-century scientific revolutions were still a long way off.

So what could Kant actually demonstrate by means of his "experiment of pure reason"? From a logical point of view, a thought experiment has a stronger result than an empirical experiment in physics or chemistry. We have to assume that Kant knew this. Level (5) of his analogy, the comparison of the "experiment of pure reason" with synthesis in chemistry, thus mainly served him by substantiating his thought experiment through associating it with the authority of natural science, and thereby affirming that his transcendental philosophy indeed gives metaphysics the "secure course" of a science. From today's point of view, the "secure course" of a science is no more and no less than a method of making inferences to the best explanation, in science as well as in metaphysics. But inferences to the best explanation do not guarantee truth, contrary to Newton's beliefs behind his first rule of philosophy (see Appendix A.2.2.2).

6.3 What Remains of the Pre-critical Cosmology?

In the course of the critical turn, Kant substantially revised the method and the foundations of his pre-critical cosmology. Nevertheless, in some crucial regards his critical philosophy continues the pre-critical system and the methods of deriving it. The analogy between Newtonian science and metaphysics, the project of integrating Newtonian physics into a system of metaphysics, and the criticism of naturalism or materialism, are the most evident features of this continuity. The most obvious difference between the pre-critical system and Kant's critical project is that, in the latter, transcendental philosophy replaces traditional ontology, in correspondence to his use of the critical method (which has been called the "transcendental" method) rather than the traditional "analytic" method defended in the *Prize Essay*. All these issues are well known, but it is worth taking a closer look at their systematic relations.

6.3.1 Analysis and Synthesis

First of all it has to be noted that the critical Kant does not reject the "analytic" method of his pre-critical philosophy. He still considers it a heuristic tool for philosophy, and even attempts to refine its usage by resorting to Pappus's original, two-stage analytic-synthetic method. In addition, he considers the analytic-synthetic method to be a logical tool, the use of which he repeatedly explains in his logic lectures. Accordingly, we have to distinguish between a logical and a heuristic meaning of the analytic-synthetic method, or its parts, in Kant's critical work.

According to the *Prize Essay*, the analytic method of philosophy consists in decomposing complex concepts into their distinct marks, and the synthetic method in obtaining complex concepts from the combination of simple concepts. Such conceptual analysis and synthesis gives rise to the logical distinction between analytic and synthetic definitions, as well as to Leibniz's distinction between analytic judgments, which are *a priori*, and synthetic judgments, which are *a posteriori*. To this distinction, the critical Kant adds his conception of synthetic judgments *a priori*.

In his logic lectures, Kant describes the analytic and synthetic methods as opposite logical procedures, following the Port-Royal logic (Arnauld and Nicole 1685, 309–310). He characterizes analysis as a regressive method of proceeding from consequences to grounds (or wholes to parts), and synthesis as a progressive method that proceeds vice versa (see Appendix A.2.3). Kant's *Logic* adds a note on the analytic method as a logic of invention or discovery, or more generally as a heuristic tool, again in accordance with the Port-Royal logic:

> Analytic method is also called the method of *invention*. Analytic method is more appropriate for the end of popularity, synthetic method for the end of scientific and systematic preparation of cognition. (§ 117, 9:291)

This heuristic feature is the bridge to the ways in which Kant still employs the analytic and synthetic method, and their combination, in his critical philosophy. He considers the synthetic method as an approach from principles and identifies it with the way in which the *CPR* proceeds. There, he derives the foundations of transcendental philosophy from the pure forms of intuition and the categories, and gives a transcendental deduction of the objective validity of the latter. We may assume that the preceding analytic step of deriving the principles of transcendental philosophy from something already given was the painstaking work of Kant's silent decade.

In contrast, the *Prolegomena* makes explicit use of the analytic method as a heuristic way of making these principles more plausible "for the end of popularity", as the *Logic* puts it—or at least this is what Kant says he aims to do, according to his remarks in the *Prolegomena* (4:263; 4:274–275). These remarks show, however, that Kant's critical method reverses the traditional Newtonian methodology, according to which analysis has to *precede* synthesis (Newton 1730, 404; see Appendix A.2.2.2). After the critical turn, Kant no longer considers the analytic method as an appropriate method of justifying metaphysical principles. He merely thinks that it might complement the deductions of the *CPR* by making their results more accessible, via a popular approach.

Kant's heuristic use of the *complete* analytic-synthetic method in the "experiment of pure reason" (see Sect. 6.2.2) also only aims at making the results of the *CPR* more accessible. In his thought experiment, which is intended to make transcendental idealism more plausible, Kant employs the complete analytic-synthetic method of Newtonian science by analogy, in order to get the readers of the *CPR* engaged with his doctrine of the cosmological antinomy. In his view, the antinomy (and in particular, its mathematical part) definitely refutes transcendental realism and leaves transcendental idealism as the only tenable position. The "experiment of pure reason" has the structure of an inference to the only possible explanation (see Sect. 6.2.2.2), which he furthermore underlines by appealing to the method of Newtonian science.

However, the heuristic value of the analytic-synthetic method is questionable. On the one hand, the "experiment of pure reason" of 1787 evokes the epistemic authority of Newtonian science and the successes of its two-staged analytic-synthetic method. On the other hand, Kant knew that a logically rigorous thought experiment is stronger than any inductive inference of Newtonian science. Indeed, he saw the need for giving an apodictic transcendental foundation to Newton's physics, which he attempted in the *MFNS* of 1786. Nevertheless, Kant may have trusted in the quasi-logical credibility of the complete analytic-synthetic proof procedures of Newtonian science, and he attempted to reproduce them by analogy in his critical philosophy. Indeed, the structure of his critical system of metaphysics also reflects this attempt (see Sect. 6.3.2).

6.3.2 The Critical Unification Project

The *CPR* seeks, after all, to lay the foundations for the Cartesian project that Kant had already pursued in his pre-critical philosophy, that is, to establish metaphysics as a science of the principles of human cognition. As he emphasizes at the beginning of the transcendental doctrine of method in the *CPR*, his metaphysical project aims at a system "of the sum total of all cognition of pure and speculative reason"; and it is not yet given with the "building materials" provided by the first *Critique*. It is the task of the architectonic chapter to sketch the plan of such a system.

Kant's critical system divides into the metaphysics of nature and the metaphysics of morals (A 841/B 869). The *CPR* provides the foundations for the metaphysics of nature. Its structure follows Wolff's and Baumgarten's metaphysics, with its subdivision into general metaphysics, i.e., ontology, and special metaphysics, further subdivided into rational pschology, cosmology, and theology. The critical Kant slightly modifies the structure of this system, and above all he substantially revises the contents of its parts. He replaces ontology by transcendental philosophy, understood as a doctrine of pure concepts and principles "that are related to objects in general, without assuming objects that **would be given** (*Ontologia*)", thus also including the speculative objects of traditional metaphysics that fall prey to the critique of the transcendental dialectic; and he subdivides the special metaphysics of nature or "**physiology** of pure reason" into "**immanent**" and "**transcendent**" metaphysics (A 845/B 873). The parts of the immanent metaphysics are rational physics and psychology; they relate to the objects of inner and outer sense, respectively, and are based on "the mere concept of matter (impenetrable lifeless extension)" and "the concept of a thinking being (in the empirically inner representation 'I think')" (A 848/B 876). Transcendent metaphysics contains rational cosmology and theology; neither of these metaphysical disciplines give rise to objective cognition, but they demonstrate that "human reason [...] is already dialectical on account of the tendency of its nature" and requires "the **preparatory** (propaedeutic) critique of reason that dares to fly with its own wings" (A 845/B 873), which the *CPR* carries out.

As a result of this critique, the cosmology part of the Wolff–Baumgarten system becomes split into rational physics, to which Kant attributes apodictic certainty, and rational cosmology, which he considers to result in the antinomy of pure reason. According to the critical system, only rational physics gives rise to objective scientific cognition, namely to the principles of mathematical physics. Kant substantiates this only remaining content of his critical special metaphysics in the *MFNS*.

The critical system finally realizes Kant's original unification project, even though he does not explicitly say so. The pre-critical writings of 1755/1756 aimed at embedding the principles of Newtonian physics in a metaphysics in Wolff's style, and the critical system finally succeeds in doing so. The young Kant wanted to settle the metaphysical debates of the eighteenth century regarding the concepts of time and space, the concept of force, the existence of atoms, and the relation of the material world to immaterial substances. Moreover, he wanted to do this *systematically*, that is (contrary to the eclecticism of his time) by a theory built on *uniform principles*. His pre-critical attempt to establish these metaphysical principles by means of the analytic method in analogy to Newtonian physics, however, failed, and in the course of the critical turn he developed the critical method of investigating the structure of human cognition and the corresponding transcendental principles of objective cognition.

Let us now see in which way Kant's critical system accomplished the pre-critical unification project. What are the common features and the distinguishing marks of the pre-critical and the critical systems? In view of the structure of Wolff's and Baumgarten's metaphysics and the principles of Newton's physics, Kant's pre-critical integrative project was a multi-layer unification problem, which needed to establish several horizontal and vertical links between general metaphysics, special metaphysics, and Newtonian science. In order to resolve it, he could take the analytic and synthetic methods of the time as a basis, in so far as they were uncontroversial. His unification project included two tasks, each of which comprised two partial tasks (see Sect. 1.3.2), disregarding here the links between cosmology, theology, and psychology:

1. *horizontal unification* at two levels:
   (i) *metaphysica specialis*: To reconcile the separate principles of physics and cosmology, in particular, the concepts of space, time, and matter debated in the Leibniz–Clarke correspondence.
   (ii) *metaphysica generalis*: To clarify the validity of the principle of sufficient reason.
2. *vertical unification* in two directions:
   (i) *analytic method*: To infer general metaphysical principles from specific theories.
   (ii) *synthetic method*: To deduce specific theories from general metaphysical principles.

The critical turn did not affect the content of this program of unification so much as the way in which Kant finally carried it out. His critical method no longer referred to the given contents of human cognition (as the pre-critical analytic method did according to the *Prize Essay*), but rather to the transcendental principles according to which such cognitive contents come about. Nevertheless, the partial tasks of the pre-critical project of unification still give rise to the structure of the critical system. They provide for an astonishing structural continuity of Kant's theory of nature despite the critical turn. *Vertical* unification now aims at linking ontology, or transcendental philosophy, with the remaining part of the special metaphysics of nature, that is, rational physics. *Horizontal* unification must still be able to settle the cosmological disputes regarding absolute space, the concept of force, atomism, and the systematic organization of the universe according to natural laws.

Kant's pre-critical cosmology had already addressed these questions. His critical system, however, provides new answers. According to the *CPR*, cosmology and rational theology only give rise to regulative principles, but not to objective knowledge about the essence of things and the sufficient reason of the universe. As a further result of his critique, Kant no longer attributes or denies absolute space to the universe, but conceives of it as the regulative idea of an ideal frame of inertial motions (*MFNS*, 4:481–482). In addition, he no longer explains matter in terms of atomism, but in terms of forces (4:532–533).

The *MFNS* accomplishes the task which Kant in the *Physical Monadology* had compared to mating griffins and horses, that is, of unifying mathematical physics or "geometry" with metaphysics or "transcendental philosophy" (1:475–476). In the *New Elucidation*, the young Kant had approached this problem by redefining the logical implications of the principle of sufficient reason and criticizing Leibniz's reading of the principle of indiscernibles, at the level of general metaphysics (see Sect. 1.4.2.2). He was aware that the horizontal unification between the parts of special metaphysics has to accord with the principles of general metaphysics, in order to achieve the vertical unification of both levels of metaphysics as well.

Kant's critical unification project, too, aims at vertical unification. The two derivation directions of vertical unification correspond to the two partial steps of the analytic method, which Kant took into account for the first time in his *Prize Essay*, albeit that here he still strictly distinguished between the synthetic method of mathematics and the analytic method of metaphysics (see Sect. 2.3.2.3). Indeed, these two methodological steps reappear in the relation between the *CPR* and the *MFNS*. Kant emphasizes that with the latter he wants to demonstrate that his general metaphysics or transcendental philosophy has an instance or model *in concreto*:

> And so a separated metaphysics of corporeal nature does excellent and indispensable service for *general* metaphysics, in that the former furnishes examples (instances *in concreto*) in which to realize the concepts and propositions of the latter (properly speaking, transcendental philosophy), that is, to give a mere form of thought sense and meaning. (4:778)

This corresponds to an *analytic* inference from the special to the general metaphysics of nature, which demonstrates that the latter applies to the spatio-temporal world. Conversely, the way in which the principles of rational physics derive from the principles of the pure understanding correspond to the *synthetic* procedure of deducing the principles of the metaphysics of corporeal nature in the *MFNS* from general metaphysics. Hence, the frequently noted fact that the *MFNS* and the *CPR* in a certain sense mutually presuppose each other, should not be taken as a methodological "inconsistency in Kant's approach", as Plaass (1965, 68) suggested, but may be considered a consequence of the fact that Kant still makes use of the analytic-synthetic method in order to demonstrate the internal coherence of his critical system.

This use aims at achieving significantly *more* than unifying the special and the general metaphysics vertically, in the two directions of inferences to the best explanation (analysis) and the complementary deductions (synthesis), as in the pre-critical system. According to the above quotation, in the critical system the unification aims at an inference from the true part of the special metaphysics of nature to the objective reality of general metaphysics, i.e., transcendental philosophy, and vice versa.

6.3.3 A New Case Against Naturalism

Another important continuity between Kant's pre-critical philosophy and the critical system concerns the critique of naturalism. The *CPR* refutes transcendental realism in favour of transcendental idealism, and one of the refuted dogmatic positions is materialism or naturalism. In the preface to the *Theory of the Heavens*, Kant had already termed this kind of dogmatism "Epicureanism". In the doctrine of the antinomies, he takes this term up again in order to characterize dogmatic naturalism in opposition to rationalism or "Platonism". In this respect, too, a direct line runs from Kant's pre-critical cosmology to the *CPR*.

The pre-critical physico-theology had provided a powerful argument against materialism and atheism. The preface to the *Theory of the Heavens* emphasizes that the atheist ascribes the actual law-like constitution of the solar system and the starry sky "to nature, which is left to itself", whereas for the defender of religion the systematic structure of the universe leads him to "become aware of the immediate hand of the highest being" (1:221). Here, the young Kant argues against naturalism or materialism that it is committed to attributing the actual systematic order of the world to blind contingency (as Epicurus did); whereas the physico-theological argument infers from this very order to its creator, in an inference to the best explanation. The collapse of Kant's pre-critical system, however, defeated his physico-theology of 1755, and thereby fell a crucial bastion against naturalism. Hence, it was no longer precluded that it is possible to explain the physical universe completely through contingent natural processes—at least insofar as the internal problems of physical cosmology were disregarded, which for the critical Kant belonged to empirical physics and not to the metaphysics of nature.

Kant does not explicitly address this problem in the *CPR*, but he constructs the cosmological antinomy as a new and more effective defence against naturalism (and as a way out of the astronomical problems covered by his pre-critical cosmogony; see De Bianchi 2013). In the course of his critical turn, he convinced himself that reason has cognitive limits, due to which it is impossible to complete our cognition of nature. The limitations of reason, which he finally states in the *CPR*, apply not only to the traditional metaphysical ideas of human freedom, an immortal soul, and the existence of God, but also to the world as a spatio-temporal totality, that is, to the physical universe.

The doctrine of the antinomy is the cornerstone of his critique of reason. According to the preface B, it is the "experiment of reason" which conclusively demonstrates the antinomic results of transcendental realism. A complete naturalistic explanation of the world would permit us to state "ultimate" natural causes for everything in the world, as well as for the universe as a whole. According to the doctrine of the antinomy, any attempt at such an explanation employs the contradictory concept of the phenomenal world as a totality and results in entangling reason in the cosmological antinomy. With the proofs of the particular versions of the cosmological antinomy, Kant intends to demonstrate that it is impossible to find the ultimate grounds of the objects of empirical knowledge, on pain of contradiction. Any attempt to give such an ultimate justification forces us to explain the spatio-temporal, phenomenal world from within and from outside at the same time, mixing "immanent" and "transcendent" proof premises, in fatal paralogisms. The proofs in all four antinomies inevitably make use of naturalistic premises, as shown in Section 5.4. In particular, the "mathematical" antinomy, i.e., the problems of the beginning of the world in space and time and the composition of matter from "ultimate" constituents, leads to conflicting naturalistic positions that both seem to be equally well justified, at least against the scientific background of Kant's time.

Kant was convinced that this insight critically restricts naturalism as well as any other dogmatic metaphysics. In my view, this is a most important albeit neglected result of the *CPR*. It concerns the way in which science gives rise to a new metaphysics, i.e., materialism or naturalism, which is also in need of critique. Passages from the *Prolegomena* and the *Progress* indeed confirm this interpretation:

> Similarly, the cosmological ideas, through the manifest inadequacy of all possible cognition of nature to satisfy reason in its rightful demands, serve to deter us from naturalism, which would have it that nature is sufficient unto itself. [...] The transcendental ideas therefore serve, if not to instruct us positively, at least to negate the impudent assertions of materialism, naturalism, and fatalism which constrict the field of reason, and in this way they serve to provide moral ideas with space outside the field of speculation [...]. (§ 60; 4:363)

> Thus the antinomy of pure reason leads inevitably back to that limiting of our knowledge, and what was previously proved in the Analytic, in dogmatic *a priori* fashion, is here likewise incontestably proved in the Dialectic, by an experiment of reason, which it performs on its own powers. (20:291)

> [...] this antinomy of pure reason, which seems necessarily to bring it to a skeptical standstill, eventually leads by way of criticism to dogmatic advances of reason, if it turns out, that is, that such a noumenon, qua thing-in-itself, is really and even by its own laws knowable, at least from a practical viewpoint, even though it is super-sensible. (20:294)

Hence, the antinomy doctrine not only leads to a drastic revision of the traditional claims of rational theology, psychology, and cosmology, but is, in addition, for Kant the decisive instance that prevents human reason from falling into any kind of dogmatism. Kant's doctrine of the antinomy indeed results in giving the cosmological idea an epistemic status that substantially differs from the other speculative ideas of reason. It is only the equivocations on which the antinomy is based that result in contradictory predications; by contrast, the paralogism of pure reason and the transcendental ideal exceed the capacity of human reason in a more harmless way. They may be subject to skepticism, but they do not give rise to any antinomy. Whereas the antinomy rules out naturalism as logically impossible, in Kant's view, the other speculative ideas only lack objectivity.

6.4 The Teleology of Pure Reason

Indeed Kant's new, critical defence against naturalism is closely related to his teleology of pure reason. In the architectonic chapter of the *CPR*, Kant distinguishes the "scholastic concept" of philosophy as a historically inherited discipline (*cognitio ex datis*) and the cosmopolitan concept (*Weltbegriff* or *conceptus cosmicus*) of philosophy. According to the latter, philosophy does not appeal to traditional dogmatic doctrines, but proceeds exclusively by following the principles of reason. Correspondingly, he claims that the system of metaphysics is generated from the idea of *legislation through human reason* (A 839–840/B 867–868). Philosophy in this cosmopolitan sense is a science of the undogmatic use of reason that aims at the "essential" ends of reason:

> From this point of view philosophy is the science of the relation of all cognition to the essential ends of human reason (*teleologia rationis humanae*), and the philosopher is not an artist of reason, but the legislator of human reason. (A 839/B 867)

Accordingly, the idea behind Kant's system of metaphysics is teleological. It concerns the way in which human reason is directed towards the *ends* of reason. In accordance with the primacy of practical philosophy in Kant's critical system, only the claims of practical reason seem to remain for these ends. Indeed, at the very end of the architectonic chapter Kant emphasizes that

> metaphysics is also the culmination of all culture of human reason [...] That as mere speculation it serves more to prevent errors than to amplify cognition does no damage to its value but rather gives it all the more dignity and authority [...], and prevents its cheerful and fruitful efforts from straying from the chief end, that of the general happiness. (A 850–851/B 878–879)

These remarks seem to suggest that according to the teleology of reason Kant subordinates the particular sciences to the ends of practical reason, and hence the special metaphysics of nature to the metaphysics of morals. A more detailed analysis of Kant's text shows, however, that the issue is not so simple.

6.4.1 The Essential Ends of Reason

Immediately before the passage quoted above, Kant makes the following enigmatic remark:[6]

> Mathematics, natural science, even the empirical knowledge of humankind, have a high value as means, for the most part to contingent but yet ultimately to necessary and essential ends of humanity, but only through the mediation of a rational cognition from mere concepts, which call it what one will, is really nothing but metaphysics. (A 850/B 878)

[6]The following analysis is based on Falkenburg (2005, 2019b).

Here, Kant considers mathematics, natural science, and empirical anthropology as means to the ends of humankind, and he distinguishes contingent ends from necessary and essential ends. From Kant's point of view, the former belong to human practice, involving technology and the applied sciences—let us say, from bridge construction to medicine. The distinction by itself suggests that Kant counts the "necessary and essential ends of humanity" as amounting to the domain of morality and ethics. We should notice, however, that he does not yet speak here of the final or ultimate purpose of reason, that is, the "chief end, that of the general happiness" (A 851/B 879), which is directly related to the moral destiny of humankind.

Indeed it is hard to see how science could directly serve the moral ends, except as contingent technical means to such ends. Then again, however, it is hard to see why for such a use of science a "mediation of a rational cognition from mere concepts" (A 850–851/B 878–879) should be necessary, i.e., metaphysics. Rather, we might read the passage as a hint to the problem that later becomes subject to the *Critique of the Power of Judgment*, namely: How do the realm of nature and the realm of freedom actually relate to each other? Or, how can theoretical reason be confident that practical reason is actually able to realize its intentions, under the conditions of nature? But in 1781, when he wrote the passage, Kant was still a long way off dealing with this problem.

In fact, the architectonics chapter itself provides information about the "necessary and essential ends of humanity", by explaining the "cosmopolitan" concept (*Weltbegriff*) of philosophy, according to which all human cognition is related to the essential ends of human reason, and the philosopher is the "legislator of human reason" (A 839/B 867).

The "essential ends" in question are therefore related to the fact that reason gives itself its own laws and does not depend on any external authority. These laws rest on the autonomy of human reason, which was an educational ideal of Enlightenment put forth most prominently by Kant in his essay *What is Enlightenment?* of 1784 (8:35–42). According to Kant's further explanations in the architectonics chapter, philosophy is divided into "propaedeutics" or critique (this includes the *CPR*), and the "system of pure reason" or metaphysics, whereby metaphysics encompasses "the whole (true as well as apparent) philosophical cognition from pure reason in systematic interconnection". Theoretical reason is obviously only legislative in the field of true cognition, which does not fall prey to criticism. The central task, or in Kant's terms the "essential end", of the *CPR*, is apparently to delimit this area precisely. In addition, Kant also explains what the "essential purposes" of reason are; they are either the "final purpose" of reason, i.e., the above-mentioned moral ends of humanity, or "subalternate ends, which necessarily belong to the former as means" (A 840/B 868).

These remarks suggest that Kant relates the particular sciences to the "essential ends of humanity" as follows. Mathematics, natural science, and empirical anthropology either serve as means for the moral destiny of man in the sense of happiness, which relates to the ideal of the supreme good and to the hopes of pure reason; or they serve as a means of secondary ends, which in turn are necessarily subordinate to this "final end". As I already noted above, a direct relationship between the particular sciences and the "ultimate purpose" of humanity is not very plausible. Thus, the particular sciences must have an indirect relation to the ends of human reason, which in turn are necessary and essentially related to the final moral end. But what ends might be involved here, for Kant?

Kant's writing on the *Conflict of the Faculties* suggests that the predominant end must be the search for truth, which among other things includes the critique of reason. However, Kant wrote this text about one and a half decades after the *CPR*, and in a completely different context (see Sect. 6.4.2). Furthermore, it makes little sense to see mathematics and the empirical sciences as in the service of the critique of reason. According to the first *Critique*, with which we are dealing here, Kant can only mean the highest ends of pure theoretical reason. According to the section *On the ultimate end of the pure use of our reason* of the chapter *The canon of pure reason*, these highest aims are the ideas of God, freedom, and immortality:

> The final aim to which in the end the speculation of reason in its transcendental use is directed concerns three objects: the freedom of the will, the immortality of the soul, and the existence of God. (A 797–798/B 825–826)

In the passage (A 850/B 878) under consideration here, Kant is seeking to explain why human reason is engaged in mathematics, natural science, and empirical anthropology. His explanations in *The canon of pure reason* indicate that it is not just for the sake of technical applications, but primarily with regard to the speculative cognitive ideals of theoretical reason, which in turn are subordinate to the moral ultimate purpose of practical reason. We have to take into account, however, that theoretical reason does not succeed in gaining certainty about its speculative cognitive objects, namely about human freedom, the immortality of the soul, and the existence of God. In this regard, pure reason is no better than the mathematical and empirical sciences, which are even less successful. Kant comments on this cognitive restriction as follows:

> It is humiliating for human reason that it accomplishes nothing in its pure use, and even requires a discipline to check its extravagances and avoid the deceptions that come from them. [...] The greatest and perhaps only utility of all philosophy of pure reason is thus only negative, namely that it does not serve for expansion, as an organon, but rather, as a discipline, serves for the determination of boundaries, and instead of discovering truth it has only the silent merit of guarding against errors. (A 795/B 823)

According to the *CPR*, God, freedom, and immortality are only objects of "doctrinal belief" (A 825–827/B 853–855). Belief in them is associated with moral convictions, but does not give rise to objective knowledge. In the section on opinions, knowledge, and believing, which precedes the architectonic chapter, Kant characterizes this "doctrinal belief" as follows:

> thus there is in merely theoretical judgments an **analogue** of practical judgments, where taking them to be true is aptly described by the word belief, and which we can call **doctrinal beliefs**. (A 825/B 853)

It is worth looking at Kant's remarks on doctrinal belief in more detail, given that they belong to the most neglected parts of Kant's philosophy. As Andrew Chignell

(2007, 324) puts it in an instructive paper on this topic, "the fact that some readers neglect the relevant portion of Kant's work [...] *invites* study". Kant writes:

> Now we must concede that the thesis of the existence of God belongs to doctrinal belief. For although with regard to theoretical knowledge of the world I have nothing at my command that necessarily presupposes this thought as the condition of my explanations of the appearances of the world, but am rather obliged to make use of my reason as if everything were mere nature, purposive unity is still so important a condition of the application of reason to nature that I cannot pass it by [...]. But I know no other condition for this unity that could serve me as a clue for the investigation of nature except insofar as I presuppose that a highest intelligence has arranged everything in accordance with the wisest ends. (A 826/B 854)

> The expression of belief is in such cases an expression of modesty from an objective point of view, but at the same time of the firmness of confidence in a subjective one. (A 827/B 855)

In this way, the traditional claims concerning metaphysical cognition are not completely devastated, but substantially weakened. The claim of objective cognition reduces to the more modest subjective claim of taking something to be true. Kant's critique urges us to be honest about the cognitive capacity of pure speculative reason.

Chignell (2007, 345–356) discusses Kant's account of theoretical or doctrinal belief with regard to Kant's discussion of the cosmological argument and the conception of a *ens realisimum* in the Transcendental Dialectic and elsewhere, as well as its relation to moral belief.[7] In addition, Kant's theory of doctrinal belief includes important remarks concerning physico-theology, which shed light on the relation between the "essential ends of reason" and the role which Kant attributes to the exact and empirical sciences at the end of the architectonic chapter. What answers can mathematics, natural science, and empirical anthropology provide to questions on a subject for which human reason cannot be at all legislative? According to my analysis above, Kant probably interprets them as supporting the "doctrinal belief" with which pure reason must content itself regarding its speculative interest in the ideas of God, freedom, and immortality.

Concerning mathematics, one might think of Plato's cave analogy. There, Plato gives arithmetic and geometry a crucial role in his theory of education, since those disciplines, due to their precision and the ways in which they idealize their objects, can help us ascend to the idea of the good. In contrast to Plato, Kant had a constructivist conception of mathematics, grounded in his theory of space and time as subjective forms of intuition. Hence, it is not obvious how to follow up this Platonist thought here. Concerning natural science, it is easier to see what Kant may have in mind; I suspect that he was still thinking of the physico-theological proof of God's existence of the *Theory of the Heavens*, according to which he inferred from the observable law-like order of nature to God as the creator of the universe. Although in the Transcendental Dialectic the physico-theological proof falls prey to

[7] Pasternack (2011, 214–215) criticizes Chignell of overrating Kant's theory of doctrinal belief, but his objection seems to be met by Kant's distinction between objective and subjective truth (i.e., mere semblance of truth), given that Kant only attributes the latter to doctrinal belief.

criticism (like all the other traditional proofs of God's existence), Kant still makes the following claim about it:

> This proof always deserves to be named with respect. It is the oldest, clearest and the most appropriate to common human reason. It enlivens the study of nature, just as it gets its existence from this study and through it receives ever renewed force. It brings in ends and aims where they would not have been discovered by our observation itself, and extends our information about nature through the guiding thread of a particular unity whose principle is outside nature. But this acquaintance also reacts upon its cause, namely the idea that occasioned it, and increases the belief in a highest author to the point where it becomes an irresistible conviction. (A 623–624/B 652–653)

This conviction can only be subjective, given that the physico-theological proof is merely plausible but not conclusive. It has no logical but only aesthetic truth (as Kant names the subjective semblance of truth in his logic lectures). Yet this is more than just nothing. Kant considers the semblance of truth to be a cognitive ideal, which he ranks among the "aesthetic perfections" of cognition (see Appendix A.3.4). According to Kant's claim quoted above, this subjective conviction regarding the truth of physico-theology can consolidate the doctrinal belief in the existence of God. Indeed his remarks on doctrinal belief in the chapter on opinions, knowing, and believing also concern physico-theology:

> Consequently, the presupposition of a wise author of the world is a condition of an aim which is [...] that of having a guide for the investigation of nature. [...] but rather even in this theoretical relation it can be said that I firmly believe in God; but in this case this belief must not strictly be called practical, but must be called a doctrinal belief, which the **theology** of nature (physico-theology) must everywhere necessarily produce. (A 826–827/B 854–855)

Of course, the physico-theological argument is not conclusive from a logical point of view. It has no more strength than any other proof of God's existence. It relies on a fallacious "dialectical" paralogism of reason, making an inference from the empirical structure of the sensible world to assumptions about the intelligible world, confusing empirical and intellectual proof premises in an unsound way.

Let me summarize the interpretation which I suggest here (based on Falkenburg 2005, 2019b). Kant emphasizes that natural science serves the "necessary and essential ends" of reason in two ways. On the one hand, it successfully uses the speculative idea of a unified, all-encompassing order of natural phenomena as a guide to broadening our objective cognition of nature. On the other, by doing so it can contribute to strengthening the subjective doctrinal belief in God as the ultimate cause of this lawful order.

Thomas Sturm disagrees with this interpretation, arguing that Kant

> does *not* present an argument for how natural science helps to determine the essential ends of practical reason. On the contrary, science must be kept free from these demands of practice.
>
> [...] more evidence for my claim that Kant demands that the sciences should be autonomous from most practical interests can be found elsewhere too. In his *Contest of the Faculties*, he famously defends the autonomy of science from the mundane pressures that come from politics and organized religion. (Sturm 2020, 24)

Concerning Kant's defence of the autonomy of science, Sturm is absolutely correct (see below, Sect. 6.4.2). However, he construes the "essential ends" of science in terms of wisdom, understood as the cognitive goal of achieving systematic unity in the special sciences (Sturm 2020, 16–18), and this is also how he interprets Kant's claim that the sciences serve these essential ends "through the mediation of a rational cognition from mere concepts" (A 850/B 878). In this way, he reduces Kant's critical metaphysical project to philosophy of science, disregarding the relation of the architectonic chapter to *The canon of pure reason*.[8] To mistake Kant's critical account of the metaphysics of nature for philosophy of science has a long tradition, but does not do justice to it. Kant admittedly remains enigmatic with regard to the "essential ends of humanity", which the sciences ultimately serve. Nevertheless his last word on them in the architectonic chapter is to relate them to the "chief end, that of the general happiness" (A 851/B 879), which before he had explained in terms of the highest good (A 809–810/B 837–838).

6.4.2 *Natural Science According to the* Conflict of the Faculties

One might feel that, compared to Mendelssohn's dictum of the "all-destroying Kant", the above results are quite a meager metaphysical result. Since Newton, Kant, and Darwin, natural science has been extremely successful in increasing our knowledge of nature without necessarily consolidating any doctrinal belief. Quite on the contrary, indeed, the increase in scientific knowledge today has given rise to an all-encompassing worldview in which Kant's ideas of God, freedom, and immortality no longer seem to have any place at all. To consider the significance of this objection, let me now briefly comment on some further aspects of Kant's views on natural science, as expressed in his writing on the *Conflict of the Faculties* of 1798.

From an institutional point of view, mathematics, natural sciences, and empirical anthropology belonged, in Kant's day, to the Faculty of Philosophy. In the quotation analyzed above (in Sect. 6.4.1), Kant is thus concerned with internal differences within the Faculty of Philosophy of his day, namely the ranking of the mathematical and empirical sciences, as well as philosophy, within what in his day was considered

[8] Even the end of the architectonic chapter, for Sturm, is to be understood in terms of the philosophy of science: "This obviously demands fundamental considerations of a nonempirical, rational kind, and it explains in part why Kant views the architectonical identification of appropriate ends as a special task of metaphysics—for metaphysics 'considers reason according to its elements and highest maxims, which must ground even the possibility of some sciences and the use of all of them.' (KrV, A 851/B 879) We would nowadays assign this task to the philosophy of science; it does not really matter what we call the discipline, however, as long as it is clear that the task described is an important one and that it requires a type of non-empirical, rational thinking about science" (Sturm 2020, 13).

the "lower" faculty. Philosophy for him has a clear primacy, insofar as it deals with the "essential ends" of human reason. At the same time, our analysis of Kant's text showed why he held the view that the "lower" faculty (in particular, natural science and philosophy) is superior to the "higher" faculties of his time, i.e., theology, jurisprudence, and medicine.

Reason is autonomous, giving itself its own laws. It does not obey authority, but is committed to the search for truth. This is not only true for philosophy, but also for the natural sciences, as Galileo's conflict with the Church had clearly shown. Galileo stood up for the truth of the Copernican system; he was not satisfied with considering it only as a useful hypothesis and as a valuable tool for predicting solar eclipses and other celestial phenomena. Kant had taken this kind of search for truth as a model for himself in the development of his critique of reason, when he described the new features of his epistemology as a "Copernican turn" and compared the Transcendental Dialectic with an "experiment of reason".

For Kant, according to the ends of reason the "lower" faculty counts as the upper one, and, within it, he supposes the following order of precedence: At the top is practical philosophy; theoretical philosophy is subordinate to it, with its principle of the search for truth; and finally, mathematics, mathematical physics, and the empirical natural sciences from biology to anthropology are subordinate to practical and theoretical philosophy. The curricula of the "upper" faculties, i.e., of theology, jurisprudence, and medicine, thus for Kant rank below them.

How can this ranking be reconciled with his connection of the "essential" ends of reason to the subjective physico-theological interpretation of the knowledge of nature, which I suggested above? From the point of view of contemporary science, a physico-theological interpretation of nature appears obsolete, even taken together with the critical limitations put on reason by Kant. In addition, this ranking seems to contradict the subordinate role that Kant himself attributed to the doctrines of the theological faculty, in comparison to the search for truth by philosophy and science. However, both impressions are misleading.

We must recognize, in particular, that Kant's critical physico-theology was progressive and controversial in his time, and his critical view of the relationship between our knowledge of nature and the limitations of reason still provides insights today. Kant's pre-critical physico-theology was not in accordance with his contemporary theological background of pietism (see Wood 1992, 395), given that it proposed a natural history of the universe that was truly revolutionary. By reconstructing the evolution of the solar system and the universe as a whole based on his nebular hypothesis, the young Kant did indeed desire to emphasize that the effects of God's hand in the world are visible in the physical universe.

But the physico-theological argument which distinguished his position from naturalism did not prevent Kant from being suspected of precisely such naturalism. Regarding the conflict with contemporary theology, in a certain sense he was still in Galileo's tradition. In particular, he had in common with Galileo a project to replace the Bible's creation narrative with a physical theory of nature, setting the Book of Nature in competition with the Bible as a testimony of the Word of God. In this regard, Kant's cosmogony is to a certain extent analogous to the view Galileo expressed in his famous *Letter to the Grand Duchess Christina of Tuscany*:

> that in discussions of physical problems we ought to begin not from the authority of scriptural passages but from sense-experiences and necessary demonstrations; for the holy Bible and the phenomena of nature proceed alike from the divine Word, the former as the dictate of the Holy Ghost and the latter as the observant executrix of God's commands. (Galileo 1615)

Despite all the differences between Kant and Galileo, the *Theory of the Heavens* proves Kant to be a critic of the Bible's creation narrative who followed Galileo, and a philosopher of Enlightenment who relied on the autonomy of reason in place of authority, as Galileo had done. Much later, in his essay *Conjectural Beginning of Human History* of 1786, Kant expanded his considerations of natural history even to empirical anthropology. He sketched a natural history of human culture, which illustrates

> the transition [...] from the guardianship of nature into the condition of freedom. (8:115)

6.5 Conclusions: Kant's Legacy

Since the publication of the *CPR*, the received view has been that Kant sought to put an end to metaphysics as a science of the ultimate questions. His attempts to unify science and philosophy, as well as to integrate the predominant opposing ways of thinking of his time, gave rise to his famous Copernican turn and his critique of reason. Many philosophers therefore regarded him as the destroyer of metaphysics; and ever since Moses Mendelssohn coined his expression "all-destroying Kant" in 1785 (Mendelssohn 1785, 3) the word "Alleszermalmer" has been common currency. However, there is a remarkable tension between this dictum and the issues discussed in this chapter, and in particular the teleology of pure reason and Kant's renewed critique of naturalism.

What runs through his life's work, from the first writings to the *opus postumum*, is his tremendous effort to unify the disparate theories of natural science and philosophy of his time into a comprehensive philosophical system, or at least to shape the secure foundations for such a system. Certainly, Kant did not see himself as a destroyer of metaphysics in the sense of smashing an old building in order to create space for a new one. Rather, he believed that the monuments built by his predecessors had all collapsed by themselves, due to their having unstable foundations. From his critical point of view of 1781, this held not only for the metaphysical systems of Descartes, Leibniz, or Wolff, but also for his own pre-critical system of 1755/1756.

With the three writings that gave rise to it, in his pre-critical system Kant had sought to present no more and no less than the foundations of a unified system of physics and metaphysics. In this way, he attempted to incorporate Newton's physics into the cosmology part of a metaphysical system in Wolff's style. And he wanted to accomplish this integration by showing how the key principles of Newton's physics could be reconciled with the basic assumptions of rationalist metaphysics about God, the world, and the soul. In this effort, he considered himself as a new Descartes, as becomes clear from the repeated rhetorical turns in the pre-critical writings. His goal was to renew philosophy in order to put an end to the metaphysical disputes of his time, as Descartes had tried a century before. Like Descartes and the inheritors of his tradition, Kant considered metaphysics to be the science of the principles of human cognition. When he defended the "analytical method" as the correct method of metaphysics in his *Prize Essay*, he meant precisely the variant of conceptual analysis he had employed as a method of justification in the *Physical Monadology*. Nevertheless, in the *Prize Essay* he emphasized that metaphysics "is without doubt the most difficult of all the things into which man has insight", and that "so far no metaphysics has ever been written" (2:283–284).

Yet unlike Descartes, his great predecessor, Kant never did work out a system of metaphysics. After the collapse of his pre-critical system, in the 1770 *Dissertation* he developed the critical theory of space and time as *a priori* forms of intuition, in a first decisive step towards the *CPR*. At that time he still believed that he could combine this with a metaphysics and cosmology in a rationalist style. Yet when he published the *CPR* in 1781, after his silent period, he had not abandoned his project of a systematic metaphysics. The architectonics chapter as well as the appendix to the Transcendental Dialectic show that he maintained his metaphysical ambitions as far as was possible. This crucial concern has long been neglected by Kant scholars. Mendelssohn's famous dictum cannot be said to be the sole cause of this neglect; the Marburg school of neo-Kantianism would later make its own contribution by considering the "fact of science" to be the sole object of Kant's theoretical philosophy, and reducing the *CPR* understood in constructivist terms to philosophy of science (beginning with Cohen 1883; see Falkenburg 2000, 310–316; Falkenburg 2020).

Kant himself also made his own contributions to these misinterpretations. The foundations of metaphysics established by the *CPR* are very modest constructions; and to this extent Mendelssohn was right: all the traditional objects of eighteenth-century metaphysics, i.e., the soul, the world, and God, do fall prey to Kant's critique. Given that they are objects of pure reason beyond the reach of experience, their concepts do not give rise to objective metaphysical cognition.

Yet this did not prevent Kant from dealing with them in the context of his metaphysics lectures, presented as part of his academic teaching responsibilities. There, he never missed the chance to explain where the speculative ideas of reason are located within a critical system of metaphysics. The notions of soul, world, and God do not belong to the "immanent" but to the "transcendent" use of reason. That is to say, they lack objective reality and are no more than hypothetical objects of a subjective doctrinal belief: but at least they give rise to regulative principles for the extension of empirical knowledge.

The doctrine of the antinomies of pure reason is crucial for Kant's critical endeavour. It seeks to show that our cognition can never be completed, and in this way Kant wanted to put an end to all forms of dogmatism, in particular to naturalism. This was the negative result of his critique. Consequently, only a transcendental philosophy (which he never worked out in detail), a rational physics (the foundations of which turned out to be untenable in twentieth-century physics), and the metaphysics of morals remained. As a further positive result, however, Kant insisted on the value of the regulative principles that result from the speculative ideas of the soul, the world, and God. He considered these to be indispensable tools of human reason. On the one hand, they serve to broaden our cognition of nature and to provide insight into the systematic unity of nature; on the other, they can support doctrinal belief, which must no longer be maintained as a dogmatic attitude, but serves the ultimate purposes of humanity in accordance with the metaphysics of morals. With his teleology of pure reason he wanted to make a bridge from the former to the latter.

This is the legacy of Kant's cosmology, which we should carefully spell out against the background of our time, despite the fact that his claim that transcendental realism leads to inevitable contradictions is questionable from today's point of view. If we apply Kant's views in the *Conflict of the Faculties* to the present role of science within society, we see that he was absolutely right in emphasizing that

> it could well happen that the last would some day be first (the lower faculty would be the higher)—not, indeed, in authority, but in counseling the authority (the government). (7:35)

What he did not anticipate was that philosophy would entirely abandon the ensemble of sciences that constituted the philosophical faculties of his day. Yet today we are observing the consequences of this withdrawal. Even foundational research in the natural sciences is under strong pressure to justify itself insofar as it serves purely theoretical interests without promising technological applications, as for example in modern physical cosmology.

Nevertheless, in the end it remains unclear what it could mean today to relate scientific knowledge to the "essential ends of humanity". Kant only gives us the hint that in order to understand this relation, metaphysics is indispensable. What exactly such an engagement with our "essential ends" would look like, more than two hundred years after the publication of the *CPR*, remains an open question. But if we refrain from asking ourselves this question, we permit all of our science, together with all of its technological consequences, to serve only arbitrary and contingent ends. Naturalism unquestionably dominates our age, boosted in particular by breakthroughs in brain research and artificial intelligence. Were Kant's question of how science relates to the "essential ends" of humanity once again to be raised in earnest, naturalism would be the dogmatic doctrine against which a modern critique of reason would have to be directed.

References

Arnauld, Antoine, and Pierre Nicole. 1685. *La logique, ou l'art de penser*. Quoted after the Engl. transl.: *Logic or the Art of Thinking*. Transl. and ed. J. V. Buroker. Cambridge University Press 1996.

Buzzoni, Marco. 2011. Kant und das Gedankenexperiment. Über eine kantische Theorie der Gedankenexperimente in den Naturwissenschaften und in der Philosophie. *Deutsche Zeitschrift für Philosophie* 59: 93–107.

Buzzoni, Marco. 2017. Kantian Accounts of Thought Experiments. In *The Routledge Companion to Thought Experiments*, ed. J.R. Brown, Y. Fehige, and M. Stuart, 327–341. New York: Routledge.

Carrier, Martin. 1990. Kants Theorie der Materie und ihre Wirkung auf die zeitgenössische Chemie. *Kant-Studien* 81: 170–210.

Chignell, Andrew. 2007. Belief in Kant. *The Philosophical Review* 116: 323–360.

Cohen, Hermann. 1883. *Das Prinzip der Infinitesimalmethode und seine Geschichte. Ein Kapitel zur Grundlegung der Erkenntniskritik*. Berlin: Duemmler. Reprint of the 4th ed. in: Werke, Bd. 1.1. Hildesheim, 1977: Olms.

De Bianchi, Silvia. 2013. The Evolution of the Sphere. Kant's Conception of Matter and the Expanding Universe. In *The Harmony of the Spheres. Kant and Herschel on the Universe and the Astronomical Phenomena*, ed. S. De Bianchi, 17–45. Newcastle upon Tyne: Cambridge Scholars Publishing.

Falkenburg, Brigitte. 2000. *Kants Kosmologie*. Frankfurt a. M.: Klostermann.

Falkenburg, Brigitte. 2005. Die Funktion der Naturwissenschaft für die Zwecke der Vernunft. In *Kant im Streit der Fakultäten*, ed. V. Gerhardt and T. Meyer, 117–133. Berlin: de Gruyter 2006.

Falkenburg, Brigitte. 2007. *Particle Metaphysics. A Critical Account of Subatomic Reality*. Heidelberg: Springer.

Falkenburg, Brigitte. 2015. How Do Quasi-Particles Exist? In *Why More is Different. Philosophical Issues in Condensed Matter Physics and Complex Systems*, ed. B. Falkenburg and M. Morrison, 227–250. Heidelberg: Springer.

Falkenburg, Brigitte. 2018a. Kant and the Scope of the Analytic Method. *Studies in History and Philosophy of Science* 71: 13–23.

Falkenburg, Brigitte. 2018b. Was beweist Kants Experiment der reinen Vernunft? In *Natur und Freiheit: Akten des XII. Internationalen Kant-Kongresses*, ed. Violetta L. Waibel et al., 641–659. Berlin: de Gruyter.

Falkenburg, Brigitte. 2019a. Mechanistic Explanations Generalized: How Far Can We Go? In *Mechanistic Explanations in Physics and Beyond*, ed. B. Falkenburg and G. Schiemann, 65–90. Cham: Springer Nature.

Falkenburg, Brigitte. 2019b. The Function of Natural Science for the Ends of Reason. In *Teleologische Reflexion in Kants Philosophie*, ed. P. Órdenes and A. Pickhan. Berlin: Springer, 93–109. Refined English version of Falkenburg (2006).

Falkenburg, Brigitte. 2020. On Method: The Fact of Science and the Distinction between Natural Science and the Humanities. In *Kant Yearbook 12: Kant and Neo-Kantianism*, ed. D. H. Heidemann. Berlin: de Gruyter, 1–31.

Ferrini, Cinzia. 2004. Heavenly Bodies, Crystals and Organisms. The Key Role of Chemical Affinity in Kant's Critical Cosmogony. In *The Kantian Legacy (1804–2004): Unsolved Problems*, ed. C. Ferrini. Napoli: Bibliopolis.

Fulkerson-Smith, B. A. 2013. Bacon's Illuminating Experiments and Kant's Experiment of Pure Reason. In *Kant und die Philosophie in weltbürgerlicher Absicht: Akten des XI. Kant-Kongresses*, ed. S. Bacin et al., Band 5. Berlin: De Gruyter, 455–466.

Fulkerson-Smith, Brett. 2013. Kant's Illuminating Experiment: On the Placement, Purpose, and Essential Procedure of the Experiment of Pure Reason in the Critique of Pure Reason. In *Society and Politics* 7, vol. 2, 62–82. http://socpol.uvvg.ro/docs/2013-1/4.%20Brett%20Fulkerson%20Smith.pdf. Access 4 May 2020.

Galileo Gallei. 1615. Letter to the Grand Duchess Christina of Tuscany. In *Modern History Sourcebook*, https://sourcebooks.fordham.edu/Halsall/mod/galileo-tuscany.asp. Access 4 May 2020.

Gloy, Karin. 1996. Kants Philosophie und das Experiment. In *Kant in der Diskussion der Moderne*, ed. G. Schönrich und Z. Katu, 64–91. Frankfurt a. M.

Guyer, Paul, and Allen W. Wood, eds. 1998. *The Critique of Pure Reason*. Cambridge: Cambridge University Press.

Heidemann, Dietmar H. 1998. *Kant und das Problem des metaphysischen Idealismus*. Kant-Studien Erg.-Hefte 131. Berlin: de Gruyter.

Herschel, William, and Johann Friedrich Gensichen. 1791. *Über den Bau des Himmels. Drei Abhandlungen aus dem Englischen übersetzt von Michael Sommer. Nebst einem authentischen Auszug aus Kants allgemeiner Naturgeschichte und Theorie des Himmels*. Königsberg: F. Nicolovius.

Hudson, John. 1992. *The History of Chemistry*. Houndmills: Macmillan.

Kalin, Martin G. 1972. Kant's transcendental arguments as Gedankenexperimente. *Kant-Studien* 63: 289–328.

Kim, Shi-Hyong. 2008. *Bacon und Kant. Ein erkenntnistheoretischer Vergleich zwischen dem "Novum Organum" und der "Kritik der reinen Vernunft"*. Kantstudien-Erg.-Hefte 156, Berlin: de Gruyter.

McNulty, Michael Bennett. 2017. What is Chemistry, for Kant? In *Kant Yearbook 9: Philosophy of Science*, ed. D. H. Heidemann and O. Motz, 85–112. Berlin: de Gruyter.

Mendelssohn, Moses. 1785. *Morgenstunden oder Vorlesungen über das Daseyn Gottes, Vorbericht*. Quoted after the repr.: *Gesammelte Schriften*, Bd. 3.2, Stuttgart-Bad Cannstatt 1974.

Newton, Isaac. 1730. *Opticks*. With a preface by I.B. Cohen, based on the 4th edition, New York. Fourth Edition, London 1930. With a Preface by I.B. Cohen, Dover 1979.

Pappus of Alexandria. 1589. *Pappi Alexandrini Mathematicae Collectiones*. Venedig. Engl. Transl.: Book 7 of the Collection. Part 1. Edited with Translation and Commentary by Alexander Jones. New York 1986.

Pasternack, Lawrence. 2011. Kant's Doctrinal Belief in God. In *Rethinking Kant*, ed. O. Thorndike, vol. 3, 211–218. Newcastle upun Tyne: Cambridge Publishing Scholars.

Plaass, Peter. 1965. *Kants Theorie der Naturwissenschaft*. Göttingen: Vandenhoeck & Ruprecht.

Stern, Robert A. 2017. Transcendental Arguments. In *The Stanford Encyclopedia of Philosophy* (Summer 2019 Edition), ed. Edward N. Zalta. https://plato.stanford.edu/archives/sum2019/entries/transcendental-arguments/

Strawson, Peter F. 1966. *The Bounds of Sense. An Essay on Kant's Critique of Pure Reason*. London: Methuen.

Sturm, Thomas. 2020. Kant on the Ends of the Sciences. *Kant-Studien* 111: 1–28.

van Benthem, Johan. 1983. *The Logic of Time: A Modal-Theoretic Investigation into the Varieties of Temporal Ontology and Temporal Discourse*. Dordrecht: Reidel.

Willaschek, Marcus. 2018. *Kant on the Sources of Metaphysics. The Dialectic of Pure Reason*. Cambridge: Cambridge University Press.

Wood, Allen W. 1992. Rational Theology, Moral Faith and Religion. In *The Cambridge Companion to Kant*. Cambridge: Cambridge University Press, ed. P. Guyer, 394–416.

Zuckert, Rachel. 2020. Attempting to Exit the Human Perspective: A Priori Experimentation in Kant's *Critique of Pure Reason*. In *Knowledge from a Human Point of View*, ed. A. Çretu and M. Massimi.

Appendix A
Systematic Background

The Appendix gives background information about the logic of cognition prior to Kant and the way in which Kant took it up in the methods of his systematic philosophy. Appendix A.1 explains the concept of a system, as understood in early modern science and philosophy, in Kant's pre-critical writings, and in his critical philosophy. Appendix A.2 focuses on the variants of the analytic-synthetic method which were in use in early modern science and philosophy. Appendix A.3 summarizes the most important principles of the logic of cognition employed by Descartes, Leibniz, and Kant.

A.1 The Concept of a System

Kant's pre-critical and critical uses of the term 'system' are based on the common usage in empirical science and Wolffianism, which itself had a long history. The claim that scientific cognition is systematic goes back to ancient mathematics and philosophy. Aristotle's predominant method was to classify concepts, which gave rise to conceptual systems of natural kinds. In contrast, Euclid's geometry had the logical structure of what we today call an axiomatic system, from which developed the 'synthetic' or axiomatic method of establishing a scientific theory in terms of definitions, postulates, axioms, and theorems (see Appendix A.2.1). In early modern science and seventeenth/eighteenth-century philosophy, systematic knowledge was generated following the method of synthesis, which was part of the traditional analytic-synthetic method (see Appendix A.2). The term 'system' for systematic knowledge came into use in the sixteenth century.

B. Falkenburg, *Kant's Cosmology: From the Pre-Critical System to the Antinomy of Pure Reason*, European Studies in Philosophy of Science 12,
https://doi.org/10.1007/978-3-030-52290-2

A.1.1 Early Modern Science

A prominent usage of the term 'system' in early modern science is found in Galileo's *Dialogue Concerning the Two Chief World Systems*. In the *Dialogue* as well as in its title, Galileo employs the term in a two-fold sense, namely ontological (system of celestial bodies: "solar system", "system of the universe") as well as theoretical (rival theoretical systems or doctrines: "Copernican system" vs. "Aristotelian system"). Hence, the expression "world system" in the title of the *Dialogue* is ambiguous, referring to the tension between the heliocentric and geocentric structures of the system of celestial bodies and connecting this to Copernicus's and Aristotle's rival theories.

A similar ambiguity can be found in the conception of Linné's system of natural kinds, which has an ontological as well as a logical sense, denoting the system of living beings as well as the corresponding conceptual classification system. This ambiguity traces back to Aristotle's logical method of classifying concepts of natural kinds.

A.1.2 Seventeenth-Century Philosophy

The epistemological concept of a system of knowledge (and in particular, of metaphysics) emerged in post-Cartesian philosophy. Descartes depicts the systematic organization of scientific knowledge in his famous comparison of the structure of philosophy with a tree, in the *Author's Letter* preceding his *Principles of Philosophy*:

> Thus philosophy as a whole is like a tree whose roots are metaphysics, whose trunk is physics, and whose branches, which issue from this trunk, are all the other sciences. These reduce themselves to three principal ones, viz. medicine, mechanics and morals [...]. (Descartes 1644, 211)

But he does not yet employ the term "system" for this organic structure. In the *Discourse* and in the *Principles of Philosophy* (Descartes 1637, 1644) he uses the term "system" only occasionally, in the sense of his philosophical doctrine. Rescher (1981) gives an instructive survey of how the usage of the term "system" developed in post-Cartesian philosophy. According to him, its use in the sense of a philosophical doctrine had become popular by the end of the seventeenth century, due to Leibniz:

> Now "system" became to be construed as a *particular approach* to a certain subject—a particular theory or doctrine about it as articulated in an organized complex of concordant hypotheses [...]. This is the sense borne in such phrases as "the system of occasional causes" or "the system of Stoic morality" [...] The prime promoter of the new usage was Leibniz. He often spoke of his own philosophy as "my (new) system" of pre-established harmony, contrasting it with various rival systems. (Rescher 1981, 117)

In parallel, there emerged the epistemic conception of a system in the sense of a structured body of knowledge. Rescher gives the following definitions:

> The root idea of a system is that of *structure* or *organization*, of integration into an orderly whole that functions as an "organic" unity. And a specifically *cognitive* system is to encompass these desiderata with respect to our knowledge. (Rescher 1981, 114–115)

According to Rescher, the use of the term "system" for the systematization of knowledge was only established after the Renaissance. In sixteenth-century theology and seventeenth-century philosophy, the epistemic meaning of the systematic account of a body of knowledge came into use:

> By the early years of the seventeenth century, the philosophers had borrowed the term "system" from the theologians, using it to stand for a synoptically comprehensive and connected treatment of a philosophical discipline: logic, rhetoric, metaphysics, ethics, etc. (Rescher 1981, 116)

He explains the consequences of this new epistemic usage as follows.

> This post-Renaissance redeployment of the term *system* had far-reaching significance. In the original (classical) sense, a system was a physical thing: a compositely structured complex. In the more recent sense, a system was an organically structured body of *knowledge*—not a mere accumulation or aggregation or compilation of miscellaneous information (like a dictionary or encyclopedia) but a functionally organized and connectedly articulated exposition of a unified discipline. (Rescher 1981, 116)

This is the meaning of system employed in Christian Wolff's *Latin logic*, in

> [...] Wolff's formula of a system as "a collection of truths duly arranged in accordance with the principles governing their connections". (Rescher 1981, 116; Wolff 1728, Sect. 889)

Rescher emphasizes that such a system of knowledge

> [...] is not just a constellation of interrelated elements, but one of elements assembled together in an "organic" unity by linking principles with a functionally ordered complex of rational interrelationships. The dual application of systems-terminology to physical and intellectual complexes thus reflects a longstanding and fundamental feature of the conception at issue. (Rescher 1981, 116)

A.1.3 Kant's Pre-critical Uses

When Kant developed his philosophy in the context of Wolffianism and in view of the current state of empirical science, he could draw upon the following four common uses of the term "system" sketched above (in Appendix A.1.2):

1. *Ontological*: A system of objects, bodies, or natural kinds.
2. *Logical*: A classification system of concepts corresponding to natural kinds.
3. *Theoretical*: A doctrine or theory based on certain principles, which distinguish it from rival theories.
4. *Cognitive*: A system of knowledge organically structured according to certain principles.

In his pre-critical philosophy, he employs all of them; yet at this stage his account of systematic knowledge is still embryonic. Several remarks in the *Only Possible Argument* show that he relates the conception of systematic knowledge to the use of the "synthetic" method *more geometrico* (see Appendix A.2.1), which he explicitly rejects for metaphysics in his *Prize Essay*.

1. *Ontological use*:
   (i) Kant's most important and innovative pre-critical account of a system concerns the celestial bodies and is modeled after the solar system. In the *Theory of the Heavens*, he defines the concept of the *systematic constitution of the universe* as a specification of a dynamically bound system, in which the motions around the center of gravity are approximately restricted to a common plane, in accordance with the phenomenological properties of the solar system:

   > In this treatise, I shall frequently use the expression of a *systematic constitution of the universe*. [...] all the planets and comets that belong to our universe constitute a system simply because they orbit around a common central body. But I take this term in a narrower meaning in that I consider the more precise relationships that have made their connections to one another regular and uniform. The orbits of the planets relate as closely as possible to a common plane, namely to the extended equatorial plane of the Sun; the deviation from this rule occurs only at the outermost border of the system, where the motions gradually cease. If, therefore, a certain number of heavenly bodies that are arranged around a common central point and move around this, are simultaneously restricted to a certain plane in such a way that they have the freedom to deviate from it to either side only as little as possible; if such deviation occurs gradually only in those that are most remote from the centre point and thus participate less in the relationships than the others: then, I say, that these bodies are related to each other in a *systematic constitution*. (1:246)

   (ii) In *On Fire* (1755), Kant uses a less elaborated concept of a system of particles in order to describe the forces between the particles of a solid. (1:373)
2. *Logical use*: In the *Plan and Announcement* of 1757, Kant sketches the plan of a classification system of physical geography. The system emerges from collecting and looking through all the empirical, historical, and scientific material he could find on geography (2:4). This is a classification system of empirical science (like Linné's system), which emerges from classifying the empirical knowledge about the earth according to the key concepts of physical geography, and which corresponds to a system of natural kinds.
3. *Theoretical use*: The use of the term "system" in the sense of a doctrine or theory is abundant in Kant's pre-critical writings. Here it concerns physical theories as well as metaphysical doctrines. Typical examples are:

   > The most recent philosophy lays down certain concepts of the essential forces of bodies, but these concepts are unacceptable. [...] According to this system, motion arises when [...] (*True Estimation*, 1:25–26)

> It is, however, precisely the agreement between my system and religion that raises my confidence to a fearless serenity in the face of all difficulties. (*Theory of the Heavens*, 1:222)

> I cannot determine exactly the borders between the system of Herr **Wright** and my own and in what ways I have merely imitated his model or have explained it further. (*Theory of the Heavens*, 1:232)

> The system advanced is the only means of all possible ones to do justice to both kinds of reason. (*Theory of the Heavens*, 1:240)

> However, the system of the universal interaction of substances [...] is certainly somewhat superior to the popular system of *physical influence* [...]. (*New Elucidation*, 1:416)

> [The system of *physical influence*] is not so much a system as indifference to all philosophical system [...]. (*New Elucidation*, 2:408)

4. *Cognitive use*: Some passages of the *Only Possible Argument* (1763) indicate that Kant relates the conception of a system of knowledge to the "synthetic" method, which he explicitly rejects for metaphysics in the *Prize Essay* (1764). The synthetic method is to establish a theory *more geometrico* in terms of definitions, postulates, axioms, and the demonstration of theorems (see Appendix A.2.1). In the *Only Possible Argument*, Kant explains that he does *not* proceed in this way. In particular, he explicitly contrasts the systematic presentation of a "formal doctrine" based on definitions, i.e., the synthetic method (see Appendix A.2.1), to his own "analytic" method, with the only goal of deriving principles on which a theory may established, once they are found.

> I have, on occasion, advanced common judgments of the understanding without giving them that form of rigour, through the art of logic, which the elements of a system ought to have. The reason for this omission has [...] been [...] the fact, indeed, that I regarded myself, not having promised a demonstration, as freed from the requirements which are legitimately made of systematic authors. (2:67)

> I am not here offering a determinate definition of the concept of God. If it were my purpose to treat the matter systematically, I should have to provide such a definition. But what I am here setting forth is intended to be an analysis which may serve as a foundation for the formal doctrine proper. (2:89)

In his *Directions in Space* of 1768 (see Sect. 3.4), Kant seems to employ still another usage of the term "system", in the mathematical sense of a coordinate system with a certain orientation:

> The same thing holds of geographical and, indeed, of our most ordinary knowledge of the position of places. Such knowledge would be of no use to us unless we could also orientate the things thus ordered, along with the entire system of their reciprocal positions, by referring them to the sides of our body. (2:379–380)

Here, however, he is dealing with the application of geometrical relations to geographical knowledge and to a system of physical bodies in relation to our own body. Hence, this use is derivative, depending on the meanings (1.–2.).

A.1.4 *Kant's Critical Accounts of a System*

In Kant's critical philosophy, the former uses of the term "system" recur. In addition, he makes the above concept of a system in the sense of a doctrine or theory more precise in logical terms, and now he applies it also to the structure of metaphysics. The *locus classicus* of Kant's critical concept of systematicity is the architectonic chapter of the *CPR*, where Kant relates the systematic unity of cognition to the structure of a science (A 436/B 463).

> By an architectonic I understand the art of systems. Since systematic unity is that which first makes ordinary cognition into science, i.e., makes a system out of a mere aggregate of it, architectonic is the doctrine of that which is scientific in our cognition in general, and therefore necessarily belongs to the doctrine of method.

A.1.4.1 The Logical Concept of a System

According to this definition, systematic cognition gives rise to scientific knowledge, which is based on a methodology and on uniform principles:

> Under the government of reason our cognitions cannot at all constitute a rhapsody but must constitute a system, in which alone they can support and advance its essential ends. I understand by a system, however, the unity of the manifold cognitions under one idea. This is the rational concept of the form of a whole, insofar as through this the domain of the manifold as well as the position of the parts with respect to each other is determined *a priori*. The scientific rational concept thus contains the end and the form of the whole that is congruent with it. (A 322/B 860)

Systematic cognition is opposed to an eclectic aggregate of knowledge consisting of "pieces that do not belong to any whole" (N. 1865, 16:141), as Kant states in his notes on logic, and as he repeatedly explains in his lectures on logic. A science, by contrast, is a cognitive system, whose structure or "architectonic" is completely determinate: "Thus, in addition to these partial cognitions, a science also contains an outline of the whole and the place of every particular cognition within this whole" (N. 1865, 16:141). The *Logic* puts it as follows:

> Cognition, as science, must be arranged in accordance with a method. For science is a whole of cognition as a system, and not merely as an aggregate. It therefore requires a systematic cognition, hence one composed in accordance with rules on which we have reflected. (§ 95, 9:139)
>
> As the doctrine of elements in logic has for its content the elements and conditions of the perfection of a cognition, so the universal doctrine of method, as the other part of logic, has to deal with the form of a science in general, or with the ways of acting so as to connect the manifold of cognition in a science. (§ 96, 9:139)
>
> The doctrine of method is supposed to expound the way for us to attain the perfection of cognition. Now one of the most essential logical perfections of cognition consists in its distinctness, thoroughness, and systematic ordering into the whole of a science. Accordingly, the doctrine of method will have principally to provide the means through which these perfections of cognition are furthered. (§ 97, 9:139–140)

> Systematic method is opposed to *fragmentary* or *rhapsodic method.* If one has thought in accordance with a method and then also expressed this method in the exposition, and if the transition from one proposition to another is distinctly presented, then one has treated a cognition systematically. If, on the other hand, one has thought according to a method but has not arranged the exposition methodically, such a method is to be called *rhapsodic*. (§ 116, 9:149)

A.1.4.2 The Scheme for Generating a System

A system of cognition is a discursive whole consisting of interconnected parts, namely of concepts and judgments. To establish it, one needs knowledge about the scope of the whole. In order to realize a system of cognition, the principles of reason according to which a system of cognition is interconnected, i.e., the idea of this whole, must therefore be *schematized*, like a pure concept of the understanding:

> For its execution the idea needs a **schema**, i.e., an essential manifold and an order of the parts determined *a priori* from the principle of the end. A schema that is not outlined in accordance with an idea, i.e., from the chief end of reason, but empirically, in accordance with aims occurring contingently (whose number one cannot know in advance), yields **technical** unity, but that which arises only in consequence of an idea (where reason provides the ends *a priori* and does not await them empirically) grounds **architectonic** unity. (*CPR*, A 834/B 862)

The architectonic scheme of a system is provided via an *analogue* of the schemata of sensibility, in analogy to the transcendental schemata of the categories which makes the categories applicable by giving them "sense and significance":

> If a cognition is to have objective reality, i.e., to be related to an object, and is to have significance and sense in that object, the object must be able to be given in some way. (A 154/B 194)

In the *CPJ*, Kant characterizes such an analogue of a schema as merely *symbolic* (§ 59, 5:351–352), in contrast to the transcendental schemata of the categories. It is not a schema proper, but the way in which it works is analogous to a transcendental schema. For its function is to provide a principle of construction, which makes the relations between the whole of a conceptual system and its parts intelligible in the manner of an extensional spatial model. Only in the chapter on the regulative use of the ideas of pure reason does Kant give a more detailed explanation of this analogue of a schema of sensibility:

> Yet although no schema can be found in intuition for the thoroughgoing systematic unity of all concepts of the understanding, an analogue of such a schema can and must be given, which is the idea of the maximum of division and unification of the understanding's cognition in one principle. (*CPR*, A 665/B 693)

For a system of natural kinds this means the idea of a maximum of conceptual unity and division, which he illustrates by means of a spatial representation of the horizons of empirical concepts (A 658–659/B 686–687). In analogy to the schematism of the pure understanding, which provides the semantic conditions for applying the pure concepts of the understanding, the symbolic schema of reason must provide the *logical conditions* that determine the architectonic structure of a system. According to § 97 of the *Logic* (see Appendix A.1.4.1), these logical conditions derive from the epistemic ideal of the *perfections of cognition*. Kant nowhere mentions this doctrine in the *CPR*, although he regularly explains it in the introductory passages of his lectures on logic (see Appendix A.3.4).

A.1.4.3 The System of Metaphysics

Kant wants to trace the principles of a scientific metaphysics back to the logical functions of reason, i.e., the system of categories, the synthetic principles of the pure understanding, and the regulative use of the ideas of pure reason. According to the transcendental dialectic, the "unity of principles is a demand of reason" (A 306/B 362). According to the architectonic chapter, a system of cognition is the "unity of the manifold cognitions under one idea" (A 832/B 860).

The perfections of cognition indeed define the disciplines into which Kant's critical metaphysics of nature is divided. In accordance with the systematic principle of legislation through reason, Kant divides metaphysics according to the speculative or practical use of reason into metaphysics of nature and metaphysics of morals. The metaphysics of nature is largely identical with metaphysics in the speculative use of reason. As such, it is not directed toward the practical ends of reason, but toward the *theoretical end of the cognition of all things that exist*. Its systematic structure is largely determined by the ideal of logical completeness explained in the lectures on logic, in which Kant ranks the conditions under which objects of cognition exist for us as being among the logical ideals of cognition. The metaphysics of nature is based on the *general laws of nature*, on the principles of the pure understanding. According to the *formal* meaning of 'nature', the laws of nature concern the first inner principle of the existence of a thing; according to the *material* meaning, they are laws that apply to nature as the sum total of appearances (*MFNS*, 4:467). The systematic principle according to which the metaphysics of nature is generated is the legislation through reason according to the laws of nature, in both senses of 'nature'. In order to determine the structure of metaphysics, Kant has to schematize this principle according to his epistemic ideal of the perfection of cognition.

In accordance with the theoretical end of reason of completing the cognition of all existing objects, the outline of metaphysics has also to satisfy the conditions of the application of the general laws of nature. These are semantic conditions, which we today would consider material. According to Kant, by contrast, they are the *a priori* conditions under which objects are given to us and to which (according to the perfections of cognition) *truth* and *certainty* correspond as necessary conditions of knowledge or ideals of cognition (9:64–72; N. 1765 to Meier 1752, § 27, 16:106).

These epistemic ideals also determine the structure of metaphysics, since philosophical knowledge aims at the real use of reason. Traditionally, metaphysics deals with objects that are not mere creations of our cognitive faculties, but are rather assumed as given and granted real existence by scholastic philosophy. The systematization of philosophical knowledge in a "scientific" metaphysics of nature must consequently aim at the classification of all objects that are represented *a priori* as given and not merely as thought. It must be based on those principles according to which we imagine objects as given representational contents, and especially on the principles of the pure understanding.

According to Kant's definition of it, metaphysics comprises not only true but also merely apparent philosophical cognition. Apparent cognition merely has the subjective illusory appearance of truth, that is, *aesthetic truth*, or semblance of truth, instead of logical truth:

> [...] the aesthetic truth.—A merely subjective truth consisting only in the agreement of cognition with the subject and the laws of sensory appearances (*Sinnen-Schein*) and is consequently nothing but general illusion (*allgemeiner Schein*). (9:39)

The latter part of metaphysics is not true cognition. Cognition in the proper sense only includes the true part of metaphysics, which refers to objects with objective reality and does not go beyond the cognitive horizon of a being with intuition and understanding. Kant's system of metaphysics, however, is also supposed to contain the systematic connections between true and apparent metaphysical cognition. The principles that generate its structure must therefore also include the dialectical inferences of reason, by means of which reason attempts to transcend its limits: the paralogism, the cosmological antinomy, and the transcendental ideal.

A.2 The Analytic-Synthetic Method

The method of analysis and synthesis was widely shared by scientists and philosophers in the age of early modern science. It dates back to ancient geometry and medicine. From ancient geometry, two opposite interpretations of Euclid's *Elements* were received, Proclus' commentary (synthetic method) and Pappus's commentary (analytic method) (see Engfer 1982, 72–89). The difference between the methods is related to the Aristotelian distinction between induction and deduction, but does not reduce to it. Proclus' synthetic method proceeds from definitions and axioms to the deduction of theorems, but the proofs also employ constructions (Engfer 1982, 73).

Pappus's analytic method also differs from mere induction, and it is more correct to call it the *analytic-synthetic* method. In early modern science, a description of it was available in the Latin translation of the Euclid commentary by Pappus (1589). Due to its origins and its reception in early modern science, the Greek terms *analysis* and *synthesis* or the Latin terms *resolution* and *composition* were used synonymously. In empirical science, the method combined inductive and deductive elements of reasoning; in geometry, constructive elements were added. Many investigations in the history of science and philosophy have focused on the original meaning of the method in geometry. Hintikka and Remes (1974) provide a well-known account of analysis and synthesis as methods of mathematical demonstration. The transformations of the method in medieval science were investigated by Crombie (1953, 1990), its seventeenth/eighteenth-century successors in Descartes,

Leibniz, and Wolff by Engfer (1982). An extensive general survey is provided by Beaney (2014).

A.2.1 Pappus's Account of Analysis and Synthesis

According to Pappus, analysis and synthesis are the complementary parts of a joint regressive–progressive method. Its "analytic" part is regressive, proceeding from something which is given (or assumed to be true) to an underlying first principle:

> Now, analysis is the path from what one is seeking, as if it were established, by way of its consequences, to something that is established by synthesis. That is to say, in analysis we assume what is sought as if it has been achieved, and look for the thing from which it follows, and again what comes before that, until by regressing in this way we come upon some one of the things that are already known, or that occupy the rank of a first principle. We call this kind of method 'analysis', as if to say *anapalin lysis* (reduction backward). (Pappus 1589, 82)

The meaning of the term "consequences" is unclear here and the usual interpretation of analysis as a regressive procedure is controversial (see Hintikka and Remes 1974, 11–14). Hintikka and Remes (1974, 8) translate "consequences" by "concomitants", whereas the regressive interpretation of analysis suggests it be understood in the sense of the premises from which something given derives. In ancient and medieval science, the method was indeed understood in this way, and it was uncontroversial that analysis is only the first part of the method. Its second, "synthetic" part is progressive or deductive, aiming at confirming the principles by deriving from them what was originally given or assumed:

> In synthesis, by reversal, we assume what was obtained last in the analysis to have been achieved already, and, setting now in natural order, as precedents, what before were following, and fitting them to each other, we attain the end of the construction of what was sought. This is what we call "synthesis". (Pappus 1589, 83)

For Pappus as well as for his successors in early modern science, it is crucial that principles are only well-established or justified if both complementary parts of the method are carried out, that is, if the analysis is completed by the corresponding synthesis. Whereas it is not obvious how the analysis was carried out in geometry (for details, see Hintikka and Remes 1974), a clear case of the "synthetic" part of Pappus's method was to establish a theory *more geometrico* in terms of definitions, postulates, axioms, theorems, and deductive proofs of the theorems.

A further important aspect of the traditional method is Pappus's distinction between "theorematic" and "problematic" analysis:

> There are two kinds of analysis: one seeks after truth, and is called "theorematic"; while the other tries to find what was demanded, and is called "problematic". (Ibid.)

The "theorematic" analysis serves to trace "what is sought" back to "something established", that is, to a true principle, under the assumption that retracing from something that is given to some premise from which it derives conserves the truth:

> In the case of the theorematic kind, we assume what is sought as a fact and true, then, advancing through its consequences, as if they are true facts according to the hypothesis, to something established, if this thing that has been established is a truth, then that which was sought will also be true, and its proof the reverse of the analysis; but if we should meet with something established to be false, then the thing that was sought too will be false. (Pappus 1589, 83–84)

As we know today, correct logical deductions from true propositions are truth-conserving, whereas the logical consequences of false propositions may either be true or false. Vice versa, the antecedents of true propositions may either be true or false. Pappus's method of analysis and synthesis, however, was not based on modern propositional logic but on Aristotelian syllogistic, according to which the class of objects to which a proposition refers should not be empty. Pappus's "problematic" analysis seems to deal with exactly this problem. It is a procedure that aims at investigating the possibility or impossibility of the given from which the method starts, and the possibility or impossibility of its principles:

> In the case of the problematic kind, we assume the proposition as something we know, then proceeding through its consequences, as if true, to something established, if the established thing is possible and obtainable, which is what mathematicians call 'given', the required thing will also be possible, and again the proof will be the reverse of the analysis; but should we meet with something established to be impossible, then the problem too will be impossible. (Ibid.)

In mathematics, "possible" means "logically possible" or "constructible", or, to put it in modern terms, logically or semantically consistent. Pappus's "problematic" analysis seems to underlie Kant's interpretation of the antinomy of pure reason and the corresponding "experiment of pure reason" in the preface to the second edition of the *CPR* (see Sect. 6.2.2.1).

A.2.2 Transformations in Early Modern Science and Philosophy

Bacon's inductive method is a variant of analysis and synthesis (see Falkenburg and Ihmig 2004; Ihmig 2004), that, however, rejects the use of mathematics and just gives rise to classification systems of empirical science. The genesis of modern physics is due to Galileo's and Newton's transformation of the method. Both combined experimental analysis with mathematical deductions. They reinterpret the traditional method in such a way that analysis becomes an inductive method of empirical science, and synthesis a method of mathematical deduction which derives or predicts phenomenological laws. Both combine causal analysis, or regress from phenomena to their causes, on the one hand, with analysis of the phenomena in the sense of decomposition, or regress from phenomena and bodies to their parts, on the other. Galileo's resolutive-compositive method traces back to Zabarella (Engfer 1982, 90–99). For Galileo, resolution includes the experimental and mathematical decomposition of phenomena, and composition includes the derivation of the observed phenomena from mathematical principles.

Hobbes's way of proceeding in *De corpore* is close to Galileo's and Newton's. His investigations do not employ mathematics, but for him the mind emerges from calculations performed by the parts of brain. Hobbes emphasizes that there is no other scientific method for him, and that the Greek and Latin expressions for the parts of the method are synonymous:

> There is [...] no method, by which we find out the causes of things, but is either compositive or resolutive, or partly compositive, and partly resolutive. And the resolutive is commonly called analytical method, as the compositive [...] synthetical. (Hobbes 1655, 199)

A.2.2.1 Descartes's Methodology

Descartes's view of the analytic-synthetic method shifted from a very complex regressive-progressive account in the unpublished *Regulae*, to the decomposition and composition of problems in the *Discours de la Méthode* (Engfer 1982, 82; Beaney 2014.) The four rules of the *Discours* suggest as a general methodology of science and philosophy (Descartes 1637, 92):

(1) to accept nothing as true which I did not clearly recognise to be so: that is to say, carefully to avoid precipitation and prejudice in judgments, and to accept in them nothing more than what was presented to my mind so clearly and distinctly that I could have no occasion to doubt it.

(2) to divide up each of the difficulties which I examined into as many parts as possible, and as seemed requisite in order that it might be resolved in the best manner possible.

(3) to carry on my reflections in due order, commencing with objects that were the most simple and easy to understand, in order to rise little by little, or by degrees, to knowledge of the most complex, assuming an order, even if a fictitious one, among those which do not follow a natural sequence relatively to one another.

(4) in all cases to make enumerations so complete and reviews so general that I should be certain of having omitted nothing.

Rule (1) is an epistemic criterion of truth. According to it, true knowledge depends on clear and distinct (or, evident) ideas. Rules (2) and (3) are Descartes's account of the traditional method of analysis and synthesis. Rule (2) proposes to decompose any problem into simple partial problems that may easily be resolved (analysis or resolution), while rule (3) proposes to put together the partial problem solutions or simple objects of knowledge in a well-defined order to proceed to more complex knowledge. Rule (4) is a heuristic principle for making our knowledge as complete as possible.

A.2.2.2 Newton's Analytic-Synthetic Method

Newton's account of induction, or analysis, stands in Pappus's tradition of the combined analytic-synthetic method (see Appendix A.2.1), though, in addition, he follows Galileo (see Engfer 1982, 100–102). For him, the inductive part of the method (analysis or *resolutio*) is the regress from the phenomena, as given, to their components and causes, as the principles sought. In order to support the regress from the given phenomena to the principles of a physical theory, Newton performed experiments and analyzed the observations. For him, the regress from the phenomena to the principles includes mathematical idealization and conceptual analysis.

Newton's account of analysis should not be confused with induction in an empiricist sense, i.e., mere empirical generalization. His philosophical background was Neoplatonism, in sharp contrast to Locke's empiricism. The *Essay Concerning Human Understanding* (Locke 1689), which appeared two years after Newton's *Principia*, begins with sharp criticism of the Neoplatonist theory of innate ideas. The empiricist view of Newton's methodology in philosophy of science is in particular due to Ernst Mach (1883, 466), who interpreted Newton's *Rules of Reasoning* of the *Principia* according to his own, empiricist view of physics. His view was influential in twentieth-century philosophy of science, but the empiricist interpretation of Newton's methodology dates back to eighteenth-century German Enlightenment philosophy, which positioned Newton's remarks about "induction" alongside Locke's philosophy. According to Engfer (1982, 100–102), however, the empiricist misunderstanding of Newton's method was also provoked by Newton himself.

Newton's methodology was very complex. A difficulty for Newton's interpreters is that he does not mention the method of analysis and synthesis in the *Principia*, but only in *Query 31* of the *Opticks*. But his remarks in *Query 31* of the *Opticks* and the *Rules of Reasoning* in the *Principia* complement each other and have both to be taken into account. The *Rules of Reasoning* correspond to the analytic step of the analytic-synthetic method explained in the *Queries* of the *Opticks* (Worrall 2000), while the *Principia*'s axiomatic approach to mathematical physics in terms of definitions, laws of motions, and mathematical deductions corresponds to the complementary synthetic step.

In the *Principia*, Newton terms his method "deduction from the phenomena" (see Achinstein 1991, 32–50). He also refers to it as "induction", presumably in Francis Bacon's tradition, giving rise to the empiricist misunderstanding mentioned above. The *Rules of Reasoning* show, however, in which aspects his "deduction from the phenomena" differs from an inductive inference in an empiricist sense. Its objective is not simply an empirical generalization from observations or data to empirical laws. His *Rules of Reasoning* have variously been considered as an inductivism to which Whewell's method of inference to the best explanation later became a rival (Achinstein 2013), a demonstrative induction (Norton 1994), and an inference or "deduction from the phenomena plus 'background knowledge" which differs from the hypothetico-deductive method (Worrall 2000). In fact it is an inductive inference governed by rules, which traces phenomena back to their causes. Falkenburg (2011), Falkenburg and Ihmig (2004), and Ihmig (2004, 2005) investigated in more detail how it fits in with the method of analysis and synthesis.

It turns out to be a variant of Galileo's resolutive-compositive method or of Pappus's regression from something given to the principles, as has been observed by Engfer (1982, 97–102), who, however, nevertheless adopts the prejudice of twentieth-century philosophy of science that Newton had a one-sided, empiricist methodological doctrine. Against Engfer (1982, 100–101) it has to be noted that Newton's fourth rule is not a verificationist principle, but a principle of theory conservation in absence of falsifying instances. The Rules are heuristic principles (or regulative principles, in Kant's later sense of the *CPR*). They guide the analytic part of the combined analytic-synthetic method in order to extend physical knowledge. They claim (Newton 1726, 794–796):

> I No more causes of natural things should be admitted than are both true and sufficient to explain their phenomena.
>
> II Therefore, the causes assigned to natural effects of the same kind must be, so far as possible, the same.
>
> III Those qualities of bodies that cannot be intended and remitted [i.e., qualities that cannot be increased or diminished] and that belong to all bodies on which experiments can be made should be taken as qualities of all bodies universally.
>
> IV In experimental philosophy, propositions gathered from phenomena by induction should be considered either exactly or very nearly true notwithstanding any contrary hypotheses, until yet other phænomena make such propositions either more exact or liable to exceptions.

Rule I is a principle of *ontological parsimony* (I neglect the truth condition). It restricts the number of supposed causes of empirical effects to a minimum. In this way, Newton wants to eliminate the assumption of causes such as occult powers which are not necessary for the explanation of phenomena. In accordance with Rule I, Rule II requires us to assign the same causes to effects of the same kind, as a principle of *analogy*. Rule III is a principle of *universality*. Only this rule is inductive in the sense of an empirical generalization, claiming that the extensive properties of bodies empirically found in all experiments are universal properties. The remark to Rule III is of particular interest, explicitly explaining that it applies to the smallest parts of bodies:

> The extension, hardness, impenetrability, mobility, and force of inertia of the whole arise from the extension, hardness, impenetrability, mobility, and force of inertia of the parts; and thus we conclude that every one of the least parts of all bodies is extended, hard, impenetrable, movable, and endowed with a force of inertia. (Newton 1726, 795–796)

It should be noted that Newton here only ranges the "force of inertia" or *vis inertiæ* among the primary qualities of bodies, but not gravity, which for him is a relational property. (Kant refers to Newton's concept of *vis inertiæ* in the *True Estimation*; see Sect. 1.1.2.) Rule III enables Newton to attribute extension, impenetrability, mobility, and inertia as primary qualities to atoms. But he emphasizes that Rule III does not permit him to conclude whether atoms exist, or whether matter is infinitely divisible:

> But it is uncertain whether those parts which have been distinguished in this way and not yet divided can be actually divided and separated from one another by the forces of nature. But if it were decided by even a single experiment that in the breaking of a hard and solid body, any undivided particle underwent division, we should conclude by the force of this third rule not only that divided particles are separable but also that undivided parts can be divided indefinitely. (Newton 1726, 796)

Rule IV is a principle of *theory conservatism.* Rules I–III give rise to "propositions collected by general induction", that is, the laws of physics deduced from the phenomena, by causal analysis and decomposition of the phenomena. Rule IV requires us to consider these laws to be exactly or approximately true as long as they are not challenged by "other phenomena", which indicate that they are incorrect or wrong; and not to consider "contrary hypotheses", i.e., rival theories, as long as a theory is not falsified.

Newton's goal is to establish the laws of physics in such a way that they are well grounded by the phenomena. His *Rules of Reasoning* aim at the empirical adequacy of mathematical physics, and the way in which he uses the term "general induction" in Rule IV was grist to the mill of empiricism. But for him, empirical adequacy is just a touchstone of truth. Rules I–III show that his method contains inductive elements in the usual sense, but is more complex. In particular, it implies causal analysis (Rules I and II) and the decomposition of wholes into their parts (Rule III), as indicated in *Query 31* of the *Opticks* (see below).

Newton's most prominent remarks on analysis and synthesis are found in *Query 31* of the *Opticks.* There, he explicitly compares the method used in natural philosophy to the corresponding method of mathematics, realigning Pappus's method of analysis (or resolution) and synthesis (or composition) in the context of physics.

> As in Mathematicks, so in Natural Philosophy, the Investigation of Difficult Things by the Method of Analysis, ought ever to precede the Method of Composition. (Newton 1730, 404)

Newton also depicts the reinterpretation of Pappus's mathematical method in terms of parts and causes of the phenomena, which traces back to Galileo and his predecessors.

> By this way of Analysis we may proceed from Compounds to Ingredients, from Effects to their Causes, and from Motions to the Forces producing them; and in general, from Effects to their Causes, and from particular Causes to more general ones, till the Argument end in the most general. This is the Method of Analysis: And the Method of Synthesis in assuming the Causes discover'd, and establish'd as Principles, and by them explaining the Phaenomena proceeding from them, and proving the Explanations. (Newton 1730, 404)

The quotation shows that for Newton the causal analysis of the phenomena has two aspects, namely the search of the ingredients or parts of a given compound and the search for the forces behind the motions. A further remark in *Query 31* of the *Opticks* makes it clear that analysis, or resolution, is the inductive method corresponding to the methodological *Rules of Reasoning* of the *Principia*:

> This Analysis consists in making Experiments and Observations, and in drawing general Conclusions from them by Induction, and admitting of no Objections against the Conclusions, but such as are taken from Experiments, or other certain Truths. For Hypotheses are not to be regarded in experimental Philosophy. (Newton 1730, 404)

The expression "general Conclusions from them by Induction" corresponds to the term "general induction" in Rule IV. The requirement of "admitting of no Objections against the Conclusions, but such as are taken from Experiments" corresponds to the principle of theory conservatism expressed by Rule IV; and the statement "For Hypotheses are not to be regarded in experimental Philosophy" to the famous dictum *hypotheses non fingo* of the *General Scholium* of *Book III* of the *Principia*:

> I have not as yet been able to deduce from phenomena the reason for these properties of gravity from phenomena, and I do not feign hypotheses. For whatever is not deduced from the phenomena must be called an hypothesis; and hypotheses, whether metaphysical or physical, or based on occult qualities, or mechanical, have no place in experimental philosophy. (Newton 1726, 943)

Roger Cotes took the method of analysis and synthesis up in his preface to the second edition of the *Principia*, in order to explain the scientific method of the founders of early modern science. His words are very close to Newton's remarks in *Query 31* of the *Opticks*:

> therefore they proceed by a twofold method, analytic and synthetic. From certain selected phenomena they deduce by analysis the forces of nature and the simpler laws of these forces, from which they then give the constitution of the rest of the phenomena by synthesis. (Cotes 1713, 386)

A.2.2.3 Conceptual Analysis

The *logical* analytic-synthetic distinction which refers to *judgments* should not be confused with the *methodological* analytic-synthetic distinction which refers to the use of Pappus's scientific method in early modern science, even though both have a common origin.

In the Port-Royal Logic (Arnauld and Nicole 1685), analysis, as an inductive method, is still interpreted in the traditional double sense of decomposition and causal analysis. The examples of decomposition given there concern the analysis of natural phenomena into their parts, but also into their properties. In this way, they make the bridge to the analysis of concepts, in particular in mathematics. Due to Leibniz's logic the analytic-synthetic method later shifted to *conceptual* analysis (see Koriako 1999, 28), e.g., in order to make confused concepts distinct (see Appendix A.3.2).

In the *Prize Essay*, Kant identifies the analytic method with conceptual analysis, i.e., the decomposition of complex concepts into their distinct marks, and the synthetic method with the construction of complex concepts by combining distinct concepts. Analytic judgments result from analyzing complex concepts into their distinct marks, while synthetic judgments result from the combination of distinct concepts. So far, the distinction between analytic and synthetic judgments is in accordance with Leibniz's logic.

A.2.3 Analysis and Synthesis in Kant's Logic

In his lectures on logic, Kant describes the analytic and synthetic methods as opposing logical procedures. The analytic method is regressive: it traces consequences back to principles (logical "subordination") or wholes to parts (logical "coordination"). The synthetic method is progressive and runs vice versa. The Blomberg logic of 1770 gives the following list of scientific methods (24:291):

> A. *analytic*
> B. *synthetic*.
> Analysis and synthesis are either as to coordination or as to subordination.
> α the *synthesis* of coordination is the combination of the parts with the whole.
> β the *analysis* of coordination is the resolution of the whole into its part.
> γ the *synthesis* of subordination is the combination of grounds with consequences.
> δ the *analysis* of subordination, however, is the derivation of the consequence from its grounds.

This classification follows the Port-Royal Logic (Arnauld and Nicole 1685, 233–234), and it still appears in most of Kant's later logic lectures. In the *Logic*, it reads as follows (§ 117, 9:291):

> *3. Analytic or synthetic method*
>
> *Analytic* is opposed to *synthetic* method. The former begins with the conditioned and grounded and proceeds to principles (*a principiatis ad principia*), while the latter goes from principles to consequences or from the simple to the composite. The former could also be called regressive, as the latter could progressive.
>
> ---
>
> *Note*. Analytic method is also called the method of *invention*. Analytic method is more appropriate for the end of popularity, synthetic method for the end of scientific and systematic preparation of cognition.

In the note, Kant again follows the Port-Royal Logic by characterizing the analytic method as a logic of invention or discovery, or more generally as a heuristic tool. This heuristic feature is the bridge to the ways in which he continues to use the analytic and synthetic method, and their combination, in his critical philosophy (see Sect. 6.3.1).

A.3 The Logic of Cognition

From Descartes to Wolff, the doctrine of ideas was based on Descartes's account of clear and distinct ideas, which Leibniz further refined. Kant's logic of cognition is based on the doctrine of the perfection of cognition in Meier (1752).

A.3.1 Descartes's Theory of Ideas

According to Descartes, ideas are perceptions or conceptions in the widest sense, from vague perceptions to the precise concepts of mathematics. Clear and distinct perceptions are evident. They satisfy the truth criterion of rule (1) of the *Discours* (Descartes 1637; see Appendix A.2.2.1), have mathematical certainty and are beyond any doubt. For him, an idea is clear if it is present to our mind, and distinct if all its attributes or other ideas included in it are clear:

> XLV. What a clear and distinct perception is.
> [...] For the knowledge upon which a certain and incontrovertible judgment can be formed, should not alone be clear but also distinct. I term that clear which is present and apparent to an attentive mind, in the same way as we assert that we see objects clearly when, being present to the regarding eye, they operate upon it with sufficient strength. But the distinct is that which is so precise and different from all other objects that it contains within itself nothing but what is clear. (Descartes 1644, 237)

Concerning the *adequacy* of ideas, Descartes distinguishes the *formal reality* and the *objective reality* of ideas. According to the *Meditations* (Descartes 1641), representations in the mind have formal reality, whereas the idea of an object is associated with objective reality. Ideas with objective reality point to an object which causes them, and which has at least as much reality as the formal reality of that idea in one's mind:

> Now it is manifest by the natural light that there must at least be as much reality in the efficient and total cause as in its effect. For, pray, whence can the effect derive its reality, if not from its cause? And in what way can this cause communicate this reality to it, unless it possessed it in itself? [...] But further, the idea of heat, or of a stone, cannot exist in me unless it has been placed within me by some cause which possesses within it at least as much reality as that which I conceive to exist in the heat or the stone. For although this cause does not transmit anything of its actual or formal reality to my idea, we must not for that reason imagine that it is necessarily a less real cause; we must remember that [since every idea is a work of the mind] its nature is such that it demands of itself no other formal reality than that which it borrows from my thought, of which it is only a mode [i.e., a manner or way of thinking]. But in order that an idea should contain some one certain objective reality rather than another, it must without doubt derive it from some cause in which there is at least as much formal reality as this idea contains of objective reality. (Descartes 1641, 162–163)

The context of these remarks is the proof of God's existence in the *Third Meditation*, where Descartes argues that his idea of an infinite and absolutely perfect being must have a cause which has at least as much objective reality as its formal reality.

A.3.2 Leibniz's Ideal of Perfect Cognition

The principles of Leibniz's epistemology are described in the essay *Meditationes de Cognitione, Veritate et Ideis*, published in 1684. Leibniz starts from Descartes's distinction between clear and distinct ideas, makes it more precise, and adds two further distinctions:

> Thus, knowledge is either obscure or clear, and again, clear knowledge is either confused or distinct, and distinct knowledge either inadequate or adequate, and adequate knowledge either symbolic or intuitive: and, indeed, if knowledge were, at the same time, both adequate and intuitive, it would be absolutely perfect. (Leibniz 1684, 23).

Then, he differentiates the ideas of cognition as follows (Leibniz 1684, 23–25).

(1) A *clear* idea can be recognized, in contradistinction to an *obscure* idea.

(2) A *distinct* idea differs from a *confused* conception or perception in having clear marks distinguishing it from other concepts or ideas. The marks of a distinct conception, such as the idea of gold, give rise to the *nominal definition* of an object in terms of a sufficient number of distinguishing marks.

(3) An *adequate* idea differs from an *inadequate* idea in having not just clear, but distinct marks. All marks of an adequate idea are distinct. The adequate idea of an object is its *complete notion* (*notio completa*), i.e., the conjunction of all predicates characterizing an object *de re* which give rise to a *real definition.*

(4) The distinct marks of an *intuitive* idea can be grasped in the mind at the same time. In contrast, a *symbolic* idea is too complex to keep all its distinct marks in mind. Therefore it is replaced by a sign that denotes composite ideas, as in algebra or arithmetic.

For Leibniz, a real definition gives the necessary and sufficient conditions for the real possibility of an object or substance. According to him, a substance has only real possibility if it is *compossible* with all the other substances within the system of pre-established harmony (Poser 1981). The distinction between intuitive and symbolic cognition comes close to Locke's distinction between *intuitive* and *demonstrative* knowledge in the *Essay* (Locke 1689, 530–588). Leibniz discusses Locke's distinction in the *Nouveaux essais* (Leibniz 1765).

The distinction between intuitive and symbolic cognition gave rise to Leibniz's project of a *characteristica universalis*. The project was a precursor of modern symbolic logic. It aimed at encoding all concepts by symbols in order to make it possible to calculate the truth of judgments. Leibniz developed it in order to compensate for the imperfection of human cognition, which in general has no access to adequate and intuitive ideas, the highest degree of knowledge. According to Leibniz, we may only have adequate knowledge of necessary *truths of reason*, but not of contingent *truths of fact*. His account of this distinction is as follows. The truths of reason are necessary, i.e., they are analytic judgments that merely state the logical consequences of the concept of an object. To them belong metaphysical principles such as the principle of sufficient reason, logical principles, and mathematical truths. The truths of fact are contingent. They are synthetic judgments adding empirical knowledge to the concept of an object, given that the complete notion of empirical objects is not accessible to human beings.

Leibniz's ideal of the *cognitio perfecta* and his attempt to come close to it in terms of the symbolic ideas of the *characteristica universalis* had substantial impact on the logic and epistemology of Wolffianism. In particular, the pre-Kantian doctrine of perfect knowledge was influenced by it, having been elaborated in Wolffianism in an eclectic or "rhapsodic" way; see the logic compendium of Meier (1752) which

Kant used as a textbook in his logic lectures, working out Meier's distinctions in a more systematic way (see Appendix A.3.4).

A.3.3 Analytic and Synthetic Judgments

The *logical* analytic-synthetic distinction which refers to *judgments* should not be confused with the *methodological* analytic-synthetic distinction which refers to Pappus's scientific method in early modern science, even if both have a common origin (see Appendix A.2).

Leibniz distinguished necessary *truths of reason* and contingent *truths of fact*. His distinction is as follows. The truths of reason are analytic judgments which state the logical consequences of the concept of an object. For him, metaphysical principles such as the principle of sufficient reason, logical principles, and mathematical truths belong to them. The truths of fact are synthetic judgments which add empirical knowledge to the concept of an object, given that the complete notion of empirical objects is not accessible to human beings. The truths of reason are *a priori*, the truths of fact are *a posteriori*. Leibniz's distinction is the background of Kant's distinction between analytic and synthetic judgments, with the well-known distinguishing mark that according to Kant's critical philosophy mathematics and metaphysics are built on *synthetic judgments a priori*.

A.3.4 Kant's Logic of Cognition

As an introduction to formal logic, Kant regularly taught a general epistemology, or logic of cognition (see *Logic*, 9:33–81). This derives from the rationalist doctrine of "scholarly", "philosophical", "rational", or "perfect" cognition, which he found in Meier's compendium of logic (Meier 1752). Its leitmotifs are given by Leibniz's ideal of perfect cognition (*cognitio perfectissima*).

According to the *Logic*, cognition is a complex representation that combines concepts and intuitions and relates both to consciousness (9:33). Kant distinguishes concepts and intuitions as follows:

> All cognitions, that is, all representations related with consciousness to an object, are either intuitions or concepts. An intuition is a *singular* representation (*repraesentatio singularis*), a concept a universal (*repraesentatio per notas communes*) or *reflected* representation (*repraesentatio discursiva*). Cognition through concepts is called *thought* (*cognitio discursiva*). (§ 1, 9:91)

While a singular representation (that is, an intuition without concept) is not a cognition, but only sensation, a universal representation is a concept without intuition, i.e., it does not denote an individual object. For the critical Kant, intuition is the faculty of individuating objects (*CPR*, A 320–1/B 377). In accordance with the distinction between concepts and intuitions, the logic of cognition is propaedeutic to the logic proper. The former is subordinated to the latter.

Kant's logic of cognition contains a doctrine of the *perfections of cognition*, on which is based the logical doctrine of method in the *Logic* (§§ 94–120, 9:139–150), and is also presupposed by the transcendental doctrine of method in the *CPR*, given that it aims at the "formal conditions of a complete system of pure reason" (A 707–708/ B 735–736).

In contrast to Kant's systematic approach, Meier's logic of cognition only contains a rhapsodic sequence of particular ideals of cognition. Meier (1752, §§ 25–29, 16:105–107) catalogues them according to

(i) extensiveness,
(ii) magnitude and importance,
(iii) truth,
(iv) distinctness, and
(v) certainty.

Kant systematizes them according to his categories as follows:

> A cognition is perfect (1.) as to quantity if it is *universal*; (2.) as to quality if it is *distinct*; (3.) as to relation if it is *true*; and finally (4.) as to modality if it is *certain*. (§ 1, 9:38)

Kant's epistemic ideal of quantitative perfection, or universality, comprises Meier's ideals of magnitude and importance, which Kant subdivides into the universality of the (i) *extension* or the (ii) *content* of cognition, in accordance with his distinction between extensive and intensive magnitudes (see below).

The "perfections" of cognition are partly constitutive principles of knowledge in general and partly regulative principles for the integration of individual cognitions in a system. Each of these perfections represents a degree of knowledge that cannot be exceeded, that is, an *epistemic supreme*, which is not necessarily achievable. Kant distinguishes *logical, aesthetic*, and *practical* ideals of cognition (9:40–41). The aesthetic and practical ideals of cognition are "subjective". The former concern the comprehensibility of cognition, the latter our intentions in the formation and practical use of theories. In particular, the "practical perfections" of philosophy concern the "cosmopolitan concept" (*Weltbegriff*) of philosophy, according to which all human cognition is ultimately related to the "essential ends" of humanity (Sect. 6.4). Apart from that, Kant's theoretical philosophy relies only on the logical and aesthetic perfections of cognition. Kant subdivides them as follows:

I. *Logical perfections of cognition:*
   They are *objective*. They are constitutive for cognition as such, in particular, for scientific cognition:

   1. *logical universality:* (9:40)
      (i) *extension:* multitude and manifoldness,
      (ii) *content:* richness of a ground in consequences;

2. *logical distinctness:* objective clarity of marks (9:61);
3. *truth:* agreement of a cognition with its object (9:50);
4. *rational (apodictic) certainty:* consciousness of necessary truth (9:71).

II. *Aesthetic perfections of cognition:*
They are *subjective*, because they underlie the "laws of sensibility" rather than the "laws of the understanding" (9:36). They only concern the comprehensibility of cognition under the conditions of intuition. With the exception of the beautiful, they are inferior to the logical perfections. They are neither necessary nor sufficient for objective knowledge, and they give rise to opinion and belief instead of knowledge. To a large extent, they are independent of the logical perfections of cognition and may be in conflict with them, given that logically perfect cognition may be incomprehensible (9:37–39).

1. *aesthetic universality:*
   (i) *content:* popularity,
   (ii) *extension:* multitude of examples;
2. *aesthetic distinctness:* examples *in concreto*;
3. *semblance of truth:* agreement of cognition with the subject and its sensations;
4. *aesthetic certainty:* confirmation by sensation or experience.

The logical ideals of cognition, which are objective, are the *formal completeness conditions* for theories and their contents. It should be noted that they are necessary, but not sufficient for objective philosophical cognition, or "rational" and objective cognition of the world. Rational cognition has only to be logically perfect. Objective rational cognition has in addition to have formal aesthetic perfection. The *Logic* characterizes the relation between logical and aesthetic perfection as follows:

> But in this effort to combine aesthetic with logical perfection in our cognitions we must not fail to attend to the following rules, namely: (1.) that logical perfection is the basis of all other perfections and hence cannot be wholly subordinated or sacrificed to any other; (2.) that one should look principally to *formal* aesthetic perfection, the agreement of a cognition with the laws of intuition, because it is just in this that the essentially beautiful, which may best be combined with logical perfection, consists; [...]. (9:38)

However, it is not only the beautiful that underlies the epistemic ideal of formal aesthetic perfection. The possibility of an objective rational cognition of the world, i.e., metaphysics, does so too. According to the *CPR*, synthetic judgments *a priori* only have objective validity if their objects can be given in pure intuition. Beyond mathematics, this holds only for rational physics, according to Kant's critical system. Hence, Kant's claim that transcendental philosophy is in need of an example *in concreto*, which is provided by the *Metaphysical Foundations of Natural Science*, is identical with the requirement of formal aesthetic perfection, and in particular aesthetic distinctness.

Indeed, a further cognitive ideal that is most important is the requirement that there be a correspondence between the subjective and the objective cognitive horizons. The *Logic* explains this issue in detail:

> In expanding our cognitions or in perfecting them as to their extensive quantity it is good to make an estimate as to how far a cognition agrees with our ends and capabilities. This reflection concerns the determination of the horizon of our cognitions, by which is to be understood the *congruence of the quantity of all cognitions with the capabilities and ends of the subject.*
>
> The horizon can be determined
>
> 1. *logically*, in accordance with the faculty or the powers of cognition in relation to the *interest of the understanding*. Here we have to pass judgment on how far we can go in our cognitions, how far we must go, and to what extent certain cognitions serve, in a logical respect, as means to various principal cognitions as our ends;
> 2. *aesthetically, in accordance with taste in relation to the interest of feeling*. He who determines his horizon aesthetically seeks to arrange science according to the taste of the *public*, i.e., to make it popular, or in general to attain only such cognitions as may be universally communicated, and in which the class of the unlearned, too, find pleasure and interest;
> 3. *practically*, in accordance with use in relation to the *interest of the will*. The practical horizon, insofar as it is determined according to the influence which a cognition has on our morality, is *pragmatic* and is of the greatest importance. (9:40–41)

The congruence between the logical and aesthetic perfections concerns the relation between transcendental philosophy and its objects. The congruence between both and the practical perfections concerns the relation between the metaphysics of nature and the metaphysics of morals, that is, the teleology of human reason (see Sect. 6.4). Kant's explanations in the Architectonic chapter indicate that the critical system of metaphysics is indeed expected to satisfy such thoroughgoing congruence.

Kant's completeness conditions, as specified by his doctrine of the perfections of knowledge, correspond to the modern requirement that a theory should be logically and semantically complete. But given that they aim at *epistemic completeness*, they are much stronger than the completeness conditions of modern logic. His idea of a system seeks to establish rational relations between the principles of a scientific theory, which from a modern axiomatic point of view are independent. These relations should derive from the systematic idea that generates the conceptual system of a science (see Appendices A.1.4.2 and A.1.4.3). For a more detailed account, see Falkenburg (2000, 358–366) and Falkenburg (2004).

References

Achinstein, Peter. 1991. *Particles and Waves. Historical Essays in the Philosophy of Science*. Oxford: Oxford University Press.

Achinstein, Peter. 2013. *Evidence and Method: Scientific Strategies of Isaac Newton and James Clerk Maxwell*. Oxford: Oxford University Press.

Arnauld, Antoine, and Nicole Pierre. 1685. *La logique, ou l'art de penser*. Quoted after the Engl. transl.: *Logic or the Art of Thinking*. Transl. and ed. J. V. Buroker. Cambridge University Press 1996.

Beaney, Michael. 2014. *"Analysis", The Stanford Encyclopedia of Philosophy* (Summer 2018 Edition), ed. Edward N. Zalta. http://plato.stanford.edu/archives/sum2018/entries/analysis/

Cotes, Roger. 1713. Editor's Preface to the Second Edition. In Newton (1726), 385–399.

Crombie, Alistair C. 1953. *Robert Grosseteste and the Origins of Experimental Science 1100–1700*. Oxford: Clarendon Press.

Crombie, Alistair C. 1990. *Science, Optics and Music in Medieval and Early Modern Thought*. London: Hambledon Press.

Descartes, René. 1637. *Discours de la méthode*. Quoted after the Engl. transl.: *Discourse on the Method of Rightly Conducting the Reason*. In *The Philosophical Works of Descartes in Two Volumes*, transl. by Elizabeth E. Haldane and G. R. T. Ross. Cambridge University Press 1934.

Descartes, René. 1641. *Meditationes*. Quoted after the Engl. transl.: *Meditations on First Philosophy*. In Descartes (1934), vol. I, 131–199.

Descartes, René. 1644. *Principia philosophiae*. Engl.: (Selections from) *The Principles of Philosophy*. In Descartes (1934), vol. I, 201–302.

Descartes, René. 1934. *The Philosophical Works of Descartes in Two Volumes*, transl. by Elizabeth E. Haldane and G. R. T. Ross. Cambridge University Press.

Engfer, Hans-Jürgen. 1982. *Philosophie als Analysis. Studien zur Entwicklung philosophischer Analysiskonzeptionen unter dem Einfluß mathematischer Methodenmodelle im 17. und frühen 18. Jahrhundert*. Stuttgart-Bad Cannstadt: Frommann-Holzboog.

Falkenburg, Brigitte. 2000. *Kants Kosmologie*. Frankfurt a. M.: Klostermann.

Falkenburg, Brigitte. 2004. Kant's Architectonic Principles for a Metaphysics of Nature. In *The Kantian Legacy (1804–2004): Unsolved Problems*, ed. by C. Ferrini, 127–153. Napoli: Bibliopolis.

Falkenburg, Brigitte. 2011. What are the Phenomena of Physics? *Synthese* 182: 149–163.

Falkenburg, Brigitte, and Karl-Norbert Ihmig. 2004. *"Hypotheses non fingo". Newtons wissenschaftliche Methodenlehre*. Report DFG FA 261/5-1 (with K.-N. Ihmig), http://ifpp.fk14.tu-dortmund.de/cms/ifpp/Medienpool/falkenburg_media/DFG_Bericht_Band1.pdf http://ifpp.fk14.tu-dortmund.de/cms/ifpp/Medienpool/falkenburg_media/DFG_Bericht_Band2.pdf

Hintikka, Jaako, and Unto Remes. 1974. *The Method of Analysis: Its Geometrical Origin and Its General Significance*. Boston Studies XXV. Dordrecht: Reidel.

Hobbes, Thomas. 1655. *De corpore*, Ch. IV. In *Thomas Hobbes: The Elements of Law, Natural and Politic*, ed. J.C. Gaskin, 193–212. Oxford University Press.

Ihmig, Karl-Norbert. 2004. Die Bedeutung der Methoden der Analyse und Synthese für Newtons Programm der Mathematisierung der Natur. In *Philosophiegeschichte und logische Analyse 7*, ed. U. Meixner and A. Newen, 91–119. Paderborn: Mentis.

Ihmig, Karl-Norbert. 2005. Newton's Program of Mathematizing Nature. In *Activity and Sign – Grounding Mathematics Education*, 242–262. New York: Springer.

Koriako, Darius. 1999. *Kants Philosophie der Mathematik: Grundlagen—Voraussetzungen—Probleme*. Hamburg: Felix Meiner.

Leibniz, Gottfried Wilhelm. 1684. Meditationes de Cognitione, Veritate et Ideis. In *Acta Eruditorum*. Repr. in: *Die philosophischen Schriften*, ed. C. I. Gerhardt, Berlin, 1875–1890, repr.: Hildesheim: Olms 1965, vol. 4, 422–426. Engl: Meditations on Knowledge, Truth, and Ideas. In *G. W. Leibniz Philosophical Essays*. Edited and Translated by Roger Ariew and Daniel Garber, 23–27. Indianapolis & Cambridge: Hackett Publishing Company.

Leibniz, Gottfried Wilhelm. 1679. De synthesi et analysi universali seu arte invendi et judicandi. In *Die philosophischen Schriften*, ed. C. I. Gerhardt, Berlin, 1875–1890, repr.: Hildesheim: Olms 1965, vol. 7, 292–298. Engl.: On universal synthesis and analysis, or the art of discovery and judgment. In *Gottfried Wilhelm Leibniz, Philosophical Papers and Letters*, ed. LĖ. Loemker, 129–134. Dordrecht: Reidel 1969.

Leibniz, Gottfried Wilhelm. 1765. *Nouveaux Essais*. Repr. in *Die philosophischen Schriften*, ed. C. I. Gerhardt, Berlin, 1875–90; reprint: Hildesheim: Olms, 1965, vol. 5. Engl.: *New Essays on Human Understanding*. Translated by Peter Remnant and Jonathan Bennett. Cambridge: Cambridge University Press, 1981.

Locke, John. 1689. *An Essay Concerning Human Understanding*. Ed. with an introduction, critical apparatus and glossary by Peter H. Nidditch. Oxford 1975. Reprinted (with corrections) Oxford 1979.

Mach, Ernst. 1883. *Die Mechanik in ihrer Entwicklung*. Leipzig: Brockhaus.

Meier, Georg Friedrich. 1752. *Auszug aus der Vernunftlehre*. Halle: Gebauer. Quoted after: *Kants gesammelte Schriften*, vol. 16.

Newton, Isaac. 1726. *Principia Mathematica Philosophiae Naturae*, 3rd ed. Quoted after: *The Principia: A New Translation* by I.B.Cohen and A.Whitman, 1999. Berkeley: University of California Press.

Newton, Isaac. 1730. *Opticks*. With a preface by I.B. Cohen, based on the 4th ed., New York. Fourth Edition, London 1930. With a Preface by I.B. Cohen, Dover 1979.

Norton, John. 1994. Science and Certainty. *Synthese* 99: 3–22.

Pappus of Alexandria. 1589. *Pappi Alexandrini Mathematicae Collectiones*. Venedig. Engl. Transl.: Book 7 of the Collection. Part 1. Edited with Translation and Commentary by Alexander Jones. New York 1986.

Poser, Hans. 1981. Gottfried Wilhelm Leibniz. In *Klassiker der Philosophie I*, ed. O. Höffe. München: Beck.

Rescher, Nicholas. 1981. Leibniz and the Concept of a System. *Studia Leibnitiana* 13(1): 114–122.

Wolff, Christian. 1728. *Philosophia rationalis sive Logica, methodo scientifica pertractata, et ad usum scientiarum. atque vitae aptata. Praemittitur discursus praeliminaris de philosophia in genere*. Frankfurt: Premlin.

Worrall, John. 2000. The Scope, Limits, and Distinctiveness of the Method of 'Deduction from the Phenomena': Some Lessons from Newton's 'Demonstrations' in Optics. *The British Journal for the Philosophy of Science* 51: 45–80.

Name Index

B. Falkenburg, *Kant's Cosmology: From the Pre-Critical System to the Antinomy of Pure Reason*, European Studies in Philosophy of Science 12,
https://doi.org/10.1007/978-3-030-52290-2

Subject Index

B. Falkenburg, *Kant's Cosmology: From the Pre-Critical System to the Antinomy of Pure Reason*, European Studies in Philosophy of Science 12,
https://doi.org/10.1007/978-3-030-52290-2

D

The manufacturer's authorised representative in the EU is Springer Nature Customer Service Centre GmbH, Europaplatz 3, 69115 Heidelberg, Germany. If you have any concerns regarding our products, please contact ProductSafety@springernature.com

Printed and bound by CPI Group (UK) Ltd, Croydon, CR0 4YY
30/09/2026
02225453-0017